D1726754

Herbert Städtke

Gasdynamic Aspects of Two-Phase Flow

Herbert Städtke

Gasdynamic Aspects of Two-Phase Flow

Hyperbolicity, Wave Propagation Phenomena, and Related Numerical Methods

WILEY-VCH Verlag GmbH & Co. KGaA

The Author

Dr. Herbert Städtke

Institute for Energy
Joint European Research Centre (JRC)
Ispra Establishment
herbert.staedtke@jrc.it

Library of Congress Card No.:
applied for

British Library Cataloguing-in-Publication Data
A catalogue record for this book is available from the British Library.

Bibliographic information published by Die Deutsche Bibliothek
Die Deutsche Bibliothek lists this publication in the Deutsche Nationalbibliografie; detailed bibliographic data is available in the Internet at <http://dnb.ddb.de>.

Printing betz-druck GmbH, Darmstadt
Binding Litges & Dopf GmbH, Heppenheim

Printed in the Federal Republic of Germany
Printed on acid-free paper

ISBN-13: 978-3-527-40578-7
ISBN-10: 3-527-40578-X

Contents

Gasdynamic Aspects of Two-Phase Flow. Herbert Städtke

ISBN: 3-527-40578-X

Preface

Many if not most flow processes of interest for engineering, environmental, and biological systems are of two-phase flow nature or include at least some two-phase flow features. It is therefore not surprising that two-phase flow has reached an enormous attention during the last decades. The great interest in two-phase flow is reflected by the large and continuously growing literature on this subject usually dispersed in various journals and conference proceedings. Books or monographs on two-phase flow are relatively rare and are mostly limited to specific two-phase conditions, flow phenomena, or to dedicated applications.

The purpose of the present monograph on "Gasdynamic Aspects of Two-phase Flow" is to provide a thorough review on wave propagation phenomena in two-component (water/air) and one-component (water/steam) media. The term "Gasdynamic Aspects" is used in a broader sense, covering not only compressibility effects such as sound waves, shock waves, and critical flow conditions rather than including also slow wave modes such as void waves or contact discontinuities propagating with the material velocity of the gas/vapor or liquid phase.

The numerical simulation of the wave propagation processes is based on a newly developed hyperbolic two-fluid model which allows an algebraic evaluation of the complete eigenspace (eigenvalues and related eigenvectors). For the numerical integration of the governing flow equations a second-order Flux Vector Splitting technique is used which allows a high resolution of local flow processes such as steep parameter gradients or flow discontinuities.

For most wave propagation processes investigated, results are also given for single-phase gas or homogeneous two-phase flow before dealing with more complex two-phase flow under heterogeneous and nonequilibrium conditions. Although the major emphasis is on the theoretical approach, experimental data are included where appropriate or available.

Most of the work presented in this book was performed at the European Commission's Joint Research Center at Ispra in Italy, and the author gratefully acknowledges substantional help and support from many colleagues. The book certainly could not have been performed in its present form without the large effort of Giovanni Franchello in transferring the hyperbolic two-phase flow model and related numerical methods into a compact and efficient computer program as is used for all the numerical test cases. The author is also grateful to Brian Worth for numerous discussions on two-phase flow modeling problems, for interpretation of numerical results, and for the careful reading of the manuscript.

Herbert Städtke

Cadrezzate, May 2006

Gasdynamic Aspects of Two-Phase Flow. Herbert Städtke

ISBN: 3-527-40578-X

List of frequently used symbols

Roman letters

a_i	sound velocity in phase i
a	mixture sound velocity
A^{int}	interfacial area
a^{int}	interfacial area concentration
C^D	drag coefficient
C^{int}	curvature of interfacial area
e	internal energy
F	external force per unit volume
f	external force per unit mass
F^{int}	interfacial force per unit volume
g	free enthalpy, or gravity
F	force per unit volume
h	enthalpy
$\mathcal{H}$	heat transfer coefficient
k, C^{vm}	virtual mass coefficient
p	pressure
$\bar{\mathbf{P}}$	pressure tensor
Q	external heat source
r	particle (bubble/droplet) radius
Re	Reynolds number
s	entropy
t	time
T	temperature
$\bar{\mathbf{T}}$	viscous stress tensor
$\vec{u}_i, u_i$	velocity for phase i
$\vec{u}, u$	mixture velocity
v	specific volume
M_i	phasic Mach number

Vectors and Matrices

$\mathbf{U}, \mathbf{V}, \mathbf{W}$	state vectors
$\mathbf{A}, \mathbf{B}$	coefficient matrices
$\mathbf{C}, \mathbf{D}, \mathbf{E}$	source term matrices
$\mathbf{F}$	flux vector
$\hat{\mathbf{F}}$	numerical flux vector
$\mathbf{G}, \mathbf{H}, \mathbf{R}$	coefficient matrices
$\mathbf{I}$	identity matrix
$\mathbf{J}, \mathbf{K}$	Jacobian matrices

Gasdynamic Aspects of Two-Phase Flow. Herbert Städtke

ISBN: 3-527-40578-X

Greek letters

α_i	volumetric concentration of phase i
β	thermal expansion
γ	compressibility
γ_i	distribution function of phase i
$\varkappa$	isentropic exponent
ρ	density, mixture density
λ_k	kth eigenvalue
σ^M	volumetric source for mass
σ^J	volumetric source for momentum
σ^E	volumetric source for energy
σ^Q	volumetric source for heat
σ^S	volumetric source for entropy
σ^A	volumetric source for interfacial area
$\mathbf{\Lambda}$	diagonal matrix of eigenvalues
$\Delta\vec{u}$, Δu	slip velocity

Subscripts

b	bubbly flow
cr	critical condition
d	droplet flow
g	gas/vapor phase
i	phase i, or computational cell i
l	liquid phase, or "left-side" value
n	normal to cell boundary segment
r	"right-side" value
s	cell boundary segment index
t	transverse to cell boundary segment
th	conditions at nozzle throat

Superscripts

ex	quantity exchanged between phases
hom	homogeneous condition
int	interfacial parameter
nc	nonconservative part
nv	nonviscous term
v	viscous term
sat	saturated condition
sub	subcooled condition
T	transpose of matrix
-1	inverse of matrix

Abbreviations

$\varrho = \alpha_g \varrho_g + \alpha_l \varrho_l$	mixture density
$\hat{\varrho} = \sqrt{\varrho_g \varrho_l + k\varrho^2}$	effective mixture density
$\varrho_s = \alpha_g \varrho_l + \alpha_l \varrho_g$	"mirrored" density
$\hat{\varrho}_s = \varrho_s + k\varrho$	effective "mirrored" density
$\hat{\varrho}_g = \varrho_g + k\varrho$	effective gas density
$\hat{\varrho}_l = \varrho_l + k\varrho$	effective liquid density
$\hat{\alpha}_g = \alpha_g \dfrac{\varrho_g \hat{\varrho}_l}{\hat{\varrho}^2}$	extended volumetric gas fraction
$\hat{\alpha}_l = \alpha_l \dfrac{\varrho_l \hat{\varrho}_g}{\hat{\varrho}^2}$	extended volumetric liquid fraction

1 Introduction

Two-phase flow is generally understood as being a simultaneous flow of two different immiscible phases separated by an infinitesimal thin interface. Phases are identified as "homogeneous" parts of the fluid for which unique local state and transport properties can be defined. In most cases, phases are simply referred to as the state of matter, e.g. gas/vapor, liquid, or solid. Typical examples are the flow of liquid carrying vapor or gas bubbles, or the flow of gas carrying liquid droplets or solid particles. However, more complex flow processes may exist where the phase distribution is less well defined.

Two-phase flow is of large relevance for many scientific/technical disciplines ranging from environmental research to the modeling of normal operation or accident conditions in nuclear, chemical, or process engineering installations. For a long time, the analysis of two-phase flow processes was limited to mostly empirical correlations or to largely simplified engineering models and, therefore, two-phase flow was considered as a rather "dirty" branch of fluid dynamics. This situation has changed significantly during the last two decades when a large effort was spent for the analysis of two-phase flow systems and for the development of related numerical simulation methods. Much of this work was stimulated by the specific requirements for the safety analysis of pressurized water reactors which, for obvious reasons, relies largely on the prediction capability of computer codes for complex two-phase flow and heat transfer processes.

Many of the present advanced models for the description of nonhomogeneous nonequilibrium two-phase flow are related to the two-fluid approach using separate mass, momentum, and energy equations for the two phases. These separate conservation equations are obtained in a volume and/or time averaging process starting from the local instantaneous conservation equations of the individual phases. In the averaging procedure important information on local flow processes is lost and, consequently, additional correlations are needed in order to close the system of equations. Most of these closure relations are of empirical nature or include some heuristic elements which cannot be deduced completely from first principles.

The correct formulation of the basic two-fluid equations and the appropriate form of the closure laws have been controversially discussed during the past, and up to now, there does not exist a commonly agreed approach. A specific concern has been that most models presently used in the large computer codes are based on governing equations having complex eigenvalues and, therefore, do not represent a mathematically "well-posed" initial-boundary value problem. Nevertheless, there seems to be a common agreement that the pure transport or Euler part of the governing system of equations should be of hyperbolic nature. The necessity

Gasdynamic Aspects of Two-Phase Flow. Herbert Städtke

ISBN: 3-527-40578-X

for the hyperbolicity of the governing equations of the two-fluid model has several aspects, including the following:

- any transient flow process might be seen as a response to perturbations manifesting themselves in wave propagation phenomena as characterized by the hyperbolic nature of the governing equations,

- nonhyperbolic models suffer from high wave-number instabilities and, therefore, require explicit damping mechanisms in the numerical algorithms with the consequence of excessive numerical diffusion and artificial viscosity effects,

- the existence of a hyperbolic system of equations is an essential condition for the application of advanced numerical methods such as Approximate Riemann Solver or Flux Vector Spitting techniques which make explicit use of the eigenstructure of the flow equations.

There have been various tentative proposals for a "hyperbolic two-fluid model" characterized by the existence of only real eigenvalues and a corresponding set of independent eigenvectors. The hyperbolicity is usually obtained by adding interfacial momentum coupling terms having time and/or spatial derivatives of governing parameters. Often, these terms cannot be deduced completely from first principles and, therefore, can be verified only indirectly. This approach might be justified as long as (1) there is a clear physical background for these additional closure terms, and (2) the effect of these terms on the predicted results is fully plausible. For all the present investigations of wave propagation processes a newly developed hyperbolic two-fluid model will be used which will be described in detail in Chapter 5.

Before dealing with complex two-phase flow conditions, it was felt worthwhile to recall a few facts about single-phase gasdynamics. This is done in Chapter 2, which also introduces the basic methodology for the characteristic analysis of the flow equations as used throughout the book.

The basic features and limitations of the two-fluid approach for two-phase flow are summarized in Chapter 3. A more detailed derivation of the corresponding balance equations for the two-fluid model is provided in Appendix A, based on the concept of a phasic distribution function and its differential form having the property of the Dirac delta function at the interface.

In Chapter 4, simplified two-phase models, based on the assumption of mechanical equilibrium (equal phase velocities) and thermal equilibrium (equal phasic temperatures), are analyzed. Although these models have only a limited value for practical applications, they are of interest as limiting cases for detailed two-phase models dealing with more complex flow conditions.

Chapter 5 is devoted to the development of an improved "hyperbolic" two-fluid model for nonhomogeneous, nonequilibrium flow conditions as forming the basis for the subsequent analysis of wave propagation phenomena in two-phase media. A specific feature of the model is the presence of explicit algebraic formulations for the complete eigenspectrum of the flow equations including eigenvalues and related right and left eigenvectors. A complete reference for the hyperbolic model is provided in Appendix B, including all relevant information on the coefficient matrices, various forms of the source term vector, right and left eigenvectors, and the characteristic form of basic flow equations.

Chapter 6 deals with the propagation and attenuation of sound waves in two-phase media. Based on the acoustic approximation a dispersion relation is derived describing the dependence of sound velocity and attenuation on the frequency of sound waves. Although the results of the dispersion analysis do not directly enter into the numerical simulation, they are of large importance for the understanding of pressure wave propagation processes in two-phase flow and the occurrence of shock waves and critical flow conditions.

Chapter 7 summarizes some basic features of numerical methods for hyperbolic conservation laws and their adaptation for two-phase flow processes.

The numerical results as presented in Chapter 9 cover a wide spectrum of typical two-phase flow phenomena at low and high Mach numbers. Where appropriate, a comparison with analytical solutions or existing experimental data is included. All results shown have been obtained with the Advanced Two-Phase Flow Module (ATFM), a computer code developed at the European Commission's Joint Research Centre in Ispra, based on the hyperbolic two-phase flow model and related numerical methods as described in Chapters 5 and 7. The basic physical modeling and applied numerical features of the ATFM code are briefly outlined in Chapter 8.

In Chapter 10 summarizing conclusions are given together with an perspective with regard to future developments for two-phase flow modeling and related numerical simulation strategies.

2 Single-Phase Gas Flow

In this chapter some basic features of compressible single-phase flows are summarized which are of particular importance for the understanding of the description of more complex behavior of two-phase flow processes in the subsequent chapters. At the same time the basic mathematical approach for the characteristic analysis of hyperbolic flow equations is introduced as will be used later for two-phase models of varying degree of modeling details. The reader might find more comprehensive information on compressible fluid dynamics in many monographs on gasdynamics as for example in the books of Shapiro [1] or Courant and Friederich [2].

2.1 Euler equations for one-dimensional flow

The flow of a compressible single-phase fluid is generally described by the conservation laws for mass, momentum, and energy as given by equations (A.26) to (A.29) in Appendix A. In the case of the absence of viscous forces and without bulk heat conduction processes in the fluid, the conservation relations reduce to the Euler equations which can be written for one-dimensional flow conditions in differential form as

$$\frac{\partial \varrho}{\partial t} + \frac{\partial}{\partial x}(\varrho u) = 0 \tag{2.1}$$

$$\frac{\partial}{\partial t}(\varrho u) + \frac{\partial}{\partial x}(\varrho u^2) + \frac{\partial p}{\partial x} = F \tag{2.2}$$

$$\frac{\partial}{\partial t}\left[\varrho\left(e + \frac{u^2}{2}\right)\right] + \frac{\partial}{\partial x}\left[\varrho u\left(h + \frac{u^2}{2}\right)\right] = Q + Fu, \tag{2.3}$$

where the source terms on the right-hand sides include the body force (gravity) $F = f\varrho$, the internal heat source $Q = q\varrho$, and the work of the body force $Fu = f\varrho u$ (Fig. 2.1).

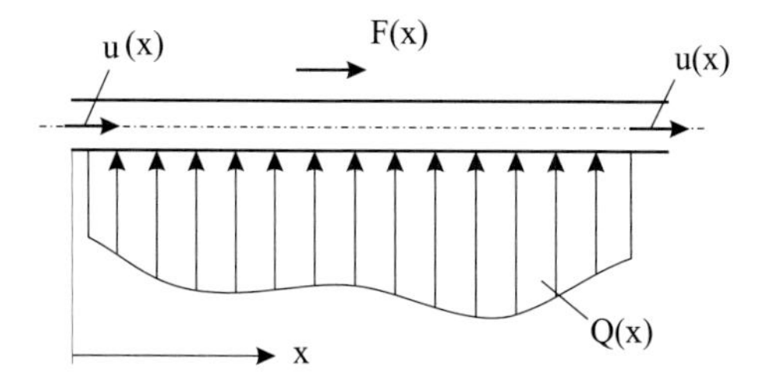

Fig. 2.1: One-dimensional flow in a duct

Gasdynamic Aspects of Two-Phase Flow. Herbert Städtke

ISBN: 3-527-40578-X

Equations (2.1) to (2.3) can also be written in a more compact vector form with the vector of conserved quantities **V**, the Flux vector **F**, and the source term vector **C**,

$$\frac{\partial \mathbf{V}}{\partial t} + \frac{\partial \mathbf{F}}{\partial x} = \mathbf{C}, \tag{2.4}$$

defined as

$$\mathbf{V} = \begin{bmatrix} \varrho \\ \varrho u \\ \varrho\left(e + \frac{u^2}{2}\right) \end{bmatrix}, \qquad \mathbf{F} = \begin{bmatrix} \varrho u \\ \varrho u^2 + p \\ \varrho u\left(h + \frac{u^2}{2}\right) \end{bmatrix}, \qquad \mathbf{C} = \begin{bmatrix} 0 \\ F \\ Q + Fu \end{bmatrix}. \tag{2.5}$$

Expanding the time and space derivatives in equations (2.1) to (2.3) the so-called primitive form of the Euler equations is obtained,

$$\frac{\partial \varrho}{\partial t} + \varrho \frac{\partial u}{\partial x} + u \frac{\partial \varrho}{\partial x} = 0 \tag{2.6}$$

$$\frac{\partial u}{\partial t} + u \frac{\partial u}{\partial x} + \frac{1}{\varrho}\frac{\partial p}{\partial x} = f \tag{2.7}$$

$$\frac{\partial e}{\partial t} + u \frac{\partial e}{\partial x} + u\left(\frac{\partial u}{\partial t} + u \frac{\partial u}{\partial x}\right) + \frac{p}{\varrho}\frac{\partial u}{\partial x} + \frac{u}{\varrho}\frac{\partial p}{\partial x} = q + f\,u. \tag{2.8}$$

With the momentum equation (2.7), the kinetic energy terms can be removed from the energy equation (2.8), resulting in

$$\frac{\partial e}{\partial t} + u \frac{\partial e}{\partial x} + \frac{p}{\varrho}\frac{\partial u}{\partial x} = q. \tag{2.9}$$

Combining equations (2.6) and (2.9), and introducing the entropy as a new state variable

$$T\delta s = \delta e - \frac{p}{\varrho^2}\delta \varrho, \tag{2.10}$$

the energy equation reduces to the simple entropy relation

$$\frac{\partial s}{\partial t} + u \frac{\partial s}{\partial x} = \frac{q}{T}. \tag{2.11}$$

The conservation equations are completed by the state equation $\varrho = \varrho(p, s)$, or in differential form

$$\delta \varrho = \left(\frac{\partial \varrho}{\partial p}\right)_s \delta p + \left(\frac{\partial \varrho}{\partial s}\right)_p \delta s. \tag{2.12}$$

With the sound velocity

$$a = \sqrt{\left(\frac{\partial p}{\partial \varrho}\right)_s}, \tag{2.13}$$

and the derivative of density with respect to entropy at constant pressure

$$\left(\frac{\partial \varrho}{\partial s}\right)_p = -\frac{\varrho \beta T}{C^p}, \tag{2.14}$$

the state equation (2.12) becomes

$$\delta \varrho = \frac{1}{a^2}\,\delta p - \frac{\varrho \beta T}{C^p}\,\delta s. \tag{2.15}$$

The sound velocity as introduced in equation (2.13) can be expressed in the most general form as

$$a = \sqrt{\frac{C^p}{\varrho \gamma C^p - T\beta^2}}, \tag{2.16}$$

with the compressibility γ, the thermal expansion β, and the specific heat at constant pressure C^p. Replacing the thermal expansion β by

$$\beta^2 = \frac{\varrho \gamma (C^p - C^v)}{T},$$

the expression for the sound velocity simplifies to

$$a = \sqrt{\frac{\varkappa}{\gamma \varrho}}, \tag{2.17}$$

with an isentropic exponent of $\varkappa = C^p/C^v$. For an *ideal gas*, the compressibility is given by $\gamma = 1/p$ and, hence, the expression for the sound speed reduces to

$$a_g = \sqrt{\varkappa_g \frac{p}{\varrho_g}}. \tag{2.18}$$

With the help of the state equation (2.15) and the entropy relationship (2.11), the mass conservation equation (2.6) yields

$$\frac{\partial p}{\partial t} + u\frac{\partial p}{\partial x} + \varrho a^2 \frac{\partial u}{\partial x} = \frac{\varrho \beta a^2}{C^p} q. \tag{2.19}$$

The modified mass conservation equation (2.19), the momentum equation (2.7), and the entropy equation (2.11) represent a system of quasi-linear partial differential equations which can be combined in the vector form

$$\frac{\partial \mathbf{U}}{\partial t} + \mathbf{G}\frac{\partial \mathbf{U}}{\partial x} = \mathbf{D}, \tag{2.20}$$

with the state vector $\mathbf{U}$, the coefficient matrix $\mathbf{G}$, and the source term vector $\mathbf{D}$,

$$\mathbf{U} = \begin{bmatrix} p \\ u \\ s \end{bmatrix}, \qquad \mathbf{G} = \begin{bmatrix} u & \varrho a^2 & 0 \\ \frac{1}{\rho} & u & 0 \\ 0 & 0 & u \end{bmatrix}, \qquad \mathbf{D} = \begin{bmatrix} \frac{\varrho \beta a^2}{C^p} q \\ f \\ \frac{q}{T} \end{bmatrix}. \tag{2.21}$$

One might notice from equations (2.21) that, apart from the source terms, the thermal and flow parameters are decoupled.

Depending on the choice of the major depending state vector, $\mathbf{U}$, different forms of the governing equations might be obtained by a similarity transformation of the equation as

$$\frac{\partial \mathbf{U}_1}{\partial t} + \mathbf{G}_1 \frac{\partial \mathbf{U}_1}{\partial x} = \mathbf{D}_1, \tag{2.22}$$

with the Jacobian matrix

$$\mathbf{J} = \frac{\partial \mathbf{U}_1}{\partial \mathbf{U}}, \tag{2.23}$$

and the new coefficient matrix and source term vector

$$\mathbf{G}_1 = \mathbf{J}\,\mathbf{G}\mathbf{J}^{-1}, \quad \mathbf{D}_1 = \mathbf{J}\,\mathbf{D}. \tag{2.24}$$

If for example the enthalpy is used instead of the entropy as major state parameter

$$h = h(s,p), \qquad \text{with} \quad \left(\frac{\partial h}{\partial s}\right)_p = T, \quad \text{and} \quad \left(\frac{\partial h}{\partial p}\right)_s = \frac{1}{\varrho},$$

the Jacobian matrix becomes

$$\mathbf{J} = \begin{bmatrix} 1 & 0 & 0 \\ 0 & 1 & 0 \\ \frac{1}{\varrho} & 0 & T \end{bmatrix}, \tag{2.25}$$

resulting in the state vector, coefficient matrix, and source term vector as

$$\mathbf{U}_1 = \begin{bmatrix} p \\ u \\ h \end{bmatrix}, \quad \mathbf{G}_1 = \begin{bmatrix} u & \varrho a^2 & 0 \\ \frac{1}{\varrho} & u & 0 \\ 0 & a^2 & u \end{bmatrix}, \quad \mathbf{D}_1 = \begin{bmatrix} \frac{\varrho \beta a^2}{C^p} q \\ f \\ \left(1 + \frac{a^2 \beta}{C^p}\right) q \end{bmatrix}. \tag{2.26}$$

For the specific case of *stationary flow*

$$\frac{\partial \mathbf{U}}{\partial t} = 0,$$

the spatial gradients become

$$\frac{\partial \mathbf{U}}{\partial x} = \mathbf{G}^{-1}\mathbf{D} \tag{2.27}$$

or more specifically

$$\left.\begin{aligned}\frac{\partial p}{\partial x} &= \frac{1}{1-M^2}\left[f\varrho - \frac{\varrho\beta}{C^p}qu\right]\\ \frac{\partial u}{\partial x} &= \frac{1}{1-M^2}\left[-f\frac{u}{a^2} + \frac{\beta}{C^p}q\right]\\ \frac{\partial s}{\partial x} &= \frac{q}{Tu}\end{aligned}\right\} . \tag{2.28}$$

From equation (2.28) it follows that the effect of heat transfer on pressure and velocity gradients linearly depends on the thermal expansion coefficient β. For the incompressible limit, with $\gamma \to 0$ and $\beta \to 0$, the Mach number approaches zero $M \to 0$ and the stationary flow in a channel of constant cross section further simplifies to

$$\frac{\partial p}{\partial x} = f\varrho, \quad \frac{\partial u}{\partial x} = 0, \quad \frac{\partial s}{\partial x} = \frac{q}{Tu}, \tag{2.29}$$

which means a constant flow velocity where the pressure and entropy (temperature) gradients are determined by body forces and external heat sources.

2.2 Quasi-one-dimensional flow in ducts of variable cross section

If the flow equations are integrated over the cross section A of a pipe or nozzle as indicated in Fig. 2.2, the following balance equations are obtained:

$$A\frac{\partial \varrho}{\partial t} + \frac{\partial}{\partial x}(\varrho u A) = 0 \tag{2.30}$$

$$A\frac{\partial}{\partial t}(\varrho u) + \frac{\partial}{\partial x}(\varrho u^2 A) + A\frac{\partial p}{\partial x} = FA \tag{2.31}$$

$$A\frac{\partial}{\partial t}\left[\varrho\left(e + \frac{u^2}{2}\right)\right] + \frac{\partial}{\partial x}\left[\varrho u A\left(e + \frac{p}{\varrho} + \frac{u^2}{2}\right)\right] = QA + FA \tag{2.32}$$

$$A\frac{\partial}{\partial t}(\varrho s) + \frac{\partial}{\partial x}(\varrho u s A) = \frac{Q}{T}A. \tag{2.33}$$

Expanding the derivative terms for the products and moving all source terms to the right-hand side of the equations results in

$$\frac{\partial \varrho}{\partial t} + \varrho\frac{\partial u}{\partial x} + u\frac{\partial \varrho}{\partial x} = -\varrho u\frac{1}{A}\frac{\partial A}{\partial x} \tag{2.34}$$

$$\frac{\partial u}{\partial t} + u\frac{\partial u}{\partial x} + \frac{1}{\varrho}\frac{\partial p}{\partial x} = f \tag{2.35}$$

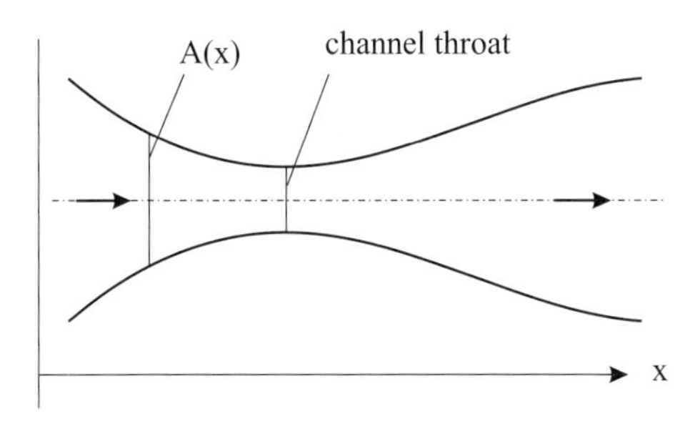

Fig. 2.2: Quasi-one-dimensional flow in a channel of variable cross section

$$\frac{\partial e}{\partial t} + u\frac{\partial e}{\partial x} + u\left(\frac{\partial u}{\partial t} + u\frac{\partial u}{\partial x}\right) + \frac{p}{\varrho}\frac{\partial u}{\partial x} + \frac{u}{\varrho}\frac{\partial p}{\partial x} = q + f\,u \tag{2.36}$$

$$\frac{\partial s}{\partial t} + u\frac{\partial s}{\partial x} = \frac{q}{T} \tag{2.37}$$

or in a compact vector form

$$\frac{\partial \mathbf{U}}{\partial t} + \mathbf{G}\frac{\partial \mathbf{U}}{\partial x} = \mathbf{D}. \tag{2.38}$$

In the case when the entropy is used as a governing state parameter, the state vector $\mathbf{U}$, the coefficient matrix $\mathbf{G}$, and the source term vector $\mathbf{D}$ are defined as

$$\mathbf{U} = \begin{bmatrix} p \\ u \\ s \end{bmatrix}, \quad \mathbf{G} = \begin{bmatrix} u & \varrho a^2 & 0 \\ \dfrac{1}{\rho} & u & 0 \\ 0 & 0 & u \end{bmatrix}, \quad \mathbf{D} = \begin{bmatrix} \dfrac{\varrho\beta a^2}{C^p}q - \dfrac{a^2\varrho u}{A}\dfrac{\partial A}{\partial x} \\ f \\ \dfrac{q}{T} \end{bmatrix}. \tag{2.39}$$

As indicated in equations (2.34) to (2.37), the only difference to the strictly one-dimensional flow is the occurrence of an additional source term related to the change of cross section in the flow direction.

For steady-state conditions the change of the state parameters in the x-direction is

$$\frac{\partial \mathbf{U}}{\partial x} = \mathbf{G}^{-1}\mathbf{D} \tag{2.40}$$

or explicitly written as

$$\left.\begin{aligned} \frac{\partial p}{\partial x} &= \frac{\varrho u^2}{1-M^2}\left[\frac{1}{A}\frac{\partial A}{\partial x} + \frac{f}{u^2} - \frac{\beta}{C^p u}q\right] \\ \frac{\partial u}{\partial x} &= \frac{-u}{1-M^2}\left[\frac{1}{A}\frac{\partial A}{\partial x} + \frac{f}{a^2} - \frac{\beta}{C^p u}q\right] \\ \frac{\partial s}{\partial x} &= \frac{q}{Tu} \end{aligned}\right\}. \tag{2.41}$$

For appropriate boundary conditions at the nozzle inlet and outlet, the system of coupled ordinary differential equations (2.41) can be solved by any standard integration techniques e.g., first or higher order Runge–Kutta method. A particular property of the system of equations is the existence of a saddle–point singularity under "critical flow" conditions for $u = a$ or $M = 1$, which is expected to occur for sufficiently low pressure values at the nozzle exit. The numerical integration is further hampered by the fact that for the general case of $f \neq 0$ or $q \neq 0$, the location of the singularity is *a priori* not known. Similar difficulties also exist for the nonequilibrium flow of gas mixtures exposed to chemical reactions or dissociation processes. A general numerical method of how to integrate through a saddle-point singularity is given by Emmanuel in [3].

2.3 Characteristic analysis of flow equations

The characteristic analysis will be based on the one-dimensional Euler equations using entropy as a major state parameter:

$$\frac{\partial \mathbf{U}}{\partial t} + \mathbf{G}\frac{\partial \mathbf{U}}{\partial x} = \mathbf{D}, \tag{2.42}$$

with the state vector $\mathbf{U}$, the coefficient matrix $\mathbf{G}$, and the source term vector $\mathbf{D}$,

$$\mathbf{U} = \begin{bmatrix} p \\ u \\ s \end{bmatrix}, \qquad \mathbf{G} = \begin{bmatrix} u & \varrho a^2 & 0 \\ \frac{1}{\rho} & u & 0 \\ 0 & 0 & u \end{bmatrix}, \qquad \mathbf{D} = \begin{bmatrix} \frac{\varrho \beta a^2}{C_p} q \\ f \\ \frac{q}{T} \end{bmatrix}. \tag{2.43}$$

The eigenvalues of $\mathbf{G}$ are the roots of the characteristic equation

$$\det(\mathbf{G} - \lambda \mathbf{I}) = \mathbf{0}, \tag{2.44}$$

which results in the characteristic velocities

$$\lambda_1 = u + a, \qquad \lambda_2 = u - a, \qquad \lambda_3 = u, \tag{2.45}$$

representing the propagation velocity of pressure waves $u \pm a$ and the material transport velocity u.

For the three different eigenvalues, a complete set of independent right eigenvectors can be obtained as

$$V_{R,1} = \begin{bmatrix} \frac{1}{2} \\ \frac{1}{2a\varrho} \\ 0 \end{bmatrix}, \qquad V_{R,2} = \begin{bmatrix} \frac{1}{2} \\ -\frac{1}{2a\varrho} \\ 0 \end{bmatrix}, \qquad V_{R,3} = \begin{bmatrix} 0 \\ 0 \\ 1 \end{bmatrix}, \tag{2.46}$$

or in the matrix form

$$\mathbf{V}_R = \begin{bmatrix} \frac{1}{2} & \frac{1}{2a\varrho} & 0 \\ \frac{1}{2} & -\frac{1}{2a\varrho} & 0 \\ 0 & 0 & 1 \end{bmatrix}. \tag{2.47}$$

Only real eigenvalues and a set of fully independent eigenvectors are the essential conditions for the existence of a hyperbolic system of equations, which represent a mathematically "well-posed" initial-boundary value problem. Since the system of equations (2.49) is hyperbolic, it can be diagonalized by the following similarity transformation:

$$\mathbf{T}^{-1}\frac{\partial \mathbf{U}}{\partial t} + \left(\mathbf{T}^{-1}\mathbf{G}\,\mathbf{T}\right)\mathbf{T}^{-1}\frac{\partial \mathbf{U}}{\partial x} = \mathbf{T}^{-1}\mathbf{D} = \mathbf{E}, \tag{2.48}$$

or

$$\mathbf{T}^{-1}\frac{\partial \mathbf{U}}{\partial t} + \mathbf{\Lambda}\,\mathbf{T}^{-1}\frac{\partial \mathbf{U}}{\partial x} = \mathbf{T}^{-1}\mathbf{D} = \mathbf{E}, \tag{2.49}$$

where

$$\mathbf{\Lambda} = \mathbf{T}^{-1}\mathbf{G}\,\mathbf{T} \tag{2.50}$$

is the diagonal matrix of the eigenvalues of the matrix $\mathbf{G}$,

$$\mathbf{\Lambda} = \begin{bmatrix} u+a & 0 & 0 \\ 0 & u-a & 0 \\ 0 & 0 & u \end{bmatrix}. \tag{2.51}$$

The column vectors of the transformation matrix $\mathbf{T}$ are the right eigenvectors of the coefficient matrix $\mathbf{G}$ given in equation (2.43),

$$\mathbf{T} = \mathbf{V}_R^T = \begin{bmatrix} \frac{1}{2} & \frac{1}{2} & 0 \\ \frac{1}{2a\varrho} & \frac{-1}{2a\varrho} & 0 \\ 0 & 0 & 1 \end{bmatrix}, \tag{2.52}$$

with the inverse representing the matrix of the left eigenvectors of $\mathbf{G}$,

$$\mathbf{T}^{-1} = \mathbf{V}_L = \begin{bmatrix} 1 & a\varrho & 0 \\ 1 & -a\varrho & 0 \\ 0 & 0 & 1 \end{bmatrix}. \tag{2.53}$$

The new source term vector in equation (2.54) is defined as

$$\mathbf{E} = \begin{bmatrix} +aF - a^2 \left(\dfrac{\partial \varrho}{\partial s}\right)_p \dfrac{q}{T} \\ \\ -aF - a^2 \left(\dfrac{\partial \varrho}{\partial s}\right)_p \dfrac{q}{T} \\ \\ \dfrac{q}{T} \end{bmatrix} . \tag{2.54}$$

Equation (2.49) represents a system of ordinary differential equations also known as the compatibility relations along the "characteristic" directions $dx/dt = \lambda_k$ in the x–t plane (Fig. 2.3)

$$\left.\begin{aligned} &\lambda_1 = u + a: && \frac{dp}{dt} + \varrho a \, \frac{du}{dt} = \left[+a\varrho f - a^2 \left(\frac{\partial \varrho}{\partial s}\right)_p \frac{q}{T} \right] \\ &\lambda_2 = u - a: && \frac{dp}{dt} - \varrho a \, \frac{du}{dt} = \left[-a\varrho f - a^2 \left(\frac{\partial \varrho}{\partial s}\right)_p \frac{q}{T} \right] \\ &\lambda_3 = u: && \frac{ds}{dt} = \frac{q}{T} \end{aligned}\right\} . \tag{2.55}$$

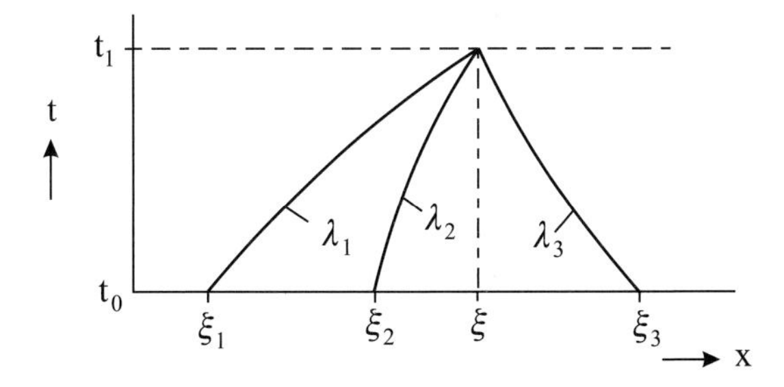

Fig. 2.3: Characteristic curves in the x–t plane

The system of equations (2.55) can be integrated along the characteristic lines from t_0 to $t_1 = t_0 + \Delta t$,

$$\left.\begin{aligned} &p(t_1, \xi) + \int_{u(t_0,\xi_1)}^{u(t_1,\xi)} \varrho a \, du = p(t_0, \xi_1) + \int_{t_0}^{t_1} \left[+a\varrho f - a^2 \left(\frac{\partial \varrho}{\partial s}\right)_p \frac{q}{T} \right] dt \\ &p(t_1, \xi) - \int_{u(t_0,\xi_3)}^{u(t_1,\xi)} \varrho a \, du = p(t_0, \xi_3) + \int_{t_0}^{t_1} \left[-a\varrho f - a^2 \left(\frac{\partial \varrho}{\partial s}\right)_p \frac{q}{T} \right] dt \\ &s(t_1, \xi) = s(t_0, \xi_2) + \int_{t_0}^{t_1} \frac{q}{T} dt. \end{aligned}\right\} . \tag{2.56}$$

Note that the "Riemann invariants" on the left-hand side of equations (2.56),

$$\mathbf{W} = \begin{bmatrix} p + \int \varrho a \, du \\ p - \int \varrho a \, du \\ s \end{bmatrix}, \tag{2.57}$$

remain constant in the case of zero source terms.

For adiabatic conditions ($q = 0$) the flow becomes isentropic $s = s_0$ and, with the assumption of an ideal state equation, the Riemann invariants simplify to

$$\mathbf{W} = \begin{bmatrix} u + \dfrac{2}{\varkappa - 1} a \\ u - \dfrac{2}{\varkappa - 1} a \\ s \end{bmatrix}, \tag{2.58}$$

and the compatibility relations (2.56) can be directly integrated resulting in the algebraic expressions

$$\left.\begin{array}{ll} \lambda_1 = u + a: & u + \dfrac{2}{\varkappa - 1} a = \text{constant} \\ \lambda_2 = u - a: & u - \dfrac{2}{\varkappa - 1} a = \text{constant} \\ \lambda_3 = u: & s = \text{constant}. \end{array}\right\}, \tag{2.59}$$

with the sound velocity a as a new state parameter.

The system of coupled ordinary differential equations (2.56) or (2.59) forms the basis of the "method of characteristics" which was widely used in the past for the solution of gas-dynamic problems. However, the integration of these equations is far from straightforward, in particular in the case of complex state equations. Furthermore, the method is difficult to implement into computer programs for two- and three-dimensional flow problems. For this reason characteristic methods have been largely replaced by finite difference or finite volume techniques.

Many numerical methods which make explicit use of the characteristic features of the flow equations are based on splitting of the coefficient matrix with respect to the individual eigenvalues. According to equation (2.50) the coefficient matrix of the one-dimensional Euler equations,

$$\frac{\partial \mathbf{U}}{\partial t} + \mathbf{G} \frac{\partial \mathbf{U}}{\partial x} = \mathbf{D}, \tag{2.60}$$

can be expressed as

$$\mathbf{G} = \mathbf{T} \, \mathbf{\Lambda} \, \mathbf{T}^{-1}. \tag{2.61}$$

with the transformation matrix $\mathbf{T}$ and the diagonal matrix of eigenvalues $\mathbf{\Lambda}$ as defined by equations (2.52) and (2.51). The split matrices are then obtained as

$$\mathbf{G}_k = \mathbf{T}\,\mathbf{\Lambda}_k\,\mathbf{T}^{-1}, \tag{2.62}$$

where the diagonal matrix $\mathbf{\Lambda}_k$ includes only the kth eigenvalue,

$$\left.\begin{array}{lll} (\Lambda_k)_{j,j} = \lambda_k & \text{for} & j = k \\ (\Lambda_k)_{j,j} = 0 & \text{for} & j \neq k \end{array}\right\}. \tag{2.63}$$

For the state vector $\mathbf{U} = \{p, u, s\}^T$ the split matrices become

$$\mathbf{G}_1 = (u+a)\begin{bmatrix} \frac{1}{2} & \frac{\varrho a}{2} & 0 \\ \frac{1}{2\varrho a} & \frac{1}{2} & 0 \\ 0 & 0 & 0 \end{bmatrix} = \lambda_1 \tilde{\mathbf{G}}_1 \tag{2.64}$$

$$\mathbf{G}_2 = (u-a)\begin{bmatrix} \frac{1}{2} & -\frac{\varrho a}{2} & 0 \\ -\frac{1}{2\varrho a} & \frac{1}{2} & 0 \\ 0 & 0 & 0 \end{bmatrix} = \lambda_2 \tilde{\mathbf{G}}_2 \tag{2.65}$$

$$\mathbf{G}_3 = u\begin{bmatrix} 0 & 0 & 0 \\ 0 & 0 & 0 \\ 0 & 0 & 1 \end{bmatrix} = \lambda_3 \tilde{\mathbf{G}}_3, \tag{2.66}$$

with the conditions

$$\sum_{k=1}^{3} \mathbf{G}_k = \mathbf{G} \quad \text{and} \quad \sum_{k=1}^{3} \tilde{\mathbf{G}}_k = \mathbf{I}. \tag{2.67}$$

For the numerical treatment of flows with embedded discontinuities like shock waves, finite volume numerical schemes are often preferred based on the conservative form as can be directly obtained from the general balance equations for mass, momentum, and energy as given by equations (2.1) to (2.4). However, the conservative form of the Euler equations can also be derived from the "primitive" form of equations (2.42)

$$\frac{\partial \mathbf{U}}{\partial t} + \mathbf{G}\frac{\partial \mathbf{U}}{\partial x} = \mathbf{D} \tag{2.68}$$

by a similarity transformation as

$$\frac{\partial \mathbf{V}}{\partial t} + \mathbf{J}\,\mathbf{G}\,\mathbf{J}^{-1}\frac{\partial \mathbf{V}}{\partial x} = \mathbf{J}\,\mathbf{D} = \mathbf{C} \tag{2.69}$$

or

$$\frac{\partial \mathbf{V}}{\partial t} + \mathbf{H}\frac{\partial \mathbf{V}}{\partial x} = \mathbf{C}, \tag{2.70}$$

with the new coefficient matrix

$$\mathbf{H} = \mathbf{J}\,\mathbf{G}\,\mathbf{J}^{-1}, \tag{2.71}$$

and the Jacobian matrix

$$\mathbf{J} = \frac{\Delta \mathbf{V}}{\Delta \mathbf{U}}. \tag{2.72}$$

With the definition of the "primitive" and the conservative state vectors $\mathbf{U}$ and $\mathbf{V}$ as given in equation (2.5) the Jacobian matrix is

$$\mathbf{J} = \frac{\partial \mathbf{V}}{\partial \mathbf{U}} = \begin{bmatrix} \dfrac{1}{a^2} & 0 & \left(\dfrac{\partial \varrho}{\partial s}\right)_p \\ \dfrac{u}{a^2} & \varrho & u\left(\dfrac{\partial \varrho}{\partial s}\right)_p \\ \dfrac{1}{a^2}\left(h + \dfrac{u^2}{2}\right) & \varrho u & \varrho T + \left(\dfrac{\partial \varrho}{\partial s}\right)_p \left(h + \dfrac{u^2}{2}\right) \end{bmatrix} \tag{2.73}$$

and the coefficient matrix $\mathbf{H}$ for the conserved state parameters becomes

$$\mathbf{H} = \begin{bmatrix} 0 & 1 & 0 \\ a^2 - u^2 & 2u & 0 \\ ua^2 - u\left(h + \dfrac{u^2}{2}\right) & \left(h + \dfrac{u^2}{2}\right) & u \end{bmatrix} + \left(\frac{\partial \varrho}{\partial s}\right)_p \frac{a^2}{T\varrho} \begin{bmatrix} 0 & 0 & 0 \\ \left(h - \dfrac{u^2}{2}\right) & u & -1 \\ u\left(h - \dfrac{u^2}{2}\right) & u^2 & -u \end{bmatrix}. \tag{2.74}$$

The thermodynamic parameters in equation (2.74) include, apart from the enthalpy h, the sound velocity and the partial derivatives of the density with respect to pressure which are defined as

$$a = \sqrt{\frac{\varkappa}{\gamma \varrho}}, \quad \left(\frac{\partial \varrho}{\partial s}\right)_p = -\frac{\beta \varrho T}{C_P}. \tag{2.75}$$

With the assumption of state equations for a perfect gas these parameters simplify to

$$a = \sqrt{\varkappa R T}, \quad \left(\frac{\partial \varrho}{\partial s}\right)_p = -\frac{\varrho}{C_P}, \tag{2.76}$$

resulting in the following form for the coefficient matrix as often found in the literature:

$$\mathbf{H} = \begin{bmatrix} 0 & 1 & 0 \\ a^2 - u^2 & 2u & 0 \\ ua^2 - u\left(h + \frac{u^2}{2}\right) & \left(h + \frac{u^2}{2}\right) & u \end{bmatrix} \\ -(\varkappa - 1)\begin{bmatrix} 0 & 0 & 0 \\ \left(h - \frac{u^2}{2}\right) & u & -1 \\ u\left(h - \frac{u^2}{2}\right) & u^2 & -u \end{bmatrix}. \tag{2.77}$$

Introducing the gradient of the flux vector, the conservative form of the flow equations (2.70) can also be written as

$$\frac{\partial \mathbf{V}}{\partial t} + \frac{\partial \mathbf{F}}{\partial x} + \left(\mathbf{H} - \frac{\partial \mathbf{F}}{\partial \mathbf{V}}\right)\frac{\partial \mathbf{V}}{\partial x} = \mathbf{C}. \tag{2.78}$$

Comparing equation (2.78) with the conservative form of the flow equations as was derived from the basic conservation principles for mass, momentum, and energy,

$$\frac{\partial \mathbf{V}}{\partial t} + \frac{\partial \mathbf{F}}{\partial x} = \mathbf{C}, \tag{2.79}$$

one obtains

$$\mathbf{H} = \frac{\partial \mathbf{F}}{\partial \mathbf{V}}. \tag{2.80}$$

This means that the coefficient matrix $\mathbf{H}$ is identical with the Jacobian matrix describing the derivative of the flux vector with respect to the vector of conserved variables. As will be shown later this is not necessarily true for the nonhomogeneous two-phase flow equations.

A remarkable property of the Euler equations (2.70) is that for ideal state equations the *homogeneity property* can be derived as

$$\mathbf{F} = \mathbf{H}\,\mathbf{V}, \tag{2.81}$$

which forms the basis for the original form of the Flux Vector Splitting technique of Steger and Warming [4] as will be described in Chapter 7.

2.4 Shock waves

A property of hyperbolic flow equations is the possibility for the existence of discontinuous solutions such as shock waves or contact discontinuities which might occur for specific flow and boundary conditions. For the one-dimensional case as schematically shown in Fig. 2.4, the parameter changes across the shock can be calculated from the general conservation equations.

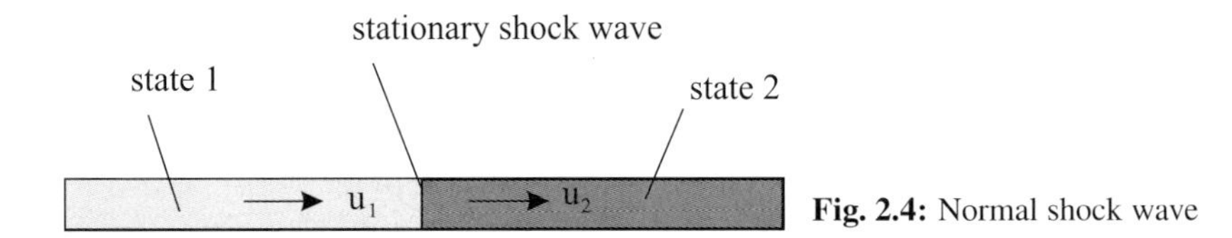

Fig. 2.4: Normal shock wave

Assuming constant steady state conditions in front and behind the shock wave, and neglecting viscosity and heat conduction effects, the mass, momentum, and energy equations (2.1) to (2.3) simplify to the jump relations

mass:

$$\varrho_1 u_1 = \varrho_2 u_2 \tag{2.82}$$

momentum:

$$\varrho_1 u_1^2 + p_1 = \varrho_2 u_2^2 + p_2 \tag{2.83}$$

energy:

$$\frac{1}{2}u_1^2 + h_1 = \frac{1}{2}u_2^2 + h_2. \tag{2.84}$$

From equations (2.82) and (2.83) the velocities u_1 and u_2 can be eliminated resulting in the Rankine–Hugoniot relation

$$h_2 - h_1 = \frac{1}{2}(p_2 - p_1)\left(\frac{1}{\varrho_1} + \frac{1}{\varrho_2}\right), \tag{2.85}$$

which includes only thermodynamic state parameters. Assuming ideal gas laws with a constant isentropic exponent $\varkappa$, the Rankine–Hugoniot relation (2.85) simplifies to

$$\frac{p_2}{p_1} = \frac{(\varkappa+1)\,\varrho_2/\varrho_1 - (\varkappa-1)}{(\varkappa+1) - (\varkappa-1)\,\varrho_2/\varrho_1} \tag{2.86}$$

or evaluated for the density ratio

$$\frac{\varrho_2}{\varrho_1} = \frac{(\varkappa+1)\, p_2/p_1 + (\varkappa-1)}{(\varkappa-1)\, p_2/p_1 + (\varkappa+1)}. \tag{2.87}$$

For weak shock waves with $p_2/p_1 \to 1$, the Rankine–Hugoniot equation (2.87) approaches the isentropic relation

$$\frac{\varrho_2}{\varrho_1} = \left(\frac{p_2}{p_1}\right)^{1/\varkappa} \tag{2.88}$$

as shown in Fig. 2.5.

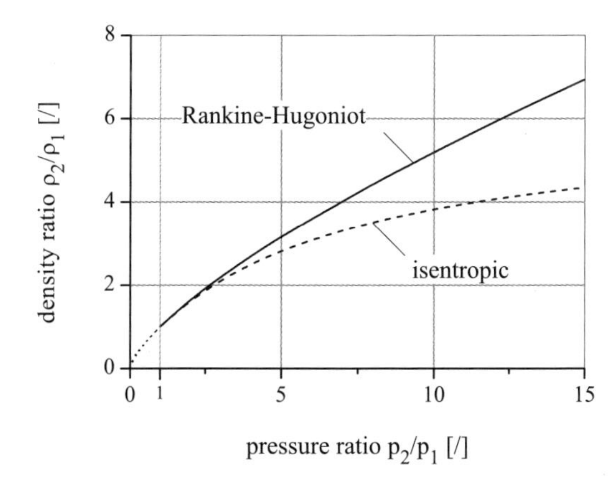

Fig. 2.5: Density ratio across the shock wave

Combining the momentum equation (2.83) and the Rankine–Hugoniot relation (2.85) a rather simple equation is obtained for the pressure ratio across the shock wave as a function of the Mach number upstream of the shock,

$$\frac{p_2}{p_1} = 1 + \frac{\varkappa+1}{2\varkappa}\left(M_1^2 - 1\right). \tag{2.89}$$

If all state parameters in front of the shock wave are known, the corresponding values on the downstream side can be calculated from equations (2.87), (2.89), and the state equations for ideal gases. This includes in particular the entropy rise across the shock,

$$\frac{s_2 - s_1}{C^v} = \ln\left[\frac{p_2}{p_1}\left(\frac{p_2/p_1\,(\varkappa-1) + (\varkappa+1)}{p_2/p_1\,(\varkappa+1) + (\varkappa-1)}\right)^{\varkappa}\right], \tag{2.90}$$

and finally the Mach number behind the shock,

$$M_2 = \frac{(\varkappa+1) + (\varkappa-1)\left(M_1^2 - 1\right)}{(\varkappa+1) + 2\varkappa\left(M_1^2 - 1\right)}. \tag{2.91}$$

The increase in the entropy indicating the dissipation of the kinetic energy across the shock wave is strongly dependent on the strength of the shock wave as shown in Fig. 2.6. From equations (2.90) and (2.91) it follows that only compression shocks ($p_2 > p_1$) are physically feasible ($s_2 > s_1$) where the flow changes from supersonic ($M_1 > 1$) to subsonic ($M_1 < 1$) conditions.

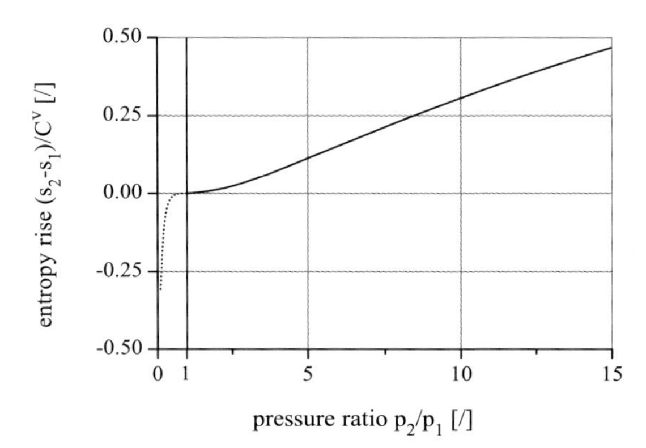

Fig. 2.6: Entropy rise across a plane shock wave

The relations for a shock wave propagating into a gas at rest as schematically shown in Fig. 2.7 can be derived from jump conditions (2.82) to (2.84) introducing a new reference system moving with constant velocity $-u_1$. For the velocity and Mach number of the shock wave one obtains

$$u_{\text{sw}} = a_1 \sqrt{1 + \frac{\varkappa + 1}{2\varkappa}\left(\frac{p_2}{p_1} - 1\right)} \tag{2.92}$$

and

$$M_{\text{sw}} = \frac{u_{\text{sw}}}{a_1} = \sqrt{1 + \frac{\varkappa + 1}{2\varkappa}\left(\frac{p_2}{p_1} - 1\right)}, \tag{2.93}$$

respectively, where the Mach number is related to the sound velocity in the undisturbed region in front of the shock wave. As can be seen from equations (2.92) and (2.93) the propagation velocity of the shock is directly related to its strength expressed by the pressure ratio p_2/p_1, with the limiting case of the sound velocity $u_{\text{sw}} = a_1$ for weak shocks $(p_2/p_1) \to 1$.

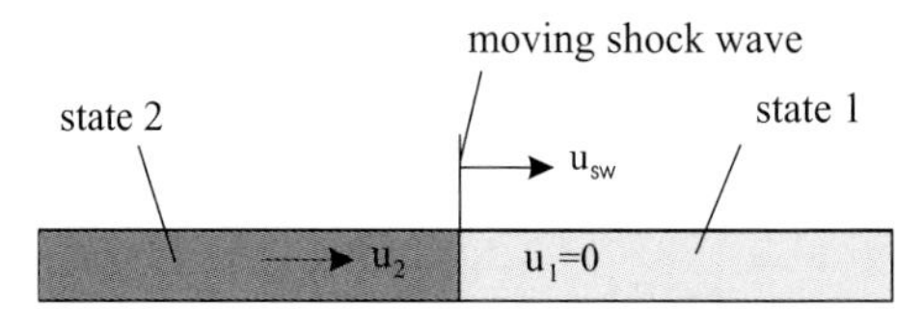

Fig. 2.7: Moving shock wave

Of further interest might be the velocity and Mach number behind the shock which can be derived as

$$u_2 = a_1 \left(\frac{p_2}{p_1} - 1 \right) \sqrt{\frac{2/\varkappa}{\frac{p_2}{p_1}(\varkappa + 1) + (\varkappa - 1)}}, \tag{2.94}$$

$$M_2 = \frac{u_2}{a_2} = \left(\frac{p_2}{p_1} - 1 \right) \sqrt{\frac{2/\varkappa}{\frac{p_2}{p_1}\left[\frac{p_2}{p_1}(\varkappa - 1) + (\varkappa + 1)\right]}}. \tag{2.95}$$

The Mach number for the shock wave propagation and the Mach number behind the moving shock are shown in Fig. 2.8. The figure indicates that for strong shock waves, supersonic flow conditions might occur behind the moving shock, however, due to the large dissipation of the kinetic energy across the shock wave, the maximum possible Mach number is limited. For an isentropic exponent of $\varkappa = 1.4$ one obtains for $p_2/p_1 \rightarrow \infty$ the maximum Mach number $M_2^{\max} = 1.889$.

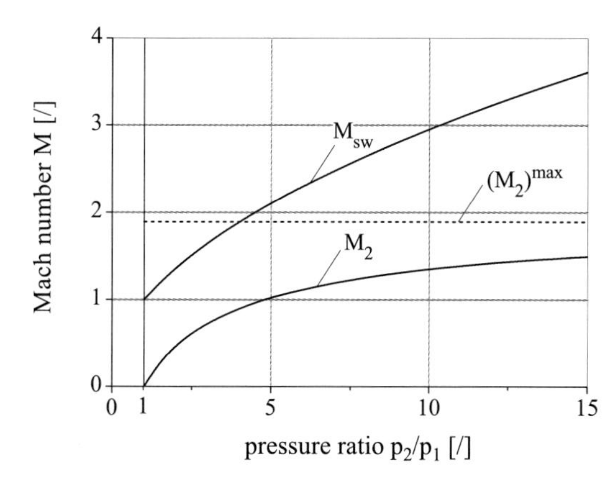

Fig. 2.8: Mach number of the moving shock wave and the Mach number behind the shock

2.5 Flow through convergent–divergent nozzles

Assuming a steady state, quasi-one-dimensional flow through a convergent–divergent nozzle, equations (2.30) to (2.33) simplify to

$$\frac{\partial}{\partial x}(\varrho u A) = 0 \tag{2.96}$$

$$\frac{\partial}{\partial x}(\varrho u^2 A) + A\frac{\partial p}{\partial x} = FA \tag{2.97}$$

$$\frac{\partial}{\partial x}\left[\varrho u A\left(h+\frac{u^2}{2}\right)\right]=QA+FA \tag{2.98}$$

$$\frac{\partial}{\partial x}(\varrho u s A)=\frac{Q}{T}A. \tag{2.99}$$

In the case of absence of external forces ($F=0$) and heat sources ($Q=0$), the mass and energy conservation equations can be immediately integrated,

$$\varrho u A=\dot{m}=\text{const} \tag{2.100}$$

$$h+\frac{u^2}{2}=h^{\text{tot}}=h_0. \tag{2.101}$$

The entropy balance equation (2.99) reduces to the isentropic condition

$$s=s_0 \tag{2.102}$$

as long as the flow is free of discontinuities.

Depending on the exit pressure value, four different flow situations can be distinguished as shown in Fig. 2.9.

(a) As long as the exit pressure is below a threshold value p_1 the flow in the nozzle remains subsonic ($M<1$) with a minimum pressure at the nozzle throat above the critical pressure value.

(b) For an exit pressure of $p_{\text{exit}}=p_1$ the sonic line is reached at the throat ($M=1$), characterized by a saddle-point singularity. Downstream of the throat the flow returns to subsonic conditions with a recovery of the pressure up to the exit values p_1.

(c) For pressure values $p_{\text{exit}}<p_1$, the flow accelerates continuously to supersonic conditions and, depending on the actual values of p_1, a shock wave is formed at the divergent part of the nozzle where the flow downstream of the shock returns to subsonic conditions.

(d) With the further decrease of the exit pressure the position of the shock waves moves further downstream and for $p_{\text{exit}}=p_3$, the shock has reached the exit plane. Any further decrease of the exit pressure then has no effect any longer on the nozzle flow.

For the case of ideal state equations the critical conditions ($M=1$) can be immediately given as a function of the upstream reservoir condition as

$$\left.\begin{aligned}
T_{\text{cr}}&=T_0\left(\frac{2}{\varkappa+1}\right)\\
p_{\text{cr}}&=p_0\left(\frac{2}{\varkappa+1}\right)^{\varkappa/(\varkappa-1)}\\
\varrho_{\text{cr}}&=\varrho_0\left(\frac{2}{\varkappa+1}\right)^{1/(\varkappa-1)}
\end{aligned}\right\}. \tag{2.103}$$

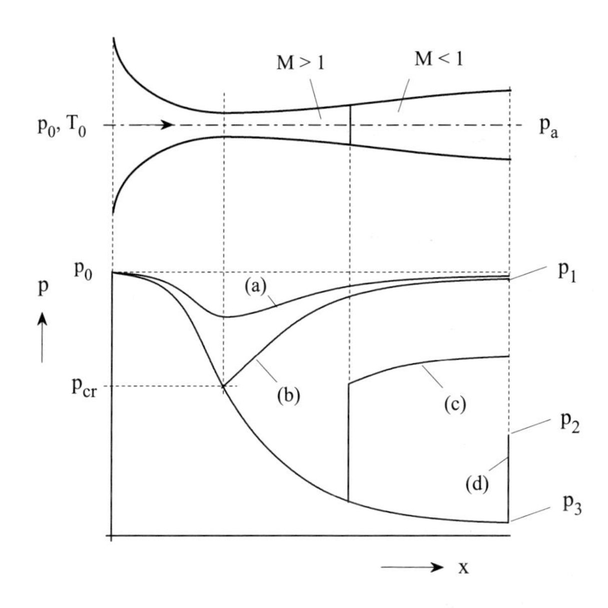

Fig. 2.9: Flow of single-phase gas through a convergent–divergent nozzle; effect of back pressure

In addition, an explicit relation between the nozzle area ratio A/A_{th} and the Mach number M can be derived as

$$A/A_{\mathrm{th}} = \frac{1}{M^2}\left[\frac{2}{\varkappa+1}\left(1+\frac{\varkappa-1}{2}M^2\right)\right]^{\frac{\varkappa+1}{\varkappa-1}}, \tag{2.104}$$

with the condition of $A/A_{\mathrm{th}} \to 1$ for $M \to 1$.

As long as the flow is isentropic (adiabatic and free of discontinuities) all flow parameters are exclusive functions of the area ratio A/A_{th} and can be iteratively calculated on the basis of the steady state flow equations (2.100) to (2.102). In the case of shock waves two isentropic regions in front of and behind the shock have to be linked by the shock relations (2.82) to (2.84). Results for pressure and Mach number as a function of the area ratio A/A_{th} are shown in Figs. 2.10 and 2.11 assuming ideal state equations and an isentropic exponent of $\varkappa = 1.4$.

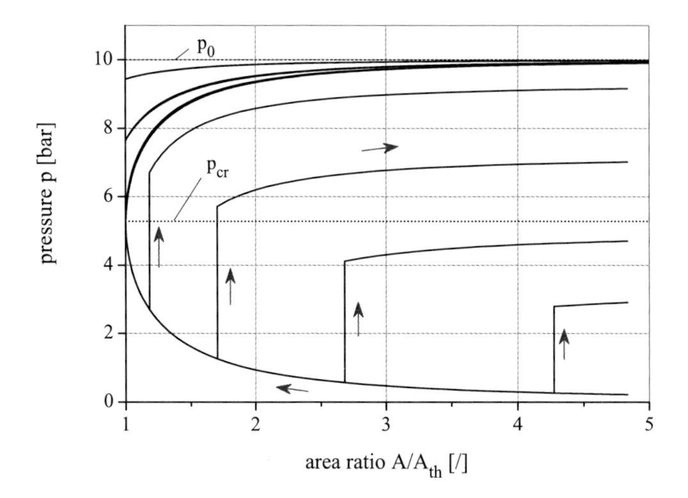

Fig. 2.10: Nozzle flow of single-phase gas, pressure as a function of the area ratio A/A_{th}, exit pressure: 9.98 bar $\geq p_{\mathrm{exit}} \geq 0.22$ bar

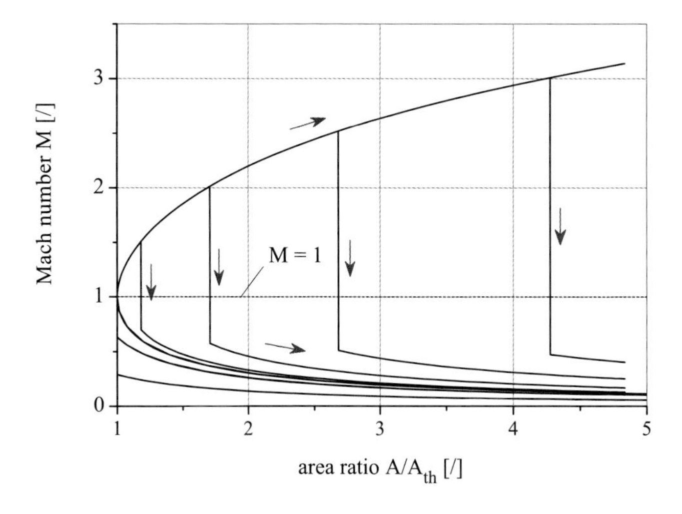

Fig. 2.11: Nozzle flow of single-phase gas, Mach number as a function of the area ratio A/A_{th}, exit pressure: 9.98 bar $\geq p_{\mathrm{exit}} \geq 0.22$ bar

A more detailed quantitative picture for the gas flow through a convergent–divergent nozzle is given in Fig. 2.12, including pressure, Mach number, temperature, and entropy distributions along the nozzle axis. The nozzle contour $A \sim x^2$ is chosen to demonstrate the distribution of flow parameters in axial direction.

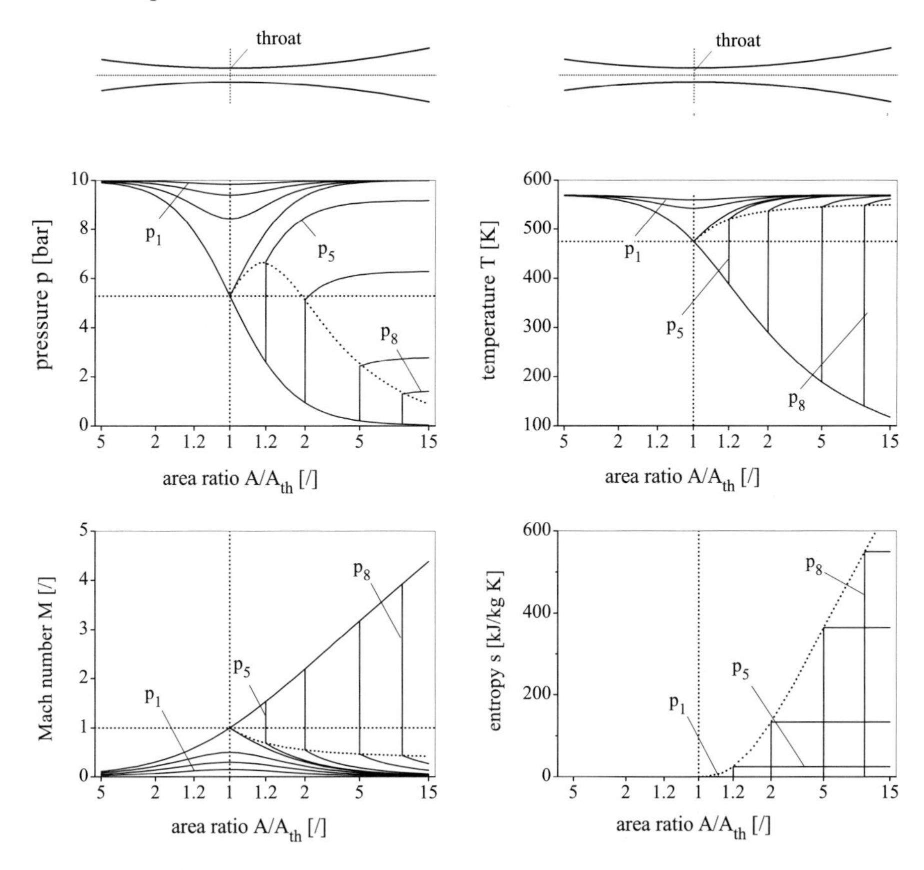

Fig. 2.12: Flow of single-phase gas through a convergent–divergent nozzle, reservoir pressure $p_0 = 10$ bar, exit pressure: $p_1 \geq p_{\rm exit} \geq p_9$, with $p_1 = 9.999$ bar, $p_5 = 9.175$ bar, $p_9 = 0.040$ bar

2.6 Shock tube

Shock tube devices have been extensively used to study shock wave propagation phenomena in compressible fluids like gases or gas–liquid two-phase mixtures. Usually a high (left) and a low (right) pressure region is separated by a diaphragm as schematically shown in Fig. 2.13. The transient is initiated by an instantaneous removal of the diaphragm. Assuming strictly one-dimensional flow conditions, the shock tube mathematically represents a "Riemann problem" where the initial flow velocities on both sides of the diaphragm have been set to zero,

$$\frac{\partial \mathbf{U}}{\partial t} + \mathbf{G}\frac{\partial \mathbf{U}}{\partial x} = 0 \tag{2.105}$$

with the initial conditions

$$\left.\begin{aligned} \mathbf{U} &= \mathbf{U}_L \quad \text{for} \quad x < x_{\text{dia}} \\ \mathbf{U} &= \mathbf{U}_R \quad \text{for} \quad x > x_{\text{dia}} \end{aligned}\right\} \tag{2.106}$$

and $p_R > p_L$.

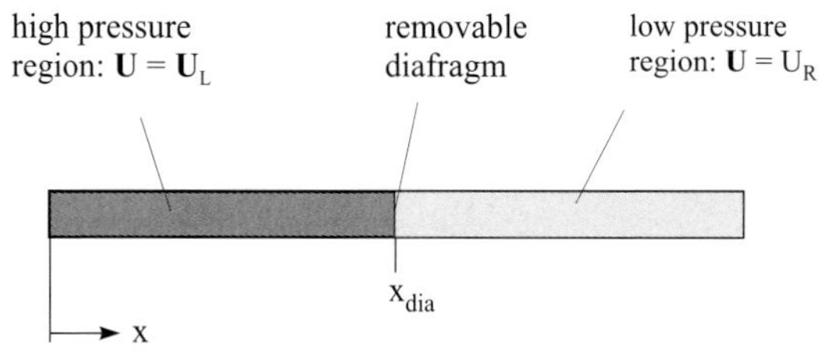

Fig. 2.13: Shock tube test problem

If viscous effects are ignored a self-similar solution exists where all parameters are only functions of the ratio x/t as illustrated in Fig. 2.14. Three different wave phenomena can be distinguished which separate the uniform regions 1, 2, 3, and 4: (1) a shock wave propagating into the low pressure region, followed by (2) a contact discontinuity traveling with subsonic velocity into the right-hand side of the pipe, and (3) a centered rarefaction wave (expansion fan) propagating into the high pressure region.

For the specific case of ideal state equations, an iterative analytical solution for the shock tube problem can be derived as will be briefly described in the following. From equation (2.94) the velocity behind the shock wave can be expressed as a function of the pressure ratio p_2/p_1 as

$$u_2 = a_1\left(\frac{p_2}{p_1} - 1\right)\sqrt{\frac{2/\varkappa}{\frac{p_2}{p_1}(\varkappa + 1) + (\varkappa - 1)}}. \tag{2.107}$$

There are two characteristics originating at the left compartment which provide information for the prediction in the region 3 which are, according to the compatibility relations (2.59),

$$\lambda = u + a: \qquad \frac{2}{\varkappa - 1}a_4 = \frac{2}{\varkappa - 1}a_3 - u_3 \tag{2.108}$$

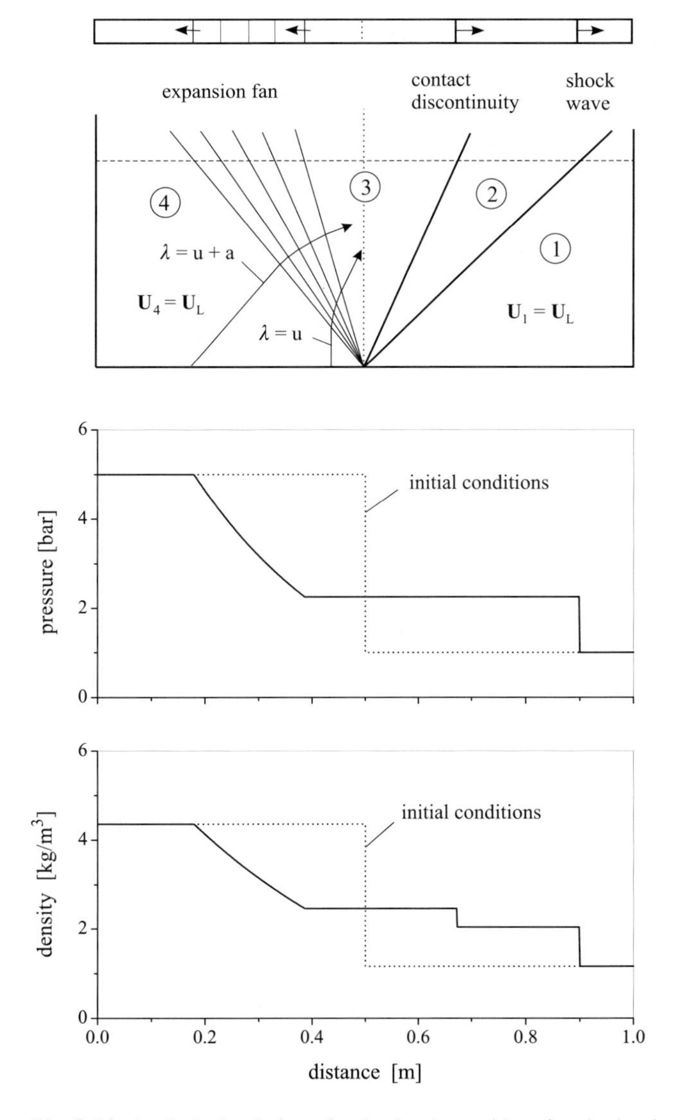

Fig. 2.14: Analytical solution of a shock tube problem for single-phase gas flow

and

$$\lambda = u : \qquad s_4 = s_3. \tag{2.109}$$

Since the velocity and pressure are constant across the contact discontinuity, $u_3 = u_2$ and $p_3 = p_2$, and due to the isentropic condition (2.109), equation (2.107) can be rewritten as

$$\frac{p_1}{p_4} = \frac{p_1}{p_2}\left[1 - \frac{1-\varkappa}{2}\left(\frac{u_2}{a_1}\right)\frac{a_1}{a_4}\right]^{\frac{2\varkappa}{\varkappa-1}} . \tag{2.110}$$

Combining equations (2.107) and (2.110) results in a relation between the pressure ratio over the diaphragm p_4/p_1 and the pressure ratio over the shock wave p_2/p,

$$\frac{p_1}{p_4} = \frac{p_1}{p_2}\left[1 - \frac{\varkappa - 1}{2\varkappa}\frac{a_1}{a_2}\frac{\left(\frac{p_1}{p_2} - 1\right)}{\sqrt{1 + \frac{\varkappa + 1}{2\varkappa}\left(\frac{p_1}{p_2} - 1\right)}}\right]^{-\frac{2\varkappa}{\varkappa - 1}}, \tag{2.111}$$

with the ratio of sound velocities

$$\frac{a_1}{a_2} = \sqrt{\frac{T_1}{T_2}} = \sqrt{\frac{p_1/p_2}{\varrho_1/\varrho_2}}. \tag{2.112}$$

From the basic shock tube equation (2.111) the shock strength p_2/p_1 can be iteratively calculated. The corresponding density and temperature ratios then follow from the Rankine-Hugoniot relation (2.87) as

$$\frac{\varrho_2}{\varrho_1} = \frac{(\varkappa + 1)\,p_2/p_1 + (\varkappa - 1)}{(\varkappa - 1)\,p_2/p_1 + (\varkappa + 1)}. \tag{2.113}$$

and

$$\frac{T_2}{T_1} = \frac{p_2/p_1}{\varrho_2/\varrho_1}.$$

If the states at 2 and 3 are known the remaining parameter distributions across the expansion fan can be easily obtained from the compatibility relations (2.59).

Due to the existing analytical solution the shock tube problem has become a standard numerical benchmark test case for the assessment of numerical methods as will be shown in Chapter 9.

2.7 Multidimensional flow conditions

For three-dimensional flow conditions the Euler equations are obtained from the general balance equations for mass, momentum, and energy (A.26) to (A.29) as given in Appendix A. Dropping the viscosity and heat conduction terms, these equations simplify to the three-dimension form of the Euler equations

$$\frac{\partial \varrho}{\partial t} + \nabla \cdot (\varrho\,\vec{u}) = 0 \tag{2.114}$$

$$\frac{\partial}{\partial t}(\varrho\,\vec{u}) + \nabla \cdot (\varrho\vec{u}\,\vec{u}) = \vec{F}_i \tag{2.115}$$

$$\frac{\partial}{\partial t}\left[\varrho\left(e + \frac{u^2}{2}\right)\right] + \nabla \cdot \left[\varrho\vec{u}\left(h + \frac{u^2}{2}\right)\right] = \vec{F} \cdot \vec{u} + Q. \tag{2.116}$$

Removing the kinetic energy term from the energy equation (2.116) and introducing the entropy as major depending variable one obtains the entropy relation

$$\frac{\partial}{\partial t}(\varrho\, s) + \nabla \cdot (\varrho\, \vec{u}\, s) = \frac{1}{T}\varrho\, Q, \tag{2.117}$$

as will be used in the following for the characteristic analysis of the flow equations. By expansion of equations (2.114), (2.115), and (2.117), the primitive form of the Euler equations is obtained which will be given here for simplicity reasons only for two-dimensional flow conditions

$$\frac{\partial \varrho}{\partial t} + u_x \frac{\partial \varrho}{\partial x} + u_y \frac{\partial \varrho}{\partial x} + \varrho \left(\frac{\partial u_x}{\partial x} + \frac{\partial u_y}{\partial y} \right) = 0 \tag{2.118}$$

$$\frac{\partial u_x}{\partial t} + u_x \frac{\partial u_x}{\partial x} + u_y \frac{\partial u_x}{\partial y} + \frac{1}{\varrho}\frac{\partial p}{\partial x} = \frac{F_x}{\varrho} = f_x \tag{2.119}$$

$$\frac{\partial u_y}{\partial t} + u_x \frac{\partial u_y}{\partial x} + u_y \frac{\partial u_y}{\partial y} + \frac{1}{\varrho}\frac{\partial p}{\partial x} = \frac{F_y}{\varrho} = f_y \tag{2.120}$$

$$\frac{\partial s}{\partial t} + u_x \frac{\partial s}{\partial x} + u_y \frac{\partial s}{\partial x} = \frac{Q}{\varrho T} = \frac{q}{T}. \tag{2.121}$$

With the state equation in expanded form

$$\delta \varrho = \left(\frac{\partial \varrho}{\partial p} \right)_s \delta p + \left(\frac{\partial \varrho}{\partial s} \right)_p \delta s, \qquad \text{respectively,} \qquad \delta \varrho = \frac{1}{a^2}\delta p + \left(\frac{\partial \varrho}{\partial s} \right)_p \delta s,$$

the density derivative terms can be removed from equation (2.118) and, hence, the the following compact vector form of the Euler equations is obtained as

$$\frac{\partial \mathbf{U}}{\partial t} + \mathbf{G}_x \frac{\partial \mathbf{U}}{\partial x} + \mathbf{G}_x \frac{\partial \mathbf{U}}{\partial x} = \mathbf{D} \tag{2.122}$$

with the state and source term vectors $\mathbf{U}$ and $\mathbf{D}$ defined as

$$\mathbf{U} = \begin{bmatrix} p \\ u_x \\ u_y \\ s \end{bmatrix}, \qquad \mathbf{D} = \begin{bmatrix} -a^2 \left(\frac{\partial \varrho}{\partial s} \right)_p \frac{q}{T} \\ f_x \\ f_y \\ \frac{q}{T} \end{bmatrix}, \tag{2.123}$$

and the coefficient matrices $\mathbf{G}_x$ and $\mathbf{G}_y$ for x- and y-directions

$$\mathbf{G}_x = \begin{bmatrix} u_x & \varrho a^2 & 0 & 0 \\ \frac{1}{\varrho} & u_x & 0 & 0 \\ 0 & 0 & u_x & 0 \\ 0 & 0 & & u_x \end{bmatrix}, \qquad \mathbf{G}_y = \begin{bmatrix} u_y & \varrho a^2 & 0 & 0 \\ \frac{1}{\varrho} & u_y & 0 & 0 \\ 0 & 0 & u_y & 0 \\ 0 & 0 & 0 & u_y \end{bmatrix}. \tag{2.124}$$

For the hyperbolic numerical schemes as described in Chapter 7, it is often required to "project" the flow equations in to an arbitrary direction of the flow field (e.g., normal to the boundary of a computational cell) as schematically shown in Fig. 2.15. The velocity components in the x-y and n-t coordinate systems are related to each other as

$$\left.\begin{aligned} u_n &= u_x n_x + u_y n_y \\ u_t &= -u_x n_y + u_y n_x \end{aligned}\right\} \tag{2.125}$$

or, respectively

$$\left.\begin{aligned} u_x &= u_n n_x - u_t n_y \\ u_y &= u_n n_y + u_t n_x \end{aligned}\right\} \tag{2.126}$$

with the x- and y-components of the unit vector $n_x = \sin(\Phi)$ and $n_x = \cos(\Phi)$.

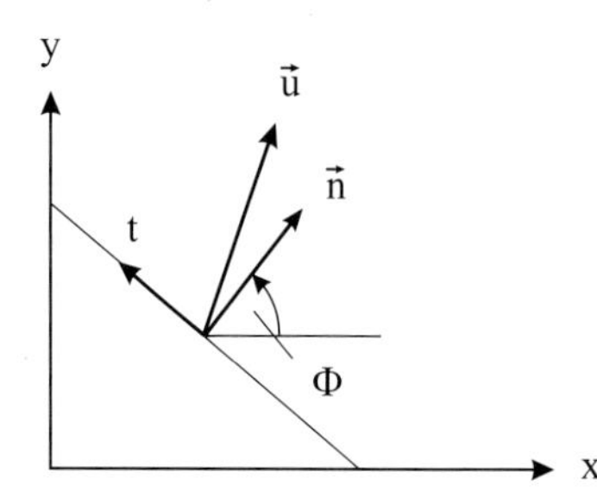

Fig. 2.15: Rotation of coordinate system

The projection of the flow equations (2.122) in $\vec{n}$-diretion then yields

$$\frac{\partial \mathbf{U}}{\partial t} + \mathbf{G}_n \frac{\partial \mathbf{U}}{\partial n}\frac{\partial n}{\partial x} + \mathbf{G}_y \frac{\partial \mathbf{U}}{\partial n}\frac{\partial n}{\partial y} = \mathbf{D} \tag{2.127}$$

or, with $\mathbf{G}_n = \mathbf{G}_x n_x + \mathbf{G}_y n_y$

$$\frac{\partial \mathbf{U}}{\partial t} + \mathbf{G}_n \frac{\partial \mathbf{U}}{\partial n} = \mathbf{D}. \tag{2.128}$$

The flow equations in the rotated coordinate system can then be obtained by a linear transformation of equation (2.128)

$$\frac{\partial \mathbf{U}}{\partial \mathbf{U}^*}\frac{\partial \mathbf{U}^*}{\partial t} + \mathbf{G}_n \frac{\partial \mathbf{U}}{\partial \mathbf{U}^*}\frac{\partial \mathbf{U}^*}{\partial n} = \mathbf{D} \tag{2.129}$$

or

$$\frac{\partial \mathbf{U}^*}{\partial t} + \underbrace{\left(\mathbf{J}\,\mathbf{G}_n \mathbf{J}^{-1}\right)}_{\mathbf{G}_n^*}\frac{\partial \mathbf{U}^*}{\partial n} = \mathbf{J}\,\mathbf{D} = \mathbf{D}^* \tag{2.130}$$

where the new state vector $\mathbf{U}^*$ and the Jacobian $\mathbf{J} = \partial \mathbf{U}^*/\partial \mathbf{U}$ are defined as

$$\mathbf{U} = \begin{bmatrix} p \\ u_n \\ u_t \\ s \end{bmatrix}, \qquad \mathbf{J} = \frac{\partial \mathbf{U}^*}{\partial \mathbf{U}} = \begin{bmatrix} 1 & 0 & 0 & 0 \\ 0 & +n_x & -n_y & 0 \\ 0 & +n_y & +n_x & 0 \\ 0 & 0 & 0 & 1 \end{bmatrix}. \tag{2.131}$$

Introducing the Jacobian matrix (2.131) the equation for the rotated coordinate system (2.130) becomes

$$\frac{\partial \mathbf{U}^*}{\partial t} + \mathbf{G}_n^* \frac{\partial \mathbf{U}^*}{\partial n} = \mathbf{D}^* \tag{2.132}$$

with the new coefficient matrix $\mathbf{G}_n^*$ and the related source term vector $\mathbf{D}^*$.

$$\mathbf{G}_n^* = \begin{bmatrix} u_n & \varrho a^2 & 0 & 0 \\ \frac{1}{\varrho} & u_n & 0 & 0 \\ 0 & 0 & u_n & 0 \\ 0 & 0 & 0 & u_n \end{bmatrix}, \qquad \mathbf{D}^* = \begin{bmatrix} -a^2 \left(\frac{\partial \varrho}{\partial s}\right)_p \frac{q}{T} \\ f_n \\ f_t \\ \frac{q}{T} \end{bmatrix}. \tag{2.133}$$

Comparing the coefficient matrices before and after the transformation of the coordinate system as given in equations (2.124) and (2.133), one immediately verifies the rotational invariance property of the Euler equations.

The eigenvalues of the coefficient matrix $\mathbf{G}_n^*$ are the characteristic velocities in $\vec{n}$ direction for the different wave modes

$$\begin{aligned} &\text{sound waves:} && \lambda_{1,2} = \vec{u}\cdot\vec{n} \pm a = u_n \pm a \\ &\text{shear waves:} && \lambda_3 = \vec{u}\cdot\vec{n} = u_n \qquad , \\ &\text{entropy waves:} && \lambda_3 = \vec{u}\cdot\vec{n} = u_n . \end{aligned} \tag{2.134}$$

and the corresponding right and left eigenvectors become

$$\mathbf{V}_R = \begin{bmatrix} \frac{1}{2} & +\frac{1}{2\varrho a} & 0 & 0 \\ \frac{1}{2} & -\frac{1}{2\varrho a} & 0 & 0 \\ 0 & 0 & 1 & 0 \\ 0 & 0 & 0 & 1 \end{bmatrix}, \qquad \mathbf{V}_L = \begin{bmatrix} 1 & +\varrho\, a & 0 & 0 \\ 1 & -\varrho\, a & 0 & 0 \\ 0 & 0 & 1 & 0 \\ 0 & 0 & 0 & 1 \end{bmatrix} \tag{2.135}$$

Apart from the shear wave, all other wave modes including pressure and entropy waves are fully equivalent to the one-dimensional case as described above. The newly appearing shear wave, propagating as the entropy wave with the material velocity, describes the transport of information on the transverse momentum.

As for the one-dimensional case flow equation (2.135) can be diagonalized, resulting in the characteristic equation

$$\mathbf{T}^{-1}\frac{\partial \mathbf{U}^*}{\partial t} + \underbrace{\mathbf{T}^{-1}\mathbf{G}^*\,\mathbf{T}}_{\mathbf{\Lambda}}\mathbf{T}^{-1}\frac{\partial \mathbf{U}^*}{\partial n} = \mathbf{T}^{-1}\mathbf{D}^* = \mathbf{E}^*, \tag{2.136}$$

or

$$\mathbf{T}^{-1}\frac{\partial \mathbf{U}^*}{\partial t} + \mathbf{\Lambda}\mathbf{T}^{-1}\frac{\partial \mathbf{U}^*}{\partial n} = \mathbf{E}^*, \tag{2.137}$$

with the transformation matrix $\mathbf{T}$, the diagonal matrix of eigenvectors $\mathbf{\Lambda}$, and the source term vector $\mathbf{E}^*$ as

$$\mathbf{T} = \mathbf{V}_R^T = \begin{bmatrix} \frac{1}{2} & \frac{1}{2} & 0 & 0 \\ +\frac{1}{2\varrho a} & -\frac{1}{2\varrho a} & 0 & 0 \\ 0 & 0 & 1 & 0 \\ 0 & 0 & 0 & 1 \end{bmatrix}, \tag{2.138}$$

and

$$\mathbf{\Lambda} = \mathbf{T}^{-1}\mathbf{G}\,\mathbf{T} = \begin{bmatrix} u_n + a & 0 & 0 & 0 \\ & u_n - a & 0 & 0 \\ 0 & 0 & u_n & 0 \\ 0 & 0 & 0 & u_n \end{bmatrix}. \tag{2.139}$$

From the characteristic form of the flow equations (2.137) the following compatibility relations can be obtained

$$
\left.
\begin{aligned}
&\textit{pressure waves} \\
&\lambda_1 = u_n + a: && \frac{dp}{dt} + \varrho a \, \frac{du}{dt} = \left[+ a \varrho f_n - a^2 \left(\frac{\partial \varrho}{\partial s} \right)_p \frac{q}{T} \right] \\
&\lambda_2 = u_n - a: && \frac{dp}{dt} - \varrho a \, \frac{du}{dt} = \left[- a \varrho f_n - a^2 \left(\frac{\partial \varrho}{\partial s} \right)_p \frac{q}{T} \right] \\
&\textit{shear waves} \\
&\lambda_3 = u_n: && \frac{du_t}{dt} = f_t \\
&\textit{entropy wave} \\
&\lambda_4 = u_n: && \frac{ds}{dt} = \frac{q}{T}
\end{aligned}
\right\} \tag{2.140}
$$

For the specific case of adiabatic flow ($q = 0$) and for the absence of external forces ($f = 0$), the the shear and entropy waves are completely decoupled from the pressure waves and the compatibility relations (2.140) simplify to

$$
\left.
\begin{aligned}
&\textit{pressure waves} \\
&\lambda_1 = u_n + a: && \frac{dp}{dt} + \varrho a \, \frac{du}{dt} = 0 \\
&\lambda_2 = u_n - a: && \frac{dp}{dt} - \varrho a \, \frac{du}{dt} = 0 \\
&\textit{shear waves} \\
&\lambda_3 = u_n: && \frac{du_t}{dt} = 0 \\
&\textit{entropy wave} \\
&\lambda_4 = u_n: && \frac{ds}{dt} = 0
\end{aligned}
\right\} . \tag{2.141}
$$

As described for the one-dimensional case, the coefficient matrix of the Euler equation $\mathbf{G}_n^*$ can be split into elementary parts

$$
\mathbf{G}_n^* = \sum_{k=1}^{4} \mathbf{G}_{n,k}^*
$$

with

$$
\mathbf{G}_{n,k}^* = \mathbf{T} \, \mathbf{\Lambda}_k \mathbf{T}^{-1},
$$

where the diagonal matrix $\mathbf{\Lambda}_k$ contains only the kth eigenvalue. The resulting split matrices are given in Appendix B.1 by equations (B.10) and (B.12).

References

[1] H. Shapiro, *The Dynamics and Thermodynamics of Compressible Flow*, Vol. 1, Ronald Press, New York, 1953.

[2] R. Courant and K.O. Friedrichs, *Supersonic Flow and Shock Waves,* Interscience Publishers, New York, 1948.

[3] G. Emmanuel, *A General Method for the Numerical Integration through a Saddle-point Singularity with Application to One-Dimensional Non-Equilibrium Nozzle Flow*, Arnold Engineering Development Center, AECD-TDR-64-29, 1G, 1964.

[4] J.L. Steger and R.R. Warming, *Flux Vector Splitting of the Inviscid Gasdynamic Equations with Application to Finite Difference Methods*, Journal of Computationel Physics, 40, 263–293, 1981.

3 Two-Fluid Model of Two-Phase Flow

In the spatial domain where either liquid or vapor/gas is present, the local flow processes can be described by the instantaneous single-phase flow equations of the corresponding phase, e.g., the Navier–Stokes equations. Together with appropriate boundary conditions for both phases at the moving interface, the two-phase flow is completely determined.

However, since such a "direct simulation" of two-phase flow is out of the scope of our present predictive capability (and will remain so, it seems, for the near future), all two-phase models of practical interest are restricted to the average flow parameters rather than local flow quantities. The corresponding "macroscopic" separate balance equations for the two phases are obtained by a space and/or time or ensemble averaging of the local instantaneous phasic flow equations, which leads to what is often referred as the "two-fluid model" of two-phase flow. There exists a large literature on the derivation two-fluid models and their applications to various two-phase flow problems. The reader might find detailed information on this subject in the more fundamental publications of Ishii [1], Boure [2], Delhaye and Achard [3], and Drew and Lahey [4]. The present form of the balance equations as is used throughout the book has been derived using the concept of distribution functions and the Dirac delta function as applied by Gray and Lee [5] for the volume averaging of multiphase flows. The detailed procedure for the derivation is given in Appendix A.

3.1 Balance equations of two fluid model of two-phase flow

With the index $i = g$ (gas or vapor), and $i = l$ (liquid), the most general form of balance equations of mass, momentum, and energy, given by equations (A.94) to (A.96) in Appendix A, can be written as

mass:

$$\frac{\partial}{\partial t}(\alpha_i \varrho_i) + \nabla \cdot (\alpha_i \varrho_i \, \vec{u}_i) = \sigma_i^M \tag{3.1}$$

momentum:

$$\frac{\partial}{\partial t}(\alpha_i \varrho_i \, \vec{u}_i) + \nabla \cdot (\alpha_i \varrho_i \, \vec{u}_i \, \vec{u}_i) + \alpha_i \nabla p_i + \left(p_i - p_i^{\mathrm{int}}\right) \nabla \alpha_i + \nabla \cdot \left(\alpha_i \bar{\mathbf{T}}_\mathbf{i}\right) = \vec{\sigma}_i^{\mathrm{J}} + \vec{F}_i \tag{3.2}$$

Gasdynamic Aspects of Two-Phase Flow. Herbert Städtke

ISBN: 3-527-40578-X

energy:

$$\frac{\partial}{\partial t}\left[\alpha_i \varrho_i \left(e_i + \frac{u_i^2}{2}\right)\right] + \nabla \cdot \left[\alpha_i \varrho_i \,\vec{u}_i \left(h_i + \frac{u_i^2}{2}\right)\right] + \nabla\left(\alpha_i\,\vec{q}_i\right) - \nabla\left(\alpha_i \bar{\mathbf{T}}_i \cdot \,\vec{u}_i\right)$$

$$+p_i^{\text{int}} \frac{\partial \alpha_i}{\partial t} = \sigma_i^E + Q_i + \,\vec{F}\cdot\vec{u}_i. \tag{3.3}$$

The source terms for mass, momentum, and energy on the right-hand side of equations (3.1) to (3.3) distinguish between the "internal" contributions describing the interfacial mass, momentum, and energy exchange between the phases,

$$\left.\begin{array}{ll} \textit{mass}: & \sigma_i^M \\ \textit{momentum}: & \vec{\sigma}_i^J = \vec{F}_i^{\text{int}} + \sigma_i^M \,\vec{u}^{\text{ex}} \\ \textit{energy}: & \sigma_i^E = \sigma_i^Q + \sigma_i^M \left(h + \dfrac{u^2}{2}\right)^{\text{ex}} + \vec{F}_i^{\text{int}} \cdot \vec{u}_i^{Fi} \end{array}\right\} \tag{3.4}$$

and the external forces such as gravity, $\vec{F}_i$, and any external heat addition Q_i.

Equations (3.1) to (3.3) are equivalent to those often found in the literature. The only difference is related to the right-hand side of the equations where the properties of the "exchanged" quantities are explicitly introduced. All the state and flow parameters on the left-hand side of the equations are considered to be volume/time averaged quantities. It is further assumed that the average of the products of the parameters is equal to the product of the averaged parameters. During the averaging process, the local volumetric concentrations of the two phases α_i, with $i = g, l$, are introduced as a measure for the composition of the two-phase mixture. The volume concentration of the gas or vapor phase α_g is also known as "void fraction".

The parameters on the r.h.s. of equations (3.1) to (3.3) include volumetric source terms representing mass, momentum, and energy transfer processes at the interface as well as external forces such as gravity and heat sources. The physical interpretation of the various terms and related parameters is given in Appendix A.

The conservation principles for mass, momentum, and energy require the following balances of the corresponding source terms:

$$\sum_{i=g,l} \sigma_i^M = 0, \qquad \sum_{i=g,l} \sigma_i^Q = 0, \qquad \sum_{i=g,l} \vec{F}_i^{\text{int}} = 0. \tag{3.5}$$

The system of equations has to be further completed by two state equations for the average quantities of the phases,

$$\varrho_i = \varrho_i(p_i, T_i) \qquad \text{and} \qquad e_i = e_i(p_i, T_i) \qquad \text{with} \quad i = g, l \tag{3.6}$$

and the constraint for the volumetric phase concentration

$$\alpha_g + \alpha_l = 1. \tag{3.7}$$

Using the momentum equation, the kinetic terms can be removed from the energy equations. Assuming further that the following thermodynamic relationship is also valid for the average quantities

$$T_i \delta s_i = \delta u_i - \frac{p_i}{\varrho_i^2} \delta \varrho_i, \tag{3.8}$$

the energy equation can be simplified by introducing the entropy as a new state variable which results in the following balance equation for the phasic entropy:

$$\frac{\partial}{\partial t}(\alpha_i \varrho_i s_i) + \nabla \cdot [\alpha_i \varrho_i \, \vec{u}_i s_i] + \frac{\nabla \cdot (\alpha_i \vec{q}_i)}{T_i} + \frac{p^{\text{int}} - p_i}{T_i}\left(\frac{\partial \alpha_i}{\partial t} + \vec{u}_i \nabla \alpha_i\right) \tag{3.9}$$

$$-\frac{\alpha_i \bar{\mathbf{T}}_i : (\nabla \cdot \vec{u}_i)}{T_i} = \sigma_i^{S,\text{int}} + \frac{Q_i}{T_i}, \tag{3.10}$$

with the internal volumetric entropy source term for phase i resulting from interfacial transfer processes,

$$\sigma_i^{S,\text{int}} = \frac{\sigma_i^Q}{T_i} + \frac{\sigma_i^M}{T_i}\left[s_i T_i + (h^{\text{ex}} - h_i) + \frac{1}{2}(u^{\text{ex}} - u_i)^2\right] + \frac{\vec{F}_i^{\text{int}}}{T_i} \cdot (\vec{u}_i^{Fi} - \vec{u}_i). \tag{3.11}$$

In agreement with the second law of thermodynamics, the overall internal entropy source has to be positive definite,

$$\sum_{i=g,l} \sigma_i^{S,\text{int}} \geq 0. \tag{3.12}$$

The entropy equation does not give any further information with respect to equations (3.1) to (3.3), nevertheless, due to the simplified form it might be worthwhile to use in some situations the entropy equation instead of the complete energy balance equation (3.3).

3.2 Single pressure two-fluid model

Considering the state equations (3.6) and the constraint for the volume fractions (3.7), there remain eight major dependent flow parameters in the six balance equations for the phasic mass, momentum, and entropy

$$\{\alpha_g, u_g, u_l, T_g, T_l, p_g, p_l, p^{\text{int}}\}.$$

There have been various attempts to complete the system of balance equations by adding additional differential or algebraic equations on a more or less heuristic way. This includes among others

- different phasic pressures resulting from static gravity heads in the case of stratified flow conditions, see for example Ardron [6];
- different interfacial pressures due to surface tension effects in the presence of an interfacial curvature, Ramshaw and Trapp [7];

- differences between the average pressure in the continuous phase and the average interfacial pressure resulting from local flow disturbances induced by moving particles (solid particles, bubbles droplets), see for example Milne-Thompson [8] for idealized flow conditions (frictionless flow around spheres);
- pressure differences resulting from inertia effects during bubble growth or collapse, which can be described for ideal conditions (e.g., spherical bubble surrounded by incompressible liquid) by Rayleigh's equation [9];
- pressure differences resulting from a "transverse" movement of the interface in the case of essentially one-dimensional flow, Ransom and Hicks [10].

Although all of these "two-pressure" models show interesting aspects of two-phase flows, they are valid only for specific flow regimes or for a limited range of flow conditions. None of them has yet reached a state of maturity for broader applications to scientific or technical problems. This is the reason why in most of the present two-fluid models of two-phase flow, the assumption of equal local pressure for the two phases is introduced, $p_g = p_l = p^{\text{int}} = p$. This seems to be justified for many technical application as long as surface tension effects can be neglected.

Assuming a single local pressure value, the balance equations (3.1) to (3.3) and (3.9) simplify to

mass:

$$\frac{\partial}{\partial t}(\alpha_i \varrho_i) + \nabla \cdot (\alpha_i \varrho_i \, \vec{u}_i) = \sigma_i^M \tag{3.13}$$

momentum:

$$\frac{\partial}{\partial t}(\alpha_i \varrho_i \, \vec{u}_i) + \nabla \cdot (\alpha_i \varrho_i \, \vec{u}_i \, \vec{u}_i) + \alpha_i \nabla p + \nabla \cdot (\alpha_i \bar{\mathbf{T}}_\mathbf{i}) = \vec{\sigma}_i^{\mathrm{J}} + \vec{F}_i \tag{3.14}$$

energy:

$$\frac{\partial}{\partial t}\left[\alpha_i \varrho_i \left(e_i + \frac{u_i^2}{2}\right)\right] + \nabla \cdot \left[\alpha_i \varrho_i \, \vec{u}_i \left(h_i + \frac{u_i^2}{2}\right)\right] + \nabla (\alpha_i \, \vec{q}_i) - \nabla (\alpha_i \bar{\mathbf{T}}_i \cdot \vec{u}_i)$$

$$+ p \frac{\partial \alpha_i}{\partial t} = \sigma_i^E + Q_i + \vec{F}_i \cdot \vec{u}_i \tag{3.15}$$

entropy:

$$\frac{\partial}{\partial t}(\alpha_i \varrho_i s_i) + \nabla \cdot [\alpha_i \varrho_i \, \vec{u}_i s_i] + \frac{\nabla \cdot (\alpha_i \vec{q}_i)}{T_i} - \frac{\alpha_i \bar{\mathbf{T}}_i : (\nabla \cdot \vec{u}_i)}{T_i} = \sigma_i^{S,\text{int}} + \frac{Q_i}{T_i}. \tag{3.16}$$

Neglecting the influence of the bulk heat conduction ($\vec{q}_i = 0$) and bulk viscous effects ($\bar{\mathbf{T}}_i = 0$) compared with the governing effects resulting from the interfacial heat and mass transfer processes, these equations can be written as *mass*:

$$\frac{\partial}{\partial t}(\alpha_i \varrho_i) + \nabla \cdot (\alpha_i \varrho_i \, \vec{u}_i) = \sigma_i^M \tag{3.17}$$

momentum:

$$\frac{\partial}{\partial t}(\alpha_i \varrho_i \, \vec{u}_i) + \nabla \cdot (\alpha_i \varrho_i \, \vec{u}_i \, \vec{u}_i) + \alpha_i \nabla p = \vec{F}_i^{\text{int}} + \sigma_i^M \, \vec{u}_i + \vec{F}_i \tag{3.18}$$

energy:

$$\frac{\partial}{\partial t}\left[\alpha_i \varrho_i \left(e_i + \frac{u_i^2}{2}\right)\right] + \nabla \cdot \left[\alpha_i \varrho_i \, \vec{u}_i \left(h_i + \frac{u_i^2}{2}\right)\right] + p\frac{\partial \alpha_i}{\partial t}$$

$$= \sigma_i^Q + \sigma_i^M \left(h + \frac{u^2}{2}\right)^{\text{ex}} + \vec{F}_i^{\text{int}} \cdot u_i^{Fi} + Q_i + \vec{F}_i \cdot \vec{u}_i \tag{3.19}$$

entropy:

$$\frac{\partial}{\partial t}(\alpha_i \varrho_i s_i) + \nabla \cdot [\alpha_i \varrho_i \, \vec{u}_i s_i] = \vec{\sigma}_i^S + \frac{Q_i}{T_i}. \tag{3.20}$$

For the specific case of *one-dimensional flows* one obtains from (3.17) to (3.20)

mass:

$$\frac{\partial}{\partial t}(\alpha_i \varrho_i) + \frac{\partial}{\partial x}(\alpha_i \varrho_i u_i) = \sigma_i^M \tag{3.21}$$

momentum:

$$\frac{\partial}{\partial t}(\alpha_i \varrho_i u_i) + \frac{\partial}{\partial x}(\alpha_i \varrho_i u_i^2) + \alpha_i \frac{\partial p}{\partial x} = \vec{\sigma}_i^{\text{J}} + F_i \tag{3.22}$$

energy:

$$\frac{\partial}{\partial t}\left[\alpha_i \varrho_i \left(e_i + \frac{u_i^2}{2}\right)\right] + \frac{\partial}{\partial x}\left[\alpha_i \varrho_i u_i \left(e_i + \frac{p}{\varrho_i} + \frac{u_i^2}{2}\right)\right] + p\frac{\partial \alpha_i}{\partial t}$$

$$= \sigma_i^E + Q_i + F_i u_i \tag{3.23}$$

entropy:

$$\frac{\partial}{\partial t}(\alpha_i \varrho_i s_i) + \frac{\partial}{\partial x}(\alpha_i \varrho_i u_i s_i) = \sigma_i^{S,\text{int}} + \frac{Q_i}{T_i}, \tag{3.24}$$

with the entropy source resulting from interfacial transfer processes,

$$\sigma_i^{S,\text{int}} = \frac{\sigma_i^M}{T_i}\left[T_i s_i + (h^{\text{ex}} - h_i) + \frac{1}{2}(u^{\text{ex}} - u_i)^2\right] + \frac{F_i^{\text{int}}}{T_i}(u^{\text{int}} - u_i). \tag{3.25}$$

In the following a slightly different definition for the entropy is used based on the expanded form of the balance equation (3.24)

$$\alpha_i \varrho_i \frac{\partial s_i}{\partial t} + \alpha_i \varrho_i u_i \frac{\partial s_i}{\partial x} = \sigma_i^S \tag{3.26}$$

with the total phasic entropy source including also external heat addition to the fluid

$$\sigma_i^S = \frac{\sigma_i^M}{T_i}\left[(h^{\text{ex}} - h_i) + \frac{1}{2}(u^{\text{ex}} - u_i)^2\right] + \frac{F_i^{\text{int}}}{T_i}(u^{\text{int}} - u_i) + \frac{Q_i}{T_i}. \tag{3.27}$$

From equation (3.27) one then obtains for the material derivative of the entropy

$$\frac{d^i s_i}{dt} = \frac{\sigma_i^S}{\alpha_i \varrho_i} \qquad \text{with} \qquad \frac{d^i s_i}{dt} = \frac{\partial s_i}{\partial t} + u_i \frac{\partial s_i}{\partial x}. \tag{3.28}$$

The set of equations (3.17) to (3.20) or their one-dimensional counterparts (3.21) to (3.24) might be seen an equivalence to the Euler equations of gasdynamics. However, even for the absence of external forces ($F_i^{\text{int}} = 0$) and external heat sources (Q_i) there remains a strong coupling effect between the balance equations resulting from the source terms describing the interfacial mass, momentum transfer between the phases.

3.3 Remarks on interfacial transfer terms

From the derivation as given in Appendix A, it is evident that the balance equations (3.17) to (3.20) of the two fluid model represent an approximation of the nonhomogeneous two-phase flow and as such, have not the same validity as the Navier Stokes or Euler equations for single-phase flow of gas or liquid. Nevertheless, two-fluid models form the basis of most present computational tools for the numerical simulation of nonequilibrium two-phases flow processes. However, any application of these models relies on a realistic modeling of the source term on the right-hand side of the equations describing the interfacial transfer processes for mass, momentum, and energy.

As shown in Appendix A by equations (A.97) to (A.98) and (A.102), the interfacial source terms can be formulated as the product of the interfacial area concentration and a corresponding (area average) flux across the interface which results in

interfacial mass transfer due to evaporation/condensation:

$$\sigma_i^M = a^{\text{int}} m_i^{\text{int}} \quad \text{with} \quad m_i^{\text{int}} = -\frac{1}{A_\xi^{\text{int}}} \int_{A_\xi^{\text{int}}} \varrho_i \left(\vec{u}_i - \vec{u}^{\text{int}}\right) \cdot \vec{n}_i^{\text{int}} \, dA \tag{3.29}$$

interfacial heat transfer:

$$\sigma_i^Q = a^{\text{int}} q_i^{\text{int}} \quad \text{with} \quad q_i^{\text{int}} = -\frac{1}{A_\xi^{\text{int}}} \int_{A_\xi^{\text{int}}} \vec{q}_i \cdot \vec{n}_i^{\text{int}} \, dA \tag{3.30}$$

interfacial forces:

$$\vec{F}_i^{\text{int}} = a^{\text{int}} f_i^{\text{int}} \quad \text{with} \quad f_i^{\text{int}} = \frac{1}{A_\xi^{\text{int}}} \int\limits_{A_\xi^{\text{int}}} \bar{\mathbf{T}}_i \cdot \vec{n}_i^{\text{int}}\, dA. \tag{3.31}$$

The negative sign for the interfacial fluxes for mass and heat in equations (3.29) and (3.30) corresponds to the definition of the interfacial normal unit vector $\vec{n}_i^{\text{int}}$ for phase i.

As indicated in equations (3.29) to (3.31) the governing parameter for the interfacial coupling processes is the interfacial area concentration a^{int} which determines the amount of contact area between the gas/vapor and liquid phases within a unit volume. The prediction of the interfacial area concentration possibly is the most crucial point for the application of the two-fluid model. The major difficulty arises from the fact that even for the same void fraction and pressure values different "flow regimes" might exist which largely differ from each other in local phase distributions and interfacial area.

Two different approaches can be distinguished for the determination of the interfacial area concentration.

(1) At present the standard method applied is a direct correlation of the interfacial area with the local flow and transport parameters,

$$a^{\text{int}} = F(\alpha_g, u_g, \varrho_g, \varrho_l, \mu_g, \mu_l, ...). \tag{3.32}$$

This is usually done on the basis of empirical "flow maps" for specific geometrical flow conditions which allows us to identify characteristic flow regimes and related phase distributions. For relatively small void fractions ($\alpha_g \lesssim 0.1$), a *bubbly flow regime* might exist where the gas or vapor phase is assumed to be distributed in the form of (not necessarily spherical) bubbles in liquid carrier media. For relatively large void fractions ($\alpha_g \gtrsim 0.9$), a *droplet flow regime* is expected where the liquid phase is distributed in the form of droplets of different sizes and shapes in gas media. The shape and size of the particle (bubble or droplets) are then estimated by semi-empirical correlations based on the stability limit for particles exposed to the external flow fields. More complex or less well-structured flow regimes and flow regime transition such as annular flow, slug-flow churn-turbulent flow, etc. are dealt with by highly empirical correlations. Often "smoothing" procedures are introduced in order to dampen the effect of nonphysical discontinuities during the transition between different flow regimes.

(2) As an alternative to the "static" modeling of the interfacial area concentration as described in (1), it has been often proposed to explicitly model the transport of the interfacial area in a "dynamic" way using a separate balance equation and related source terms [11]. As shown in Appendix A, such a transport equation can be easily derived on the basis of a phasic distribution function γ_i and a related Dirac delta function $\nabla\gamma_i$ at the moving interphase,

$$\frac{\partial a_i}{\partial t} + \nabla \cdot \left(a^{\text{int}} \vec{u}^{\,\text{int}}\right) = \sigma^A \tag{3.33}$$

where the source term σ^A describes the creation (or destruction) of the interfacial area due to pressure changes (expansion, compression), phase change (evaporation, condensation), particle break-up or coalescence, and flow regime transitions. The method allows a more physically based evolution of the interfacial area and avoids unrealistic discontinuities as are often present in the "static" approach. Nevertheless, as for all volume averaged equations, there is a need for additional modeling of the corresponding transport velocity $\vec{u}^{\,\mathrm{int}}$ and the related source term σ^A.

In accordance to the principles of nonequilibrium thermodynamics, the interfacial fluxes for mass and heat are assumed to be linear functions of "driving forces" as arising from the local deviations from thermal or mechanical equilibrium between the phases. This results in the following interfacial source terms for the mass and heat as

$$\sigma_g^M = -\sigma_l^M = \frac{a^{\mathrm{int}}}{\Delta h^s}(\mathcal{H}_g^q \Delta T_g + \mathcal{H}_l^q \Delta T_l) \tag{3.34}$$

and

$$\sigma_g^Q = -\sigma_l^Q = a^{\mathrm{int}} \mathcal{H}^q \left(T_l - T_g\right), \tag{3.35}$$

with the transfer coefficients $\mathcal{H}_g^q$, $\mathcal{H}_l^q$, and $\mathcal{H}^q$ on the gas $(i = g)$ and the liquid $(i = l)$ side of the interface. For the driving temperature differences in the expression for the evaporation rate (3.34), often only the metastable contributions are taken into account,

$$\Delta T_g = \min\left(T_g - T^{\mathrm{sat}}, 0\right), \qquad \Delta T_l = \max\left(T_l - T^{\mathrm{sat}}, 0\right). \tag{3.36}$$

The major contribution to the interfacial forces results from the interfacial friction, $F_i^{\mathrm{int}} \approx F_i^{\mathrm{v}}$, which can be described by a resistance law of the form

$$F_l^{\mathrm{v}} = -F_g^{\mathrm{v}} = a^{\mathrm{int}} C^D \rho^{\mathrm{ref}} \left|\vec{u}_g - \vec{u}_l\right| \left(\vec{u}_g - \vec{u}_l\right), \tag{3.37}$$

with the "drag" coefficient C^D and the reference density ρ^{ref}.

For the case of well-structured flow conditions such as mono-dispersed bubbly or droplet flows existing physically based correlations can be applied for the interfacial drag coefficient C^D and heat transfer coefficients $\mathcal{H}_g^q$, $\mathcal{H}_l^q$, and $\mathcal{H}^q$ as introduced in equations (3.34) to (3.37). For the case of more complex flow regimes highly empirical correlations are often used which require some model calibration through sensitivity studies in comparison with existing experimental data. The way how the interfacial source terms are modeled in the ATFM code is described in some detail in Chapter 8.

References

[1] M. Ishii, *Thermo-Fluid Dynamics of Two-Phase Flow*, Eyrolles, Paris, 1975.

[2] J.A. Boure, *On a Unified Presentation of the Non-Equilibrium Two-Phase Flow Models in Non-Equilibrium Two-Phase Flows*, Proceedings of the ASME Symposium, New York, 1975.

[3] J.M. Delhaye and J.L. Achard, *On the Averaging Operators Introduced in Two Phase Flow Modelling* , Proceedings of a CSNI Specialist Meeting in Transient Two-Phase Flow, Toronto, 1976.

[4] D.A. Drew and R. T. Lahey Jr., *Application of General Constitutive Principle to the Derivation of Multidimensional Two phase Flow Equations*, International Journal of Multiphase Flow, 5, 243–264, 1979.

[5] W.G. Gray and P.C.Y. Lee, *On the Theorems for Local Volume Averaging of Multiphase Systems*, International Journal of Multiphase Flow, 3, 333–340, 1977.

[6] K.H. Ardron, *One-Dimensional Two-Fluid Equations for Horizontal Stratified Two-Phase Flow*, International Journal of Multiphase Flow, 6, 299–304, 1980.

[7] J.D. Ramshaw and J.A. Trapp, *Characteristics, Stability and Short-Wavelength Phenomena in Two-Phase Flow Equation Systems*, Nuclear Science and Engineering, 66, 93–102, 1978.

[8] L.M. Milne-Thompson, *Theoretical Hydrodynamics*, Macmilan, 5th edition, New York, 1968.

[9] J.W.S. Rayleigh, cited in H. Lamb, *Hydrodynamics*, Dover, New York, 1945, p. 122.

[10] V.H. Ransom and D.L. Hicks, *Hyperbolic Two-Pressure Models for Two-Phase Flow*, Journal of Computational Physics, 53, 124–151, 1984.

[11] G. Kacomustafaogullari and M. Ishii, *Modelling of Interfacial Area Concentration in a Two-Phase Flow System*, International Journal of Heat and Mass Transfer, 38 (3), 481–493, 1995.

4 Simplified Two-Phase Flow Models

Multiphase flow processes are normally governed by deviations from mechanical and thermal equilibrium between the phases. These nonequilibrium effects are a result of the generally large differences in the state and transport properties for the two phases and the finite rates for the interfacial transfer processes. The assumption of homogeneous flow (equal local average phase velocities) and thermal equilibrium between the phases (equal local average temperatures) represents a large simplification and, therefore, can be considered only as a rough approximation of the real flow process. Nevertheless, homogeneous two-phase flow models can be seen as special limiting cases for more detailed approaches and, for this reason, they are very helpful for the understanding of more complex conditions.

4.1 Homogeneous equilibrium model

The homogeneous equilibrium model for two-phase flow is based on the assumption of infinite transfer processes between the phases for mass, momentum, and energy, which results in equal local (average) flow velocities and equal local (average) temperatures for the two phases. This allows us to describe the two-phase mixture as a pseudo fluid and, apart from more complex state equations, all well-known properties of the gasdynamics remain (at least) qualitatively valid. The homogeneous equilibrium model also does not require any further description of interfacial coupling conditions since all transfer processes are implicitly determined by the assumption of mechanical and thermal equilibrium between the phases.

The one-dimensional homogeneous equilibrium two-phase flow is fully determined by the three conservation equations for mass, momentum, and energy. With the assumption of equal local flow velocities

$$u_g = u_l = u \tag{4.1}$$

and equal local pressure and temperatures for the two phases

$$\left.\begin{aligned} p_g &= p_l = p \\ T_g &= T_l = T \end{aligned}\right\}, \tag{4.2}$$

the conservation equations for the homogeneous equilibrium two-phase flow can be written as

mass:

$$\frac{\partial \varrho}{\partial t} + \frac{\partial}{\partial x}(\varrho u) = 0 \tag{4.3}$$

Gasdynamic Aspects of Two-Phase Flow. Herbert Städtke

ISBN: 3-527-40578-X

momentum:

$$\frac{\partial}{\partial t}(\varrho u) + \frac{\partial}{\partial x}(\varrho u^2) + \frac{\partial p}{\partial x} = F \tag{4.4}$$

energy:

$$\frac{\partial}{\partial t}\left[\varrho\left(e + \frac{u^2}{2}\right)\right] + \frac{\partial}{\partial x}\left[\varrho u\left(e + \frac{p}{\varrho} + \frac{u^2}{2}\right)\right] = Q + F\,u. \tag{4.5}$$

In these equations, mixture state properties have been used which are defined as

$$\varrho = \frac{1}{X_g/\varrho_g + X_l/\varrho_l} \tag{4.6}$$

or

$$\varrho = \alpha_g\varrho_g + \alpha_l\varrho_l \tag{4.7}$$

respectively, and

$$\left.\begin{aligned} e &= X_g e_g + X_l e_l \\ h &= X_g h_g + X_l h_l \\ s &= X_g s_g + X_l s_l . \end{aligned}\right\} \tag{4.8}$$

The vapor and liquid mass fractions are defined as

$$\left.\begin{aligned} X_g &= \frac{\alpha_g\varrho_g}{\alpha_g\varrho_g + \alpha_l\varrho_l} \\ X_l &= \frac{\alpha_l\varrho_l}{\alpha_g\varrho_g + \alpha_l\varrho_l} \end{aligned}\right\} \tag{4.9}$$

with the condition $X_g + X_l = 1$. The mass fraction vapor is also known as vapor quality.

Due to the assumption of a complete equilibrium between the phases, all state variables of the two phases are functions of only the pressure and temperature,

$$\{\varrho_g, \varrho_l, u_g, u_l, s_g, s_l\} = f(p, T).$$

For the mixture quantities, the gas/vapor mass fraction appears as a third independent parameter

$$\{\varrho, e, h, s\} = f(p, T, X_g).$$

With the assumption of a complete mechanical and thermal equilibrium between the phases, the basic thermodynamic relationships (Maxwell relations) as known for a single-phase fluid remain valid also for the whole two-phase mixture. This means in particular for the mixture entropy

$$T\delta s = \delta e - \frac{p}{\varrho^2}\delta\varrho, \tag{4.10}$$

for the state equation of the two-phase mixture

$$\varrho = f(p, s) \tag{4.11}$$

and for the homogeneous equilibrium sound velocity of the two-phase mixture

$$a = \sqrt{\left(\frac{\partial p}{\partial \varrho}\right)_s}. \tag{4.12}$$

Introducing the mixture entropy as a new state variable and replacing the density derivatives in the mass conservation equation by derivatives with respect to pressure and entropy yields the following "primitive" form of the conservation equations

mass:

$$\frac{1}{a^2}\frac{\partial p}{\partial t} + \frac{1}{a^2}u\frac{\partial p}{\partial x} + \varrho\frac{\partial u}{\partial x} = -\left(\frac{\partial \varrho}{\partial s}\right)_p \frac{Q}{\varrho T} \tag{4.13}$$

momentum:

$$\varrho\frac{\partial u}{\partial t} + \varrho u\frac{\partial u}{\partial x} + \frac{\partial p}{\partial x} = F \tag{4.14}$$

entropy:

$$\varrho T\frac{\partial s}{\partial t} + \varrho u T\frac{\partial s}{\partial x} = Q. \tag{4.15}$$

Equations (4.13) to (4.15) are identical with the corresponding conservation equations for single-phase gas flows. For this reason, all derivations made in Section 2 for single-phase flow remain at least qualitatively valid also for the homogeneous equilibrium two-phase flow. This means in particular that the equations (4.13) to (4.15) form a hyperbolic set of equations which can be combined in the vector form as

$$\frac{\partial \mathbf{U}}{\partial t} + \mathbf{G}\frac{\partial \mathbf{U}}{\partial x} = \mathbf{C}, \tag{4.16}$$

with the state vector $\mathbf{U}$, the coefficient matrix $\mathbf{G}$, and the source term vector $\mathbf{D}$,

$$\mathbf{U} = \begin{bmatrix} p \\ u \\ s \end{bmatrix}, \qquad \mathbf{G} = \begin{bmatrix} u & \varrho a^2 & 0 \\ \frac{1}{\rho} & u & 0 \\ 0 & 0 & u \end{bmatrix}, \qquad \mathbf{D} = \begin{bmatrix} -a^2\left(\frac{\partial \varrho}{\partial s}\right)_p \frac{q}{T} \\ f \\ \frac{q}{T} \end{bmatrix}. \tag{4.17}$$

The eigenvalues of the matrix $\mathbf{G}$ are the characteristic velocities

$$\lambda_1 = u + a, \qquad \lambda_2 = u - a, \qquad \lambda_3 = u \tag{4.18}$$

and the corresponding matrices containing the right and left eigenvectors are

$$\mathbf{V}_R = \begin{bmatrix} \frac{1}{2} & \frac{1}{2a\varrho} & 0 \\ \frac{1}{2} & -\frac{1}{2a\varrho} & 0 \\ 0 & 0 & 1 \end{bmatrix}, \qquad \text{and} \qquad \mathbf{V}_L = \begin{bmatrix} 1 & a\varrho & 0 \\ 1 & -a\varrho & 0 \\ 0 & 0 & 1 \end{bmatrix}. \tag{4.19}$$

As for single-phase gas, the basic system of equations (4.17) can be transferred into characteristic form

$$\mathbf{T}^{-1}\frac{\partial \mathbf{U}}{\partial t} + \mathbf{\Lambda}\,\mathbf{T}^{-1}\frac{\partial \mathbf{U}}{\partial x} = \mathbf{T}^{-1}\mathbf{D} = \mathbf{E}, \tag{4.20}$$

resulting in the compatibility relations

$$\left.\begin{aligned} \lambda_1 = u + a: &\qquad \frac{dp}{dt} + \varrho a\,\frac{du}{dt} = \left[+a\varrho f - a^2 \left(\frac{\partial \varrho}{\partial s}\right)_p \frac{q}{T} \right] \\ \lambda_2 = u - a: &\qquad \frac{dp}{dt} - \varrho a\,\frac{du}{dt} = \left[-a\varrho f - a^2 \left(\frac{\partial \varrho}{\partial s}\right)_p \frac{q}{T} \right] \\ \lambda_3 = u: &\qquad \frac{ds}{dt} = \frac{q}{T} \end{aligned}\right\}. \tag{4.21}$$

For further evaluation using the state equation for the two-phase mixture, two distinct cases have to be considered: (1) two-component two-phase mixtures without mass exchange between the phases (no evaporation or condensation) and (2) one-component two-phase flow with saturation conditions ($T_g = T_l = T^{\mathrm{sat}}$).

4.1.1 Two-component two-phase flow

For the two-component homogeneous flow conditions without mass exchange the following expressions for the equilibrium sound velocity can be derived from basic thermodynamic relationships (4.12) as

$$a = \frac{1}{\varrho}\sqrt{\frac{1}{\left[\frac{X_g\gamma_g}{\varrho_g} + \frac{X_l\gamma_l}{\varrho_l}\right] - \frac{T}{\bar{C}^p}\left[\frac{X_g\beta_g}{\varrho_g} + \frac{X_l\beta_l}{\varrho_l}\right]^2}}, \tag{4.22}$$

with the specific heat at constant pressure of the two-phase mixture

$$\bar{C}^p = X_g C_g^p + X_l C_l^p. \tag{4.23}$$

Using the volume-averaged quantities for the two-phase compressibility γ and, the thermal expansion β,

$$\left.\begin{aligned} \gamma &= \alpha_g \gamma_g + \alpha_g \gamma_g \\ \beta &= \alpha_g \beta_g + \alpha_g \beta_g \end{aligned}\right\} \tag{4.24}$$

expression (4.22) can be further simplified to

$$a = \sqrt{\frac{\bar{C}^p}{\varrho \gamma \bar{C}^p - T\beta^2}}. \tag{4.25}$$

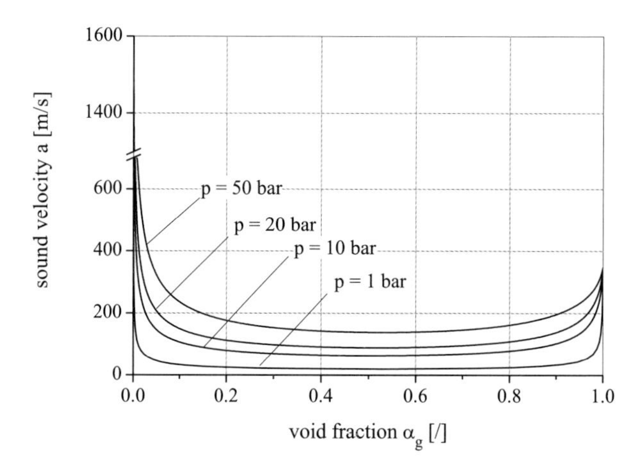

Fig. 4.1: Homogeneous equilibrium sound velocity in *two-phase water/air* media for various pressure values; fluid temperature $T = 300$ K

The single-phase sound velocities for gas and liquid phases can be obtained from equation (4.22) as limiting cases for vanishing gas or liquid content respectively as

$$X_g \to 1: \quad a_1 = \sqrt{\frac{C_g^p}{\gamma_g \varrho_g C_g^p - T\beta_g^2}} = a_g \tag{4.26}$$

$$X_l \to 1: \quad a_2 = \sqrt{\frac{C_l^p}{\gamma_l \varrho_l C_l^p - T\beta_l^2}} = a_l. \tag{4.27}$$

The behavior of the equilibrium sound velocity as a function of void fraction is shown Fig. 4.1 for a water/air mixture at different pressure values. The figure indicates that the sound velocities can be considerably smaller than the limiting values for single-phase liquid or vapor.

For not too high system pressure values, the state equations for the gas and liquid phase might be further simplified assuming the state equations for a perfect gas and neglecting the effect of the thermal expansion for the liquid phase ($\beta_l = 0$). With these assumptions, the basic thermodynamic mixture quantities for internal energy, enthalpy, and entropy reduce to

$$\left.\begin{aligned} e &= e_0 + \bar{C}^v(T - T_0) \\ h &= h_0 + \bar{C}^p(T - T_0) + \frac{X_l}{\varrho_l}(p - p_0) \\ s &= s_0 + \bar{C}^p \ln\left(\frac{T}{T_0}\right) - \bar{R}\ln\left(\frac{p}{p_0}\right) \end{aligned}\right\} \tag{4.28}$$

where the specific heats for the two-phase mixture at constant pressure $\bar{C}^p$ and constant volume $\bar{C}^v$ are defined as:

$$\left.\begin{aligned} \bar{C}^p &= X_g C_g^p + X_l C_l^p \\ \bar{C}^v &= X_g C_g^v + X_l C_l^v \end{aligned}\right\} \tag{4.29}$$

and the corresponding "gas constant" becomes

$$\bar{R} = \bar{C}^p - \bar{C}^v = X_g R_g. \tag{4.30}$$

For isentropic state changes ($s = s_0$) temperature and gas density become functions of the pressure

$$\frac{T}{T_0} = \left(\frac{p}{p_0}\right)^{\frac{\bar{\varkappa} - 1}{\bar{\varkappa}}} \quad \text{and} \quad \frac{\varrho_g}{\varrho_{g,0}} = \left(\frac{p}{p_0}\right)^{\frac{\bar{\varkappa}_g - 1}{\bar{\varkappa}_g}}, \tag{4.31}$$

with the polytropic exponent for the gas phase

$$\bar{\varkappa}_g = \frac{\bar{C}^p}{\bar{C}^p - \bar{R}_g}. \tag{4.32}$$

Under these conditions the expression for the two-phase sound velocity (4.22) simplifies to

$$a = \sqrt{\frac{1}{\dfrac{\alpha_g \varrho}{\varrho_g \bar{a}_g^2} + \dfrac{\alpha_l \varrho}{\varrho_l \bar{a}_l^2}}}, \tag{4.33}$$

with

$$\bar{a}_g^2 = \bar{\varkappa}_g \frac{p}{\varrho_g} \quad \text{and} \quad \bar{a}_l^2 = \frac{1}{\varrho_l \gamma_l}. \tag{4.34}$$

As for the single-phase gas flow an iterative algebraic procedure can be derived for the steady state flow in channels of variable cross section when applying the simplified state equations for the water/air mixture as were given by equations (4.28) to (4.32). As an example results for the flow through convergent–divergent nozzles are shown in Fig. 4.2. The initial gas mass

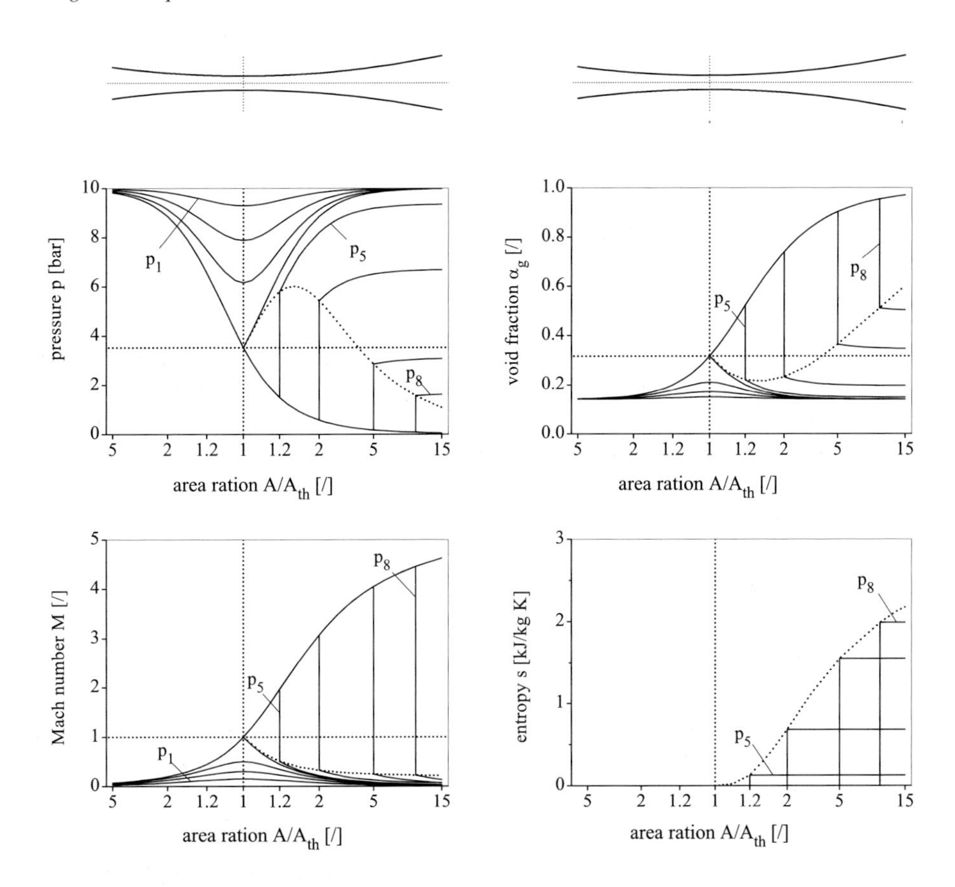

Fig. 4.2: Flow of *water/air mixture* through convergent–divergent nozzle; gas content X_0 = 0.001, reservoir pressure: $p_0 = 10$ bar, exit pressure: $p_1 \geq p_{\mathrm{exit}} \geq p_9$, $p_1 = 0.99$ bar, $p_9 = 0.04$ bar

fraction of $X_{g,0} = 0.001$ equivalent to a reservoir void fraction of $\alpha_{g,0} = 0.17$. The nozzle geometry is the same as was used for the pure gas case in Fig. 2.12.

As shown in Fig. 4.2, the two-component two-phase flow behaves qualitatively like the pure gas flow, however, larger quantitative differences occur with increasing liquid content with respect to critical pressure, flow velocities, or mass flow rates.

For many applications with not too small values for the void fraction, the compressibility of the liquid phase can be neglected and, together with the simplified state equations (4.28) to (4.32), the expression for the sound velocity (4.33) can be further simplified to

$$a = \sqrt{\bar{\varkappa} \frac{p}{\alpha_g \varrho}}. \tag{4.35}$$

For these conditions the Riemann invariants as given by equation (4.21) can be integrated resulting in

$$\mathbf{W} = \begin{bmatrix} u + \dfrac{2}{\varkappa - 1} a \\ \\ u - \dfrac{2}{\varkappa - 1} a \\ \\ s \end{bmatrix}, \tag{4.36}$$

with the mixture velocity u and sound velocity a as defined by equation (4.35). Similar to the pure gas case as presented in Chapter 2, an algebraic solution for the shock tube problem can be derived. As an example for a water–air shock tube the pipe geometry and the initial conditions are specified as given in Fig. 4.3.

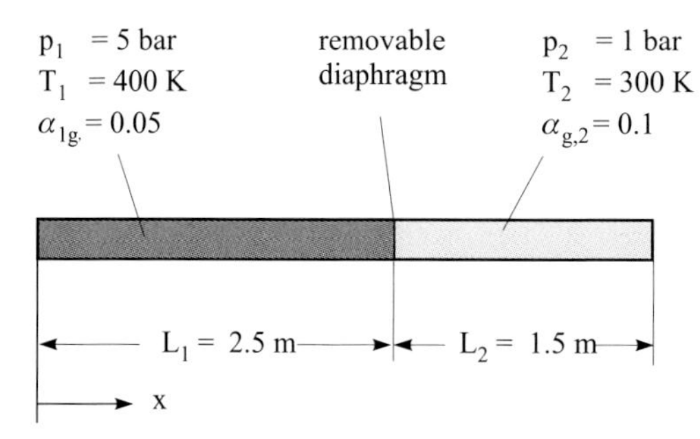

Fig. 4.3: Shock tube problem for *two-phase water/air* mixture; geometry and initial conditions

The calculated results for the governing flow parameters as given in Fig. 4.4 show a behavior which is qualitatively very similar to that obtained for the gas case. This includes in particular the propagation of a discontinuous shock wave and a contact surface into the right low pressure region of the tube, and the continuous expansion wave penetrating into the left high pressure region. However, there are also some large quantitative differences. Compared to the pure gas case, the wave velocities are drastically reduced, which becomes especially evident for the contact discontinuity. These differences originate from the reduced two-phase sound velocity and from the relatively high density of the two-phase mixture and related strong inertia effects.

As will be shown in Chapter 9, the solution obtained for homogeneous equilibrium conditions represents (at least for the present initial conditions) a reasonable approximation for the general case of nonhomogeneous nonequilibrium flow.

4.1.2 One-component two-phase flow

In one-component two-phase media undergoing phase transitions (evaporation or condensation), the thermal equilibrium assumption results in a strict coupling of pressure and temperature as given by the saturation condition $p = p^{\text{sat}}(T)$ as long as $0 \leq X_g \leq 1$. As a consequence, all thermodynamic quantities for liquid and vapor under saturation condition

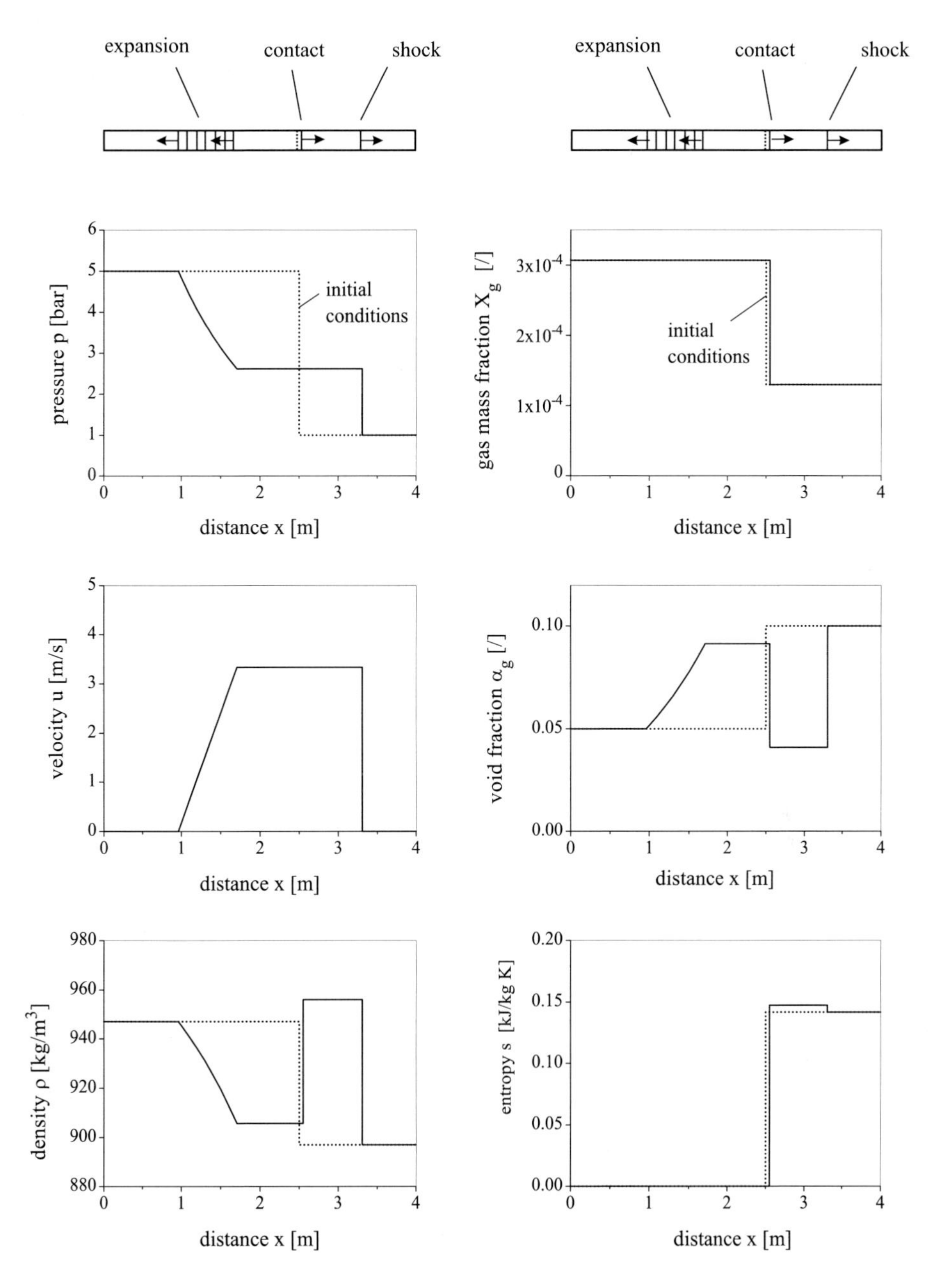

Fig. 4.4: Shock tube problem for *two-phase water/air* mixture; parameter distributions at $t = 15$ ms, analytical solution for homogeneous equilibrium flow conditions

become functions of only temperature or pressure, respectively,

$$\{\varrho_i, e_i, h_i, s_i\} = f(T) \qquad \text{for} \quad i = g, l. \tag{4.37}$$

For pressure values considerably below the critical point, and using the simplified state equations for liquid and vapor as introduced by equations (4.28) and (4.30), a relation for the saturated conditions can be obtained from the identity $\Delta s = s_g - s_l = (h_g - h_l)/T$,

$$\ln\left(\frac{p^{\text{sat}}}{p_0}\right) = \frac{\Delta s_0}{R_g}\left(1 - \frac{1}{T/T_0}\right) - \frac{C_l - C_g^p}{R_g}\ln\left(\frac{T}{T_0}\right) + \frac{\varrho_{g,0}}{\varrho_l}\frac{p^{\text{sat}}/p_0 - 1}{T/T_0}, \tag{4.38}$$

with the known reference state at $T = T_0$ and $p = p_0$. The saturation pressure p_s can then be obtained iteratively, starting with a first guess value

$$\left(\frac{p^{\text{sat}}}{p_0}\right)_n = \exp\left[\frac{\Delta s_0}{R_g}\left(1 - \frac{1}{T/T_0}\right) - \frac{C_l - C_g^p}{R_g}\ln\left(\frac{T}{T_0}\right) + \frac{\varrho_{g,0}}{\varrho_l}\frac{p^{\text{sat}}/p_0 - 1}{T/T_0}\right]. \tag{4.39}$$

With an appropriate choice for C_l, C_g^p, and R_g, a good approximation of the saturation curve can be found as shown in Fig. 4.5 for a water/steam mixture.

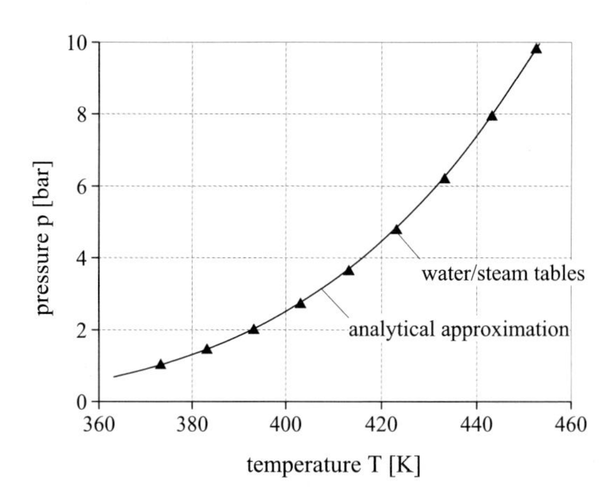

Fig. 4.5: Saturation pressure as a function of temperature; comparison of analytical approximation with steam table data

For the equilibrium sound velocity defined in (4.12) one obtains for one-component two-phase media

$$a = \frac{1}{\varrho}\sqrt{\frac{1}{\left[\dfrac{X_g\gamma_g}{\varrho_g} + \dfrac{X_l\gamma_l}{\varrho_l}\right] - 2\left[\dfrac{X_g\beta_g}{\varrho_g} + \dfrac{X_l\beta_l}{\varrho_l}\right]\dfrac{1/\varrho_g - 1/\varrho_l}{s_g - s_l} + \dfrac{\bar{C}^p}{T}\dfrac{(1/\varrho_g - 1/\varrho_l)^2}{(s_g - s_l)^2}}}, \tag{4.40}$$

with the limiting cases:

$$X_g \to 1: \qquad a_1 = \sqrt{\frac{1}{\gamma_g \varrho_g - 2\dfrac{\beta_g}{\varrho_g}\dfrac{1/\varrho_g - 1/\varrho_l}{s_g - s_l} + \dfrac{C_g^p\, \varrho_g^2}{T}\dfrac{(1/\varrho_g - 1/\varrho_l)^2}{(s_g - s_l)^2}}} \neq a_g \quad (4.41)$$

$$X_l \to 1: \qquad a_2 = \sqrt{\frac{1}{\gamma_l \varrho_l C_l^p - 2\dfrac{\beta_l}{\varrho_l}\dfrac{1/\varrho_g - 1/\varrho_l}{s_g - s_l} + \dfrac{C_l^p\, \varrho_l^2}{T}\dfrac{(1/\varrho_g - 1/\varrho_l)^2}{(s_g - s_l)^2}}} \neq a_l. \quad (4.42)$$

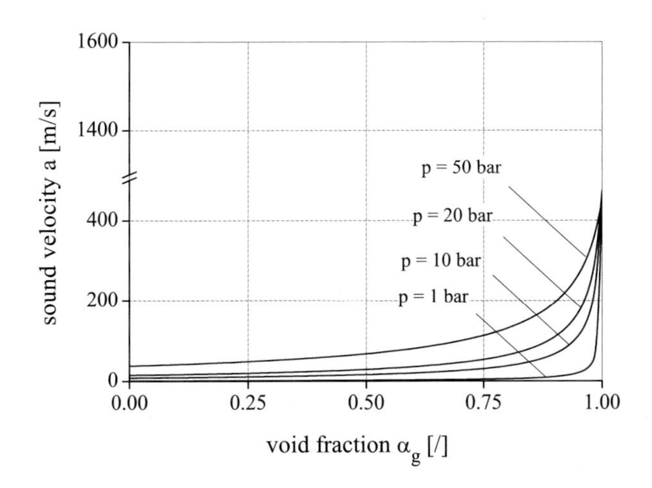

Fig. 4.6: Homogeneous equilibrium sound velocity in *two-phase water/air* media for various pressure values; fluid temperature $T = 300$ K

Introducing the volume averaged quantities for the mixture compressibility γ and the thermal expansion β as defined by equation (4.24), the equilibrium sound velocity can also be written in a more compact form as

$$a = \sqrt{\frac{1}{\gamma\, \varrho - 2\, \beta\, \varrho \dfrac{1/\varrho_g - 1/\varrho_l}{s_g - s_l} + \varrho^2 \dfrac{\bar{C}^p}{T}\dfrac{(\,1/\varrho_g - \,1/\varrho_l\,)^2}{(s_g - s_l)^2}}}. \quad (4.43)$$

Calculated values for the equilibrium sound velocity as a function of void fraction are shown in Fig. 4.6 for a saturated water/vapor mixture at atmospheric pressure. The figure indicates two interesting aspects: the thermal equilibrium assumption results in extreme low values for the sound velocity and (2) strong discontinuities for the transition between single-phase liquid or vapor and two-phase conditions which is certainly in contrast to experimental data.

The drastic change in sound velocity when crossing the saturation line between liquid and two-phase region or vapor and two-phase conditions, respectively, is illustrated in Figs. 4.7 and 4.8.

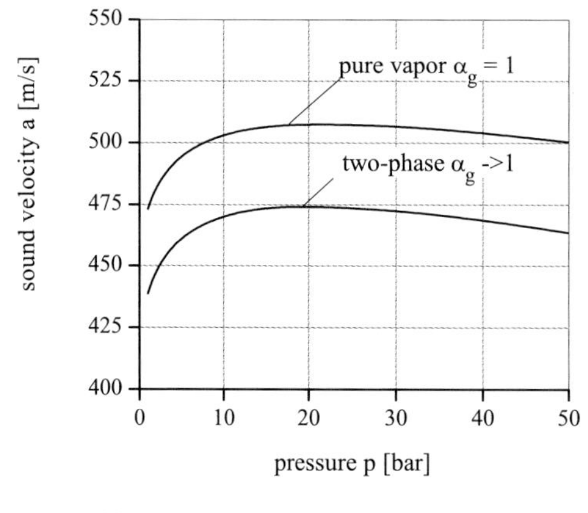

Fig. 4.7: Discontinuity of sound velocity at "right" saturation line; *water/steam mixture* at pressure $p = 1$ bar

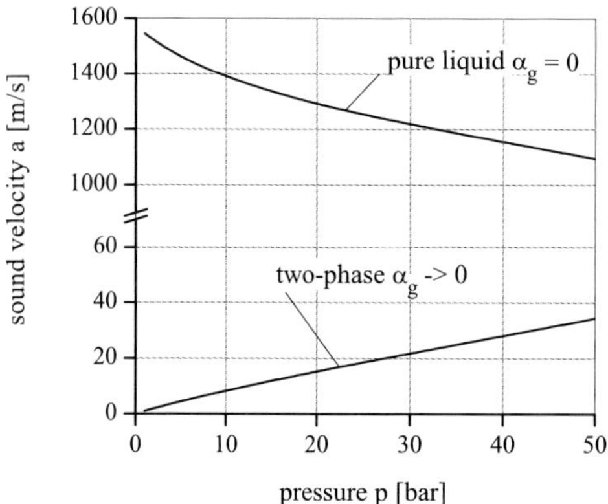

Fig. 4.8: Discontinuity of sound velocity at "left" saturation line; *water/steam mixture* at pressure $p = 1$ bar

The discontinuous behavior of the sound velocity during the transition between single and two-phase flow conditions has a strong effect for the one-component two-phase flow through convergent–divergent nozzles. This is shown in Fig. 4.9 for a water/steam mixture with a reservoir pressure of p_0 =10 bar and various values for the subcooling $\Delta T^{\mathrm{sub}} = T_0 - T^{\mathrm{sat}}(p_0)$. For subcooled reservoir conditions an important parameter is the flow velocity when crossing the saturation line at $p = p^{\mathrm{sat}}(p_0)$. Assuming constant liquid density this value can be obtained from the Bernoulli equation as

$$u^{\mathrm{sat}} = \sqrt{\frac{2\,[p_0 - p^{\mathrm{sat}}(T_0)]}{\varrho_l}}. \tag{4.44}$$

From Fig. 4.9, the following cases can be distinguished:

1. For saturated upstream reservoir conditions ($T_0 = T^{\mathrm{sat}}$) and low vapor content ($0 \leq X_g \leq \epsilon$), the onset of evaporation occurs in the divergent section of the nozzle and critical conditions at the nozzle throat are obtained as $(X_g)_{\mathrm{cr}} > 0$, and $u_{\mathrm{cr}} = a_{\mathrm{cr}}$.

2. For pure liquid ($X_g = 0$) with slightly subcooled conditions ($T_0 < T^{\mathrm{sat}}$), the crossing of the saturation line occurs in the divergent section of the nozzle with $u^{\mathrm{sat}} < a_{2,}$ and the critical conditions reached are as for case (1) at the nozzle throat with $(X_g)_{\mathrm{cr}} > 0$ and $u_{\mathrm{cr}} = a_{\mathrm{cr}}$.

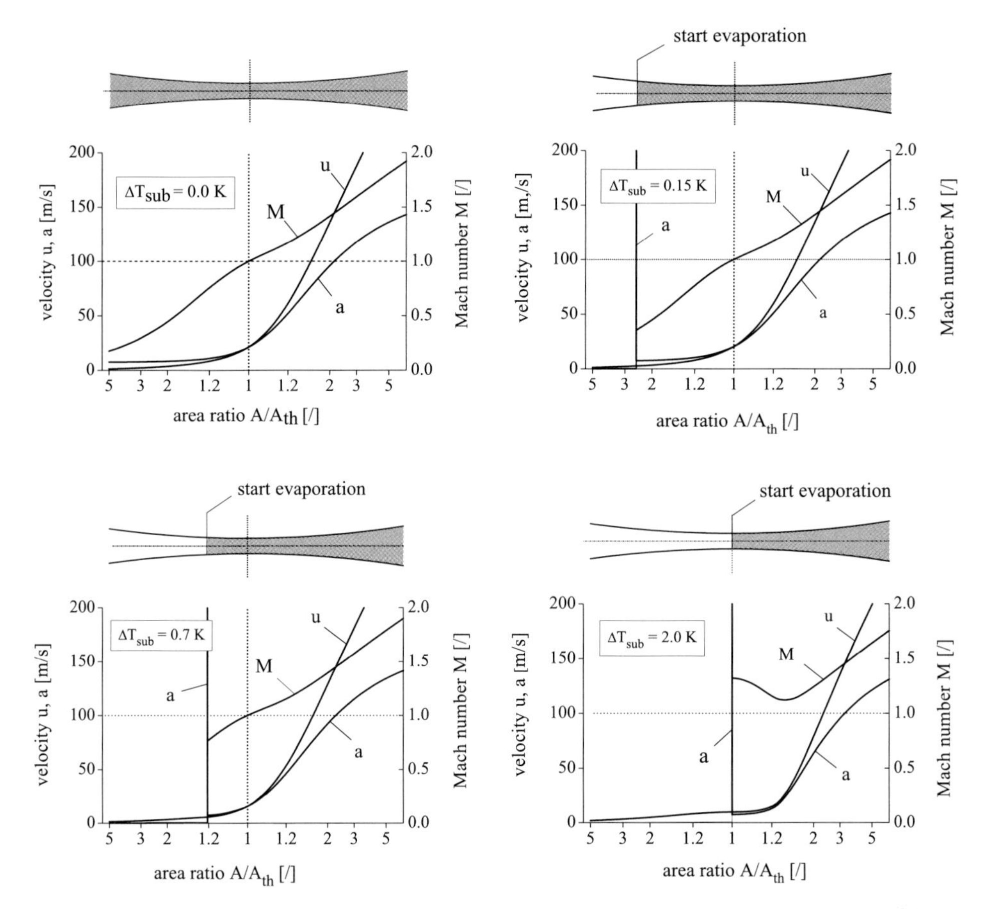

Fig. 4.9: Critical flow of *water/steam mixtures* through convergent–divergent nozzle; effect of subcooling at the upstream reservoir

3. With increasing degree of subcooling, the location where the saturation line is reached moves further downstream, and if it has reached the nozzle throat, the velocity at the saturation line is equal with the limiting value for the two-phase sound velocity $u^{\mathrm{sat}} = a_2$. In this case the location for the onset of evaporation becomes identical with the critical conditions at the nozzle throat defined as $(X_g)_{\mathrm{cr}} = 0$, $p_{\mathrm{cr}} = p^{\mathrm{sat}}(T_0)$, $u_{\mathrm{cr}} = u^{\mathrm{sat}} = (a_2)_{\mathrm{cr}}$.

4. A further increase of the degree of subcooling will not have any effect on the location for the onset of evaporation and the critical state remains as in the case 3 with $(X_g)_{\mathrm{cr}} = 0$, $p_{\mathrm{cr}} = p^{\mathrm{sat}}(T_0)$, $u_{\mathrm{cr}} = u^{\mathrm{sat}}$. Due to $u^{\mathrm{sat}} > a_2$, the Mach number at the nozzle throat changes discontinuously from subsonic ($M < 0$) to supersonic conditions ($M > 0$). For these conditions, the critical (maximum) mass flow can be explicitly given as

$$m_{\mathrm{cr}} = \varrho_l u^{\mathrm{sat}} A_{\mathrm{th}} \tag{4.45}$$

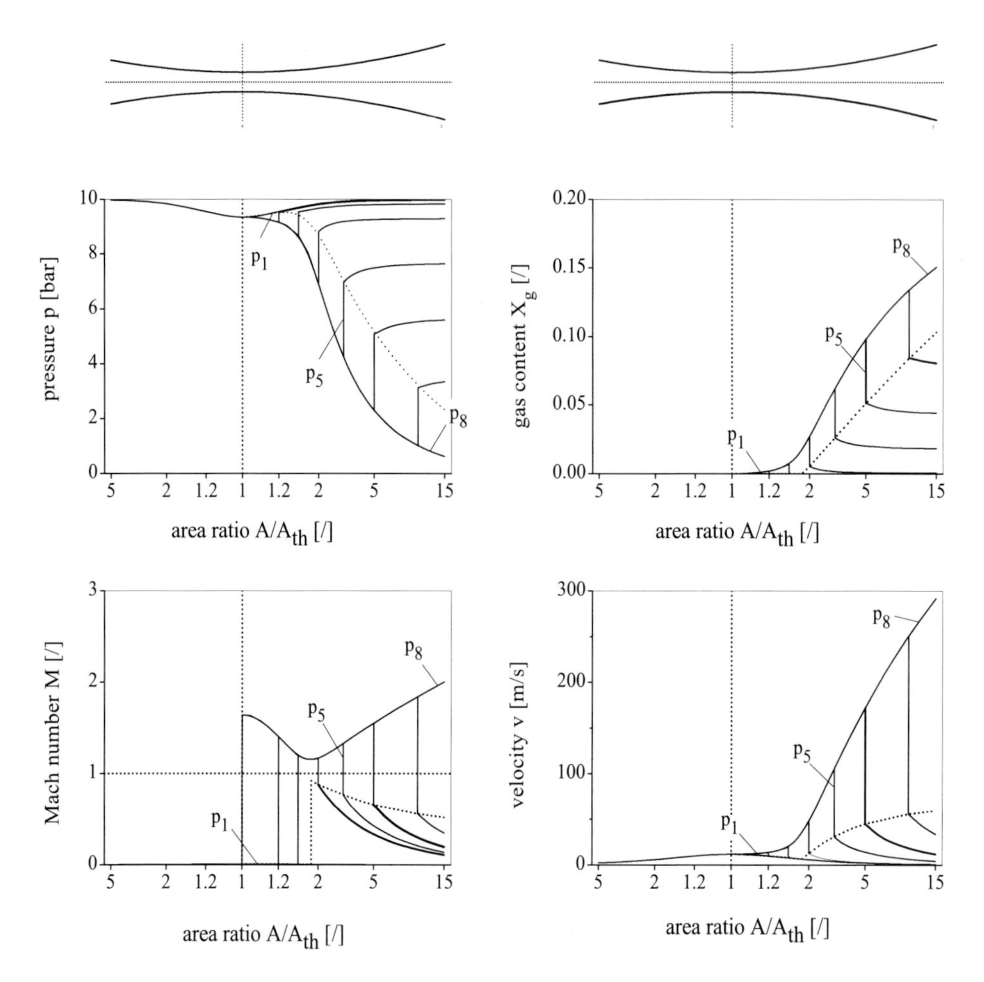

Fig. 4.10: Flow of *initially subcooled water* through a convergent–divergent nozzle; reservoir conditions: $p_0 = 10$ bar, $\Delta T^{\text{sub}} = 3$ K; exit pressure: $p_1 \geq p^{\text{exit}} \geq p_8$, $p_1 = 0.997$ bar, $p_5 = 4.243$ bar, $p_8 = 0.620$ bar

or with equation (4.44)

$$m_{\text{cr}} = \sqrt{2\varrho_l \left[p_0 - p^{\text{sat}}(T_0)\right]}A_{\text{th}}. \tag{4.46}$$

A more detailed picture of the critical flow of water/steam mixtures through convergent–divergent nozzles is given in Fig. 4.10 for subcooled reservoir conditions $\Delta T^{\text{sub}} = 3$ K and a pressure $p_0 = 10$ bar. For these conditions, the evaporation starts at the nozzle throat as was shown in Fig. 4.9. The free parameter is the pressure at the exit plane A_{exit}.

4.2 Homogeneous nonequilibrium two-phase flow

A step toward a more realistic description of two-phase flow processes is to consider deviations from the thermal equilibrium between the phases. This means to allow for differences in the local phase temperature and, in the case of one-component flows, for deviations from the state of saturation. As a consequence, additional correlations describing the mass and energy transfer process between the two phases have to be provided.

The homogeneous thermal nonequilibrium two-phase flow is described by two separate mass equations, two separate energy equations, and one momentum equation for the whole two-phase mixture. Removing the kinetic terms from the energy equations and introducing the entropy as a new state variable, the conservation equations can be written for one-dimensional flow conditions as

mass:

$$\frac{\partial}{\partial t}(\alpha_g \varrho_g) + \frac{\partial}{\partial x}(\alpha_g \varrho_g u) = \sigma_g^M \tag{4.47}$$

$$\frac{\partial}{\partial t}(\alpha_l \varrho_l) + \frac{\partial}{\partial x}(\alpha_l \varrho_l u) = \sigma_l^M \tag{4.48}$$

$$\text{with} \quad \sigma_g^M + \sigma_l^M = 0$$

momentum:

$$\frac{\partial}{\partial t}(\varrho u) + \frac{\partial}{\partial t}(\varrho u^2) + \frac{\partial p}{\partial x} = F \tag{4.49}$$

entropy:

$$\frac{\partial}{\partial t}(\alpha_g \varrho_g s_g) + \frac{\partial}{\partial x}(\alpha_g \varrho_g u s_g) = \sigma_g^M s_g + \frac{\sigma_g^M}{T_g}(h^{\mathrm{ex}} - h_g) + \frac{\sigma_g^Q}{T_g} + \frac{Q_g}{T_g} \tag{4.50}$$

$$\frac{\partial}{\partial t}(\alpha l \varrho_l\, s_l) + \frac{\partial}{\partial x}(\alpha_l \varrho_l u s_l) = \sigma_l^M s_l + \frac{\sigma_l^M}{T_l}(h^{\mathrm{ex}} - h_l) + \frac{\sigma_l^Q}{T_l} + \frac{Q_l}{T_l} \tag{4.51}$$

$$\text{with} \quad \sigma_g^Q + \sigma_l^Q = 0.$$

The external heat source in the entropy equations has been partitioned between a vapor and a liquid part $Q = Q_g + Q_l$. The mixture density in the momentum equation (4.49) remains the same as for the equilibrium conditions as $\varrho = \alpha_g \varrho_g + \alpha_l \varrho_l$.

The densities of the two phases are determined by state equations of the form $\varrho_i = f(p, s_i)$, or

$$\delta \varrho_i = \frac{1}{a_i^2}\,\delta p + \left(\frac{\partial \varrho_i}{\partial s_i}\right)_p \delta s_i. \tag{4.52}$$

Expanding the time and space derivatives, the conservation equations become

mass:

$$\frac{\alpha_g}{a_g^2}\left(\frac{\partial p}{\partial t}+u\frac{\partial p}{\partial x}\right)+\varrho_g\left(\frac{\partial \alpha_g}{\partial t}+u\frac{\partial \alpha_g}{\partial x}\right)+\alpha_g\varrho_g\frac{\partial u}{\partial x}$$
$$+\alpha_g\left(\frac{\partial \varrho_g}{\partial s_g}\right)_p\left(\frac{\partial s_g}{\partial t}+u\frac{\partial s_g}{\partial x}\right)=\sigma_g^M \tag{4.53}$$

$$\frac{\alpha_l}{a_l^2}\left(\frac{\partial p}{\partial t}+u\frac{\partial p}{\partial x}\right)+\varrho_l\left(\frac{\partial \alpha_l}{\partial t}+u\frac{\partial \alpha_l}{\partial x}\right)+\alpha_l\varrho_l\frac{\partial u}{\partial x}+$$
$$\alpha_l\left(\frac{\partial \varrho_l}{\partial s_l}\right)_p\left(\frac{\partial s_l}{\partial t}+u\frac{\partial s_l}{\partial x}\right)=\sigma_l^M \tag{4.54}$$

momentum:

$$\varrho\frac{\partial u}{\partial t}+\varrho u\frac{\partial u}{\partial x}+\frac{\partial p}{\partial x}=F \tag{4.55}$$

entropy:

$$\alpha_g\varrho_g\frac{\partial s_g}{\partial t}+\alpha_g\varrho_g u_g\frac{\partial s_g}{\partial x}=\left(h^{\text{ex}}-h_g\right)\frac{\sigma_g^M}{T_g}+\frac{\sigma_g^Q}{T_g}+\frac{Q_g}{T_g}=\sigma_g^S \tag{4.56}$$

$$\alpha_l\varrho_l\frac{\partial s_l}{\partial t}+\alpha_l\varrho_l u_l\frac{\partial s_l}{\partial x}=\left(h^{\text{ex}}-h_l\right)\frac{\sigma_l^M}{T_l}+\frac{\sigma_l^Q}{T_l}+\frac{Q_l}{T_l}=\sigma_l^S. \tag{4.57}$$

Equations (4.53) to (4.57) can be arranged in a vector form as

$$\mathbf{A}\frac{\partial \mathbf{U}}{\partial t}+\mathbf{B}\frac{\partial \mathbf{U}}{\partial x}=\mathbf{C}, \tag{4.58}$$

with the state and source term vectors $\mathbf{U}$ and $\mathbf{C}$, respectively,

$$\mathbf{U}=\begin{bmatrix} p \\ u \\ \alpha_g \\ s_g \\ s_l \end{bmatrix}, \qquad \mathbf{C}=\begin{bmatrix} \sigma_g^M \\ \sigma_l^M \\ F \\ \sigma_g^M\left(h^{\text{ex}}-h_g\right)+\dfrac{\sigma_g^M}{T_g}+\dfrac{Q_g}{T_g} \\ \sigma_l^M\left(h^{\text{ex}}-h_l\right)+\dfrac{\sigma_l^M}{T_l}+\dfrac{Q_l}{T_l} \end{bmatrix}$$

and the coefficient matrices $\mathbf{A}$ and $\mathbf{B}$ as given in Tables B.18 and B.21 in Appendix B.

Multiplying the vector equation (4.46) with the inverse matrix $\mathbf{A}^{-1}$ yields

$$\frac{\partial \mathbf{U}}{\partial t} + \mathbf{G}\frac{\partial \mathbf{U}}{\partial x} = \mathbf{D}, \tag{4.59}$$

with the coefficient matrix $\mathbf{G} = \mathbf{A}^{-1}\mathbf{B}$

$$\mathbf{G} = \begin{bmatrix} u & \dfrac{\varrho_g \varrho_l a_g^2 a_l^2}{\alpha_g \varrho_l a_l^2 + \alpha_l \varrho_g a_g^2} & 0 & 0 & 0 \\ \dfrac{1}{\varrho} & u & 0 & 0 & 0 \\ 0 & -\dfrac{\alpha_g \alpha_l \left(\varrho_l a_l^2 - \varrho_g a_g^2\right)}{\alpha_g \varrho_l a_l^2 + \alpha_l \varrho_g a_g^2} & u & 0 & 0 \\ 0 & 0 & 0 & u & 0 \\ 0 & 0 & 0 & 0 & u \end{bmatrix} \tag{4.60}$$

and the new source vector $\mathbf{D}$ given in Table B.17 in Appendix B. The eigenvalues of the coefficient matrix $\mathbf{G}$ are determined by the characteristic equation

$$\det\left(\mathbf{G} - \lambda \mathbf{I}\right) = 0, \tag{4.61}$$

which results in

$$\lambda_1 = u + a, \qquad \lambda_2 = u - a, \qquad \lambda_{3,4,5} = u, \tag{4.62}$$

with the sound velocity of the two-phase mixture as

$$a = \sqrt{\frac{1}{\dfrac{\alpha_g \varrho}{a_g^2 \varrho_g} + \dfrac{\alpha_l \varrho}{a_l^2 \varrho_l}}}. \tag{4.63}$$

The effect of the thermal nonequlibrium on the two-phase sound velocity is shown in Fig. 4.11 for a water/vapor mixture at 10 bar.

As can be seen from this figure, the inclusion of thermal nonequilibrium effects avoids the discontinuities as were typical of the equilibrium conditions and, in addition, more realistic values are calculated for small vapor contents.

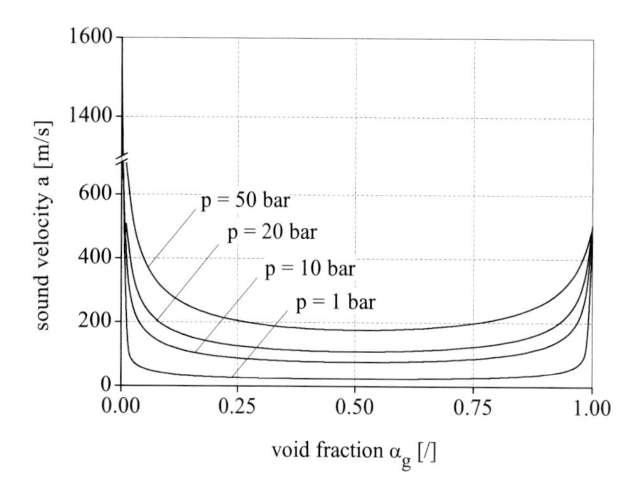

Fig. 4.11: Two-phase homogeneous sound velocity, *water/steam mixture* under saturated conditions at p = 10 bar

Introducing the sound velocity a, the coefficient matrix $\mathbf{G}$ can be further simplified to

$$\mathbf{G} = \begin{bmatrix} u & \varrho a^2 & 0 & 0 & 0 \\ \frac{1}{\varrho} & u & 0 & 0 & 0 \\ 0 & -\alpha_g \alpha_l \left(\frac{\varrho a^2}{\varrho_g a_g^2} - \frac{\varrho a^2}{\varrho_l a_l^2} \right) & u & 0 & 0 \\ 0 & 0 & 0 & u & 0 \\ 0 & 0 & 0 & 0 & u \end{bmatrix} . \tag{4.64}$$

Despite the fact that three of the five eigenvalues are identical, a complete set of independent eigenvectors exists which is the condition for the hyperbolic character of the governing equations. These eigenvectors are

pressure wave: $\lambda_1 = u + a$

$$V_{R,1} = \left[a\varrho \, , \, 1 \, , \, -\alpha_g \alpha_l \left(\frac{a\varrho}{\varrho_g a_g^2} - \frac{a\varrho}{\varrho_l a_l^2} \right) \, , \, 0 \, , \, 0 \right] \tag{4.65}$$

pressure wave: $\lambda_2 = u - a$

$$V_{R,2} = \left[a\varrho \, , \, 1 \, , \, \alpha_g \alpha_l \left(\frac{a\varrho}{\varrho_g a_g^2} - \frac{a\varrho}{\varrho_l a_l^2} \right) \, , \, 0 \, , \, 0 \right] \tag{4.66}$$

void wave: $\lambda_3 = u$

$$V_{R,3} = [\,0\,,\,0\,,\,1\,,\,0\,,\,0\,] \tag{4.67}$$

entropy wave $\lambda_4 = u$

$$V_{R,4} = [\,0\,,\,,0\,,\,0\,,\,1\,,\,0\,] \tag{4.68}$$

entropy wave $\lambda_5 = u$

$$V_{R,5} = [\,0\,,\,0\,,\,0\,,\,0\,,\,1\,]\,. \tag{4.69}$$

Since the matrix $\mathbf{G}$ has a set of independent eigenvectors, there exists a matrix $\mathbf{T}$ which allows a similarity transformation

$$\mathbf{T}^{-1}\frac{\partial \mathbf{U}}{\partial t} + \mathbf{\Lambda}\,\mathbf{T}^{-1}\frac{\partial \mathbf{U}}{\partial x} = \mathbf{T}^{-1}\mathbf{D}, \tag{4.70}$$

where

$$\mathbf{\Lambda} = \mathbf{T}^{-1}\mathbf{G}\,\mathbf{T} \tag{4.71}$$

is the diagonal matrix of the eigenvalues

$$\mathbf{\Lambda} = \begin{bmatrix} u+a & 0 & 0 & 0 & 0 \\ 0 & u-a & 0 & 0 & 0 \\ 0 & 0 & u & 0 & 0 \\ 0 & 0 & 0 & u & 0 \\ 0 & 0 & 0 & 0 & u \end{bmatrix}. \tag{4.72}$$

The columns of the transformation matrix $\mathbf{T}$ are the right eigenvectors of $\mathbf{G}$

$$\mathbf{T} = \begin{bmatrix} a\varrho & -a\varrho & 0 & 0 & 0 \\ 1 & 1 & 0 & 0 & 0 \\ -\alpha_g\alpha_l\left(\dfrac{a\varrho}{\varrho_g a_g^2} - \dfrac{a\varrho}{\varrho_l a_l^2}\right) & \alpha_g\alpha_l\left(\dfrac{a\varrho}{\varrho_g a_g^2} - \dfrac{a\varrho}{\varrho_l a_l^2}\right) & 1 & 0 & 0 \\ 0 & 0 & 0 & 1 & 0 \\ 0 & 0 & 0 & 0 & 1 \end{bmatrix} \tag{4.73}$$

with the inverse

$$\mathbf{T}^{-1} = \begin{bmatrix} \dfrac{1}{2a\varrho} & \dfrac{1}{2} & 0 & 0 & 0 \\ -\dfrac{1}{2a\varrho} & \dfrac{1}{2} & 0 & 0 & 0 \\ \alpha_g\alpha_l\left(\dfrac{1}{\varrho_g a_g^2} - \dfrac{1}{\varrho_l a_l^2}\right) & 0 & 1 & 0 & 0 \\ 0 & 0 & 0 & 1 & 0 \\ 0 & 0 & 0 & 0 & 1 \end{bmatrix}. \tag{4.74}$$

As already explained for the case of single-phase gas flows, the coefficient matrix $\mathbf{G}$ can be split into elementary parts related to each of the eigenvalues

$$\mathbf{G} = \sum_{k=1}^{5} \mathbf{G}_k, \tag{4.75}$$

with $\mathbf{G}_k = \mathbf{T}\,\mathbf{\Lambda}_k\mathbf{T}^{-1}$, where the diagonal matrix $\mathbf{\Lambda}_k$ includes only the kth eigenvalue. All the corresponding split matrices are given by Tables B.18 to B.21 in Appendix B.

With the diagonal matrix given in equation (4.72), the transformed system of equations can be written as

$$\mathbf{T}^{-1}\frac{\partial \mathbf{U}}{\partial t} + \mathbf{\Lambda}\,\mathbf{T}^{-1}\frac{\partial \mathbf{U}}{\partial x} = \mathbf{T}^{-1}\mathbf{D}, \tag{4.76}$$

which results in the following compatibility relations for the "characteristic" directions $dx/dt = \lambda_k$:

pressure wave: $\dfrac{dx}{dt} = \lambda_1 = u + a$

$$\frac{dp}{dt} + \varrho a\,\frac{du}{dt} = a\,F + \varrho a^2\left(\frac{1}{\varrho_g} - \frac{1}{\varrho_l}\right)\sigma_g^M - \frac{\varrho a^2}{\varrho_g^2}\left(\frac{\partial \varrho_g}{\partial s_g}\right)_p \sigma_g^S - \frac{\varrho a^2}{\varrho_l^2}\left(\frac{\partial \varrho_l}{\partial s_l}\right)_p \sigma_l^S \tag{4.77}$$

pressure wave: $\dfrac{dx}{dt} = \lambda_1 = u - a$

$$\frac{dp}{dt} + \varrho a\,\frac{du}{dt} = a\,F - \varrho a^2\left(\frac{1}{\varrho_g} - \frac{1}{\varrho_l}\right)\sigma_g^M - \frac{\varrho a^2}{\varrho_g^2}\left(\frac{\partial \varrho_g}{\partial s_g}\right)_p \sigma_g^S - \frac{\varrho a^2}{\varrho_l^2}\left(\frac{\partial \varrho_l}{\partial s_l}\right)_p \sigma_l^S \tag{4.78}$$

void wave: $\dfrac{dx}{dt} = \lambda_3 = u$

$$\alpha_g\alpha_l\left(\frac{1}{\varrho_g a_g^2} - \frac{1}{\varrho_l a_l^2}\right)\frac{dp}{dt} + \frac{d\alpha_g}{dt} = \frac{\varrho}{\varrho_g\varrho_l}\sigma_g^M - \frac{\alpha_l}{\varrho_g^2}\left(\frac{\partial \varrho_g}{\partial s_g}\right)_p \sigma_g^S + \frac{\alpha_g}{\varrho_l^2}\left(\frac{\partial \varrho_l}{\partial s_l}\right)_p \sigma_l^S \tag{4.79}$$

entropy wave: $\dfrac{dx}{dt} = \lambda_4 = u$

$$\frac{ds}{dt} = \frac{1}{\alpha_g \varrho_g} \sigma_g^S \tag{4.80}$$

entropy wave: $\dfrac{dx}{dt} = \lambda_5 = u$

$$\frac{ds}{dt} = \frac{1}{\alpha_l \varrho_l} \sigma_l^S . \tag{4.81}$$

The compatibility relations (4.77) to (4.81) give some insight into the wave propagation phenomena. The first two eigenvalues $\lambda_{1,2}$ correspond, as in the gasdynamic case, to pressure/density fluctuations propagating with the velocity of sound relative to the flow of the two-phase mixture. The last two eigenmodes, $\lambda_{4,5}$, describe entropy (or temperature) fluctuations propagating with the material transport velocity of the mixture. There exists no equivalence in gasdynamics for the third eigenvalue λ_3. This eigenmode describes void/pressure changes propagating as the entropy (temperature) fluctuations with the material transport velocity u. In the case of incompressible phases $(a_g \to \infty,\ a_l \to \infty)$, the third eigenmode reduces to a pure void wave

$$\frac{d\alpha_g}{dt} = \frac{\varrho}{\varrho_g \varrho_l} \sigma_g^M - \frac{\alpha_l}{\varrho_g^2} \left(\frac{\partial \varrho_g}{\partial s_g} \right)_p \sigma_g^S + \frac{\alpha_g}{\varrho_l^2} \left(\frac{\partial \varrho_l}{\partial s_l} \right)_p \sigma_l^S . \tag{4.82}$$

The system of ordinary differential equations (4.77) to (4.81) has been used by Ferch [3] as the basis for a "wave tracing algorithm" for homogeneous nonequilibrium transient flow. The author reports that this technique which appears to be close to the classical method of characteristics has largely reduced the numerical diffusion effects as compared with standard finite difference schemes. However, it was also observed that numerical difficulties can arise for transitions between single- and two-phase conditions due to the large sensitivity of the sound velocity a with respect to the volumetric gas fraction α_g for very low or very high gas volume fractions.

4.3 Wallis model

For nonhomogeneous flow conditions $(u_g \neq u_l)$ a closed system of equations can be obtained assuming equal local pressure values for the two phases:

$$p_g = p_g^{\text{int}} = p_l^{\text{int}} = p_l = p.$$

If it is further assumed that the source terms on the right-hand sides of the balance equations are exclusive algebraic functions of state and flow parameters, the single pressure model leads to what is often referred to as the "Wallis model" of nonhomogeneous two-phase flow [1].

mass:

$$\frac{\partial}{\partial t}(\alpha_g \varrho_g) + \frac{\partial}{\partial x}(\alpha_g \varrho_g u_g) = \sigma_g^M \tag{4.83}$$

$$\frac{\partial}{\partial t}(\alpha_l \varrho_l) + \frac{\partial}{\partial x}(\alpha_l \varrho_l u_l) = \sigma_l^M \tag{4.84}$$

momentum:

$$\frac{\partial}{\partial t}(\alpha_g \varrho_g u_g) + \frac{\partial}{\partial x}(\alpha_g \varrho_g u_g) + \alpha_g \frac{\partial p}{\partial x} = F_g^{\text{int}} + F_g + \sigma_g^M u^{\text{ex}} \tag{4.85}$$

$$\frac{\partial}{\partial t}(\alpha_l \varrho_l u_l) + \frac{\partial}{\partial x}(\alpha_l \varrho_l u_l) + \alpha_l \frac{\partial p}{\partial x} = F_l^{\text{int}} + F_l + \sigma_l^M u^{\text{ex}} \tag{4.86}$$

energy:

$$\frac{\partial}{\partial t}\left[\alpha_g \varrho_g (u_g + \frac{u_g^2}{2})\right] + \frac{\partial}{\partial x}\left[\alpha_g \varrho_g u_g (u_g + \frac{p_g}{\varrho_g} + \frac{u_g^2}{2})\right] + p\frac{\partial \alpha_g}{\partial t}$$

$$= \sigma_g^Q + Q_g + F_g^{\text{int}} u^{\text{int}} + F_g u_g + \sigma_g^M \left[h^{\text{ex}} + \frac{(u^{\text{ex}})^2}{2}\right] \tag{4.87}$$

$$\frac{\partial}{\partial t}\left[\alpha_l \varrho_l (u_l + \frac{u_l^2}{2})\right] + \frac{\partial}{\partial x}\left[\alpha_l \varrho_l u_l (u_l + \frac{p_l}{\varrho_l} + \frac{u_l^2}{2})\right] + p\frac{\partial \alpha_l}{\partial t}$$

$$= \sigma_f^Q + Q_l + F_l^{\text{int}} u^{\text{int}} + F_l u_l + \sigma_l^M \left[h^{\text{ex}} + \frac{(u^{\text{ex}})^2}{2}\right] \tag{4.88}$$

entropy:

$$\frac{\partial}{\partial t}(\alpha_g \varrho_g s_g) + \frac{\partial}{\partial x}(\alpha_g \varrho_g u_g s_g) = \frac{\sigma_g^Q}{T_g} + \frac{Q_g}{T_g} + \frac{F_g^{\text{int}}}{T_g}(u^{\text{int}} - u_g)$$

$$+\frac{\sigma_g^M}{T_g}\left[h^{\text{ex}} - h_g + \frac{1}{2}(u^{\text{ex}} - u_g)^2\right] + \sigma_g^M s_g \tag{4.89}$$

$$\frac{\partial}{\partial t}(\alpha_l \varrho_l s_l) + \frac{\partial}{\partial x}(\alpha_l \varrho_l u_l s_l) = \frac{\sigma_f^Q}{T_l} + \frac{Q_l}{T_l} + \frac{F_l^{\text{int}}}{T_l}(u^{\text{int}} - u_l)$$

$$+\frac{\sigma_l^M}{T_l}\left[h^{\text{ex}} - h_l + \frac{1}{2}(u^{\text{ex}} - u_l)^2\right] + \sigma_l^M s_l \tag{4.90}$$

With the state equations $\varrho_i = f(p, s_i)$, or in the differential form

$$\delta\varrho_i = \frac{1}{a_i^2}\,\delta p + \left(\frac{\partial \varrho_i}{\partial s_i}\right)_p \delta s_i, \tag{4.91}$$

the expansion of the time and space derivative terms of equations (4.83) to (4.89) yields the nonconservative forms of the mass, momentum, and entropy equations

mass:

$$\frac{\alpha_g}{a_g^2}\left(\frac{\partial p}{\partial t} + u_g\frac{\partial p}{\partial x}\right) + \varrho_g\left(\frac{\partial \alpha_g}{\partial t} + u_g\frac{\partial \alpha_g}{\partial x}\right)$$
$$+\alpha_g\left(\frac{\partial \varrho_g}{\partial s_g}\right)_p\left(\frac{\partial s_g}{\partial t} + u_g\frac{\partial s_g}{\partial x}\right) + \alpha_g\varrho_g\frac{\partial u}{\partial x} = \sigma_g^M \tag{4.92}$$

$$\frac{\alpha_l}{a_f^2}\left(\frac{\partial p}{\partial t} + u_l\frac{\partial p}{\partial x}\right) + \varrho_l\left(\frac{\partial \alpha_l}{\partial t} + u_l\frac{\partial \alpha_l}{\partial x}\right)$$
$$+\alpha_l\left(\frac{\partial \varrho_l}{\partial s_l}\right)_p\left(\frac{\partial s_l}{\partial t} + u_l\frac{\partial s_l}{\partial x}\right) + \alpha_l\varrho_l\frac{\partial u}{\partial x} = \sigma_l^M \tag{4.93}$$

momentum:

$$\alpha_g\varrho_g\left(\frac{\partial u_g}{\partial t} + u_g\frac{\partial u_g}{\partial x}\right) + \alpha_g\frac{\partial p}{\partial x} = F_g^{\text{int}} + \sigma_g^M(u^{\text{ex}} - u_g) + F_g \tag{4.94}$$

$$\alpha_l\varrho_l\left(\frac{\partial u_l}{\partial t} + u_l\frac{\partial u_l}{\partial x}\right) + \alpha_l\frac{\partial p}{\partial x} = F_l^{\text{int}} + \sigma_l^M(u^{\text{ex}} - u_l) + F_l \tag{4.95}$$

entropy:

$$\alpha_g\varrho_g\left(\frac{\partial s_g}{\partial t} + u_g\frac{\partial s_g}{\partial x}\right) = \sigma^S \tag{4.96}$$

$$\alpha_l\varrho_l\left(\frac{\partial s_l}{\partial t} + u_l\frac{\partial s_l}{\partial x}\right) = \sigma^S \tag{4.97}$$

with the phasic entropy source terms

$$\sigma_g^S = \left[h^{\text{ex}} - h_g + \frac{1}{2}(u^{\text{ex}} - u_g)^2\right]\frac{\sigma_g^M}{T_g} + \frac{\sigma_g^Q}{T_g} + (u^{\text{int}} - u_g)\frac{F_g^{\text{int}}}{T_g} + \frac{Q_g}{T_g} \tag{4.98}$$

$$\sigma_l^S = \left[h^{\text{ex}} - h_l + \frac{1}{2}(u^{\text{ex}} - u_l)^2 \right] \frac{\sigma_l^M}{T_l} + \frac{\sigma_l^Q}{T_l} + (u^{\text{int}} - u_l)\frac{F_l^{\text{int}}}{T_l} + \frac{Q_l}{T_l}. \tag{4.99}$$

The six balance equations (4.92) to (4.97) can be combined in the vector form

$$\mathbf{A}\frac{\partial \mathbf{U}}{\partial t} + \mathbf{B}\frac{\partial \mathbf{U}}{\partial x} = \mathbf{C}, \tag{4.100}$$

with the state and source vectors $\mathbf{U}$ and $\mathbf{C}$, respectively,

$$\mathbf{U} = \begin{bmatrix} p \\ u_g \\ u_l \\ \alpha \\ s_g \\ s_l \end{bmatrix}, \qquad \mathbf{C} = \begin{bmatrix} \sigma_g^M \\ \sigma_l^M \\ F_g^{\text{int}} + \sigma_g^M (u^{\text{ex}} - u_g) + F_g \\ F_l^{\text{int}} + \sigma_l^M (u^{\text{ex}} - u_l) + F_l \\ \sigma_g^S \\ \sigma_l^S \end{bmatrix} \tag{4.101}$$

and the matrices $\mathbf{A}$ and $\mathbf{B}$ given in Tables B.23 and B.24 in Appendix B. Multiplying equation (4.100) with $\mathbf{A}^{-1}$ provides the even more compact form

$$\frac{\partial \mathbf{U}}{\partial t} + \mathbf{G}\frac{\partial \mathbf{U}}{\partial x} = \mathbf{D}, \tag{4.102}$$

with the coefficient matrix

$$\mathbf{G} = \begin{bmatrix} u_g - \alpha_l \Delta u \dfrac{\varrho_g a_0^2}{\varrho_s a_l^2} & \alpha_g a_0^2 \dfrac{\varrho_g \varrho_l}{\varrho_s} & \alpha_l a_0^2 \dfrac{\varrho_g \varrho_l}{\varrho_s} & a_0^2 \Delta u \dfrac{\varrho_g \varrho_l}{\varrho_s} & 0 & 0 \\ \dfrac{1}{\varrho_g} & u_g & 0 & 0 & 0 & 0 \\ \dfrac{1}{\varrho_l} & 0 & u_l & 0 & 0 & 0 \\ \alpha_g \alpha_l \dfrac{\Delta u}{\varrho_s} \dfrac{a_0^2}{a_g^2 a_l^2} & \alpha_g \alpha_l \dfrac{\varrho_g}{\varrho_s} \dfrac{a_0^2}{a_l^2} & -\alpha_g \alpha_l \dfrac{\varrho_l}{\varrho_s} \dfrac{a_0^2}{a_g^2} & u_l + \alpha_l \Delta u \dfrac{\varrho_g a_0^2}{\varrho_s a_l^2} & 0 & 0 \\ 0 & 0 & 0 & 0 & u_g & 0 \\ 0 & 0 & 0 & 0 & 0 & u_l \end{bmatrix}, \tag{4.103}$$

and the abbreviations

$$\left.\begin{aligned} \varrho_s &= \alpha_g \varrho_l + \alpha_l \varrho_g \\ a_0^2 &= \frac{\alpha_g \varrho_l + \alpha_l \varrho_g}{\dfrac{\alpha_g \varrho_l}{a_g^2} + \dfrac{\alpha_l \varrho_g}{a_l^2}} \end{aligned}\right\} . \tag{4.104}$$

The eigenvalues of the system of equations (4.100) or (4.102) are the roots of the characteristic equation

$$f(\lambda) = \det\left(\mathbf{B} - \lambda\,\mathbf{A}\right) = 0 \tag{4.105}$$

or

$$f(\lambda) = \det\left(\mathbf{G} - \lambda \mathbf{I}\right) = 0. \tag{4.106}$$

Two of the eigenvalues can be obtained immediately from the entropy relations

$$\lambda_5 = u_g \qquad \lambda_6 = u_l. \tag{4.107}$$

The remaining eigenvalues λ_1 to λ_4 cannot be expressed in a closed algebraic form; however, it can be shown that two of them are complex conjugate,

$$\lambda_{1,2} = u_R \pm i\,\Delta\, u_I, \tag{4.108}$$

with the real part

$$u_l \le u_R \le u_g \qquad \text{for} \qquad u_l \le u_g. \tag{4.109}$$

If the first two eigenvalues are removed from the characteristic equation (4.105), a fourth-order equation $f^\star(\lambda) = 0$ remains,

$$f^\star = \frac{f}{(u_g - \lambda)(u_l - \lambda)}, \tag{4.110}$$

or in detail

$$f^\star = \alpha_l \varrho_g \left[1 - \frac{(\lambda - u_l)^2}{a_l^2}\right](\lambda - u_g)^2 + \alpha_g \varrho_l \left[1 - \frac{(\lambda - u_{gl})^2}{a_g^2}\right](\lambda - u_l)^2. \tag{4.111}$$

As an example, the characteristic function $f^*(\lambda)$ is shown in Figs. 4.12 and 4.13 for a saturated water/steam mixture at $p = 10$ bar and a volumetric vapor fraction of $\alpha_g = 0.25$. The phasic flow velocities are $u_g = 100$ m/s and $u_l = 50$ m/s respectively. The figure indicates that there exist only two further real eigenvalues which can be interpreted as propagation velocity of of pressure waves $\lambda_{3,4} = u \pm a$ with the mixture flow velocity u and mixture sound velocity a for which where no algebraic expressions can be derived.

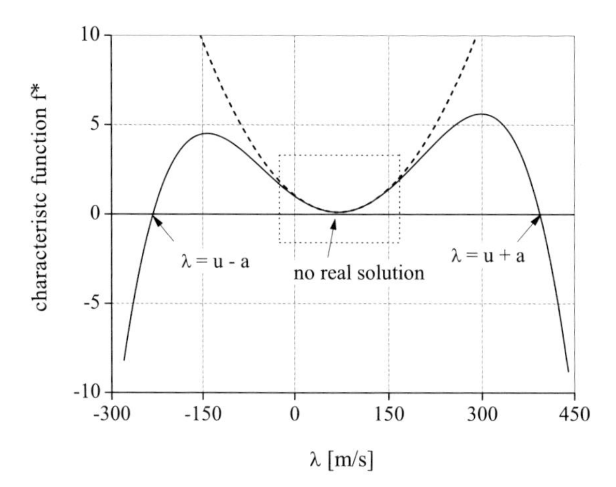

Fig. 4.12: Characteristic function of Wallis model, straight line: compressible phases, dashed line: incompressible phases

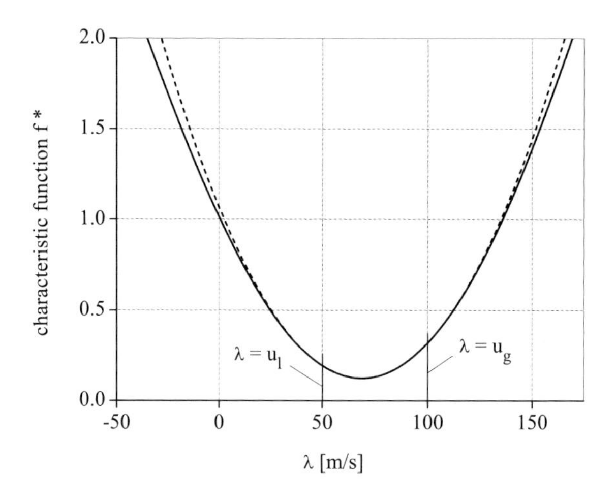

Fig. 4.13: Characteristic function of Wallis model, straight line: compressible phases, dashed line: incompressible phases

The characteristic function has the interesting property that $f^\star(\lambda)$ becomes independent of the phasic sound velocities for $\lambda = u_g$ and $\lambda = u_l$:

$$\left.\begin{array}{ll} \lambda = u_g : & f^\star = \alpha_g \varrho_l (u_g - u_l)^2 \\ \lambda = u_l : & f^\star = \alpha_l \varrho_g (u_g - u_l)^2 \end{array}\right\} . \tag{4.112}$$

For the specific case of incompressible vapor/gas and liquid, $a_g \to \infty$, $a_l \to \infty$, the characteristic function f^* reduces to the quadratic expression

$$f^{\star}(\lambda) = \alpha_l \varrho_g (\lambda - u_g)^2 + \alpha_g \varrho_l (\lambda - u_l)^2 \tag{4.113}$$

which, as shown in Figs. 4.12 and 4.13, represents a good approximation for small characteristic velocities $|\lambda| < a$. As a solution of the corresponding characteristic equation $f^{\star}(\lambda) = 0$ one obtains for the complex conjugate eigenvalues

$$\lambda_{1,2} = \frac{\alpha_g \varrho_l u_l + \alpha_l \varrho_g u_g}{\alpha_g \varrho_l + \alpha_l \varrho_g} \pm i \sqrt{\frac{\alpha_g \alpha_l \varrho_g \varrho_l}{\alpha_g \varrho_l + \alpha_l \varrho_g} (u_g - u_l)^2} \, .$$

Exclusively real eigenvalues are obtained only for the specific case of equal phasic velocities $u_g = u_l = u$

$$\left.\begin{aligned} \lambda_{1,2} &= u \\ \lambda_{3,4} &= u \pm a \\ \lambda_{5,6} &= u \end{aligned}\right\} \tag{4.114}$$

with the "sound velocity"

$$a = a_0 = \sqrt{\frac{\alpha_g \varrho_l + \alpha_l \varrho_g}{\dfrac{\alpha_g \varrho_l}{a_g^2} + \dfrac{\alpha_l \varrho_g}{a_l^2}}} \, . \tag{4.115}$$

The fact that the "Wallis model" has not a complete set of real eigenvalues has the following consequences:

- the model does not represent a "well-posed" initial-boundary value problem;
- the system of equations cannot be transposed into the characteristic form and, therefore, all numerical techniques developed for fully hyperbolic systems of equations cannot be applied;
- the model does not realistically describe pressure wave phenomena, and for this reason, it is not able to provide realistic critical flow predictions;
- high wave number (or short wave length) instabilities require specific damping mechanisms in order to obtain stable numerical results .

Despite these disadvantages, the Wallis model is still the basis for many of today's transient two-phase flow computer codes. The usually applied numerical solution methods, based on staggered grid and donor cell techniques, provide a sufficient amount of numerical diffusion that stable results can be obtained for many transient two-phase flow conditions. However, this is compromised by the severe inaccuracy in predicting local flow quantities, especially in the presence of large density or void fraction gradients.

References

[1] Graham B. Wallis, *One-dimensional Two-phase Flow,* McGraw-Hill, New York, 1969.

[2] W.G. Mathers et al., *On Finite Difference Solutions to the Transient Flow-Boiling Equations,* Specialist Meeting on Transient Two-Phase Flow, Toronto, Ontario, August 1976.

[3] R.L. Ferch, *Method of Characteristics for Non-Equilibrium Transient Two-Phase Flow,* International Journal of Multiphase Flow, 5, 265–279, 1979.

[4] P. Romstedt, *A Split-Matrix Method for the Numerical Solution of Two-Phase Flow Equations,* Nuclear Science and Engineering 104, 1–9, 1990.

5 A Hyperbolic Model for Two-Phase Flow

This chapter provides a detailed description of a hyperbolic two-phase flow model as was developed at the Joint Research Center Ispra for the numerical simulation of nonhomogeneous, nonequilibrium two-phase flow conditions. The specific features of the model include the capability for a fully algebraic evaluation of the eigenspectrum with regard to eigenvalues and related right- and left-side eigenvectors. The model forms the basis of the Advanced Two-Phase Flow Module (ATFM) as described in Chapter 8. Some background information on the development of the model and its application to first test cases is given by Städtke et al. in [1] and [2].

5.1 One-dimensional flow

The present hyperbolic two-fluid model is based on the single pressure two-fluid model as described Chapter 3 and Appendix A. With the restriction to one-dimensional two-phase flow the balance equations for mass, momentum, and energy (entropy) (3.21) to (3.24) become for $i = g$ (gas), l (liquid)

mass:

$$\frac{\partial}{\partial t}(\alpha_i \varrho_i) + \frac{\partial}{\partial x}(\alpha_i \varrho_i u_i) = \sigma_i^M \tag{5.1}$$

momentum:

$$\frac{\partial}{\partial t}(\alpha_i \varrho_i u_i) + \frac{\partial}{\partial x}(\alpha_i \varrho_i u_i^2) + \alpha_i \frac{\partial p}{\partial x} = F_i^{\text{int}} + \sigma_i^M u^{\text{ex}} + F_i \tag{5.2}$$

energy:

$$\frac{\partial}{\partial t}\left[\alpha_i \varrho_i \left(e_i + \frac{u_i^2}{2}\right)\right] + \frac{\partial}{\partial x}\left[\alpha_i \varrho_i u_i \left(h_i + \frac{u_i^2}{2}\right)\right] + p\frac{\partial \alpha_i}{\partial t}$$

$$= \sigma_i^Q + \sigma_i^M \left[h^{\text{ex}} + \frac{(u^{\text{ex}})^2}{2}\right] + F_i^{\text{int}} u^{\text{int}} + Q_i + F_i u_i \tag{5.3}$$

entropy:

$$\frac{\partial}{\partial t}(\alpha_i \varrho_i s_i) + \frac{\partial}{\partial x}(\alpha_i \varrho_i u_i s_i) = \sigma_i^{S,\text{int}} + \frac{Q_i}{T_i}, \tag{5.4}$$

Gasdynamic Aspects of Two-Phase Flow. Herbert Städtke

ISBN: 3-527-40578-X

with the internal entropy source resulting from interfacial transfer processes,

$$\sigma_i^{S,\text{int}} = \frac{\sigma_i^M}{T_i}\left[T_i s_i + (h^{\text{ex}} - h_i) + \frac{1}{2}(u^{\text{ex}} - u_i)^2\right] + \frac{F_i^{\text{int}}}{T_i}(u^{\text{int}} - u_i). \tag{5.5}$$

5.1.1 Interfacial momentum coupling terms

As shown for the Wallis model the occurrence of complex eigenvalues in the six-equation two-fluid model is linked with the presence of nonhomogeneous flow conditions $u_g \neq u_l$. This indicates that the assumption of exclusively algebraic terms for the interfacial drag results in an incomplete formulation for the interfacial momentum coupling. Apart from the non-hyperbolic character of the flow equations, there also exists strong evidence that the use of exclusively algebraic formulations for the interfacial forces leads to the calculation of incorrect values for the sound velocity and consequently erroneous critical flow rates.

In order to obtain a more general expression for the interfacial momentum coupling, the interfacial forces are split into a viscous part, F^{v}, and a nonviscous part, F^{nv},

$$\left.\begin{aligned} F_g^{\text{int}} &= F_g^{\text{v}} + F_g^{\text{nv}} \\ F_l^{\text{int}} &= F_l^{\text{v}} + F_l^{\text{nv}} \end{aligned}\right\}, \tag{5.6}$$

where the nonviscous part is assumed to include only time and space derivative terms. Apart from F^{nv}, all the other source terms on the right-hand side of the conservation equations are further assumed to be algebraic functions of flow and state parameters of the two phases.

Although there is a common agreement about the need for the derivative terms in the interfacial momentum coupling expressions, there seems to be at present no way to deduce these terms completely from basic principles. The following considerations are therefore not free of some "heuristic" elements.

In order to obtain a more complete form for the nonviscous interfacial forces, the following guidelines have been postulated:

- the nonviscous interfacial terms should at least qualitatively include our present knowledge about the virtual mass effects;
- the additional terms should provide only a moderate modification to the basic Wallis model of two-phase flow;
- since so far no specific assumption have been made for the state equations, the formulation should be "symmetrically" with regard to g (gas or vapor) and l (liquid);
- the nonviscous interfacial forces should not affect the sum of the momentum equations;
- the nonviscous interfacial terms should not contribute to the entropy source for the individual phases;
- the coefficient matrix should have only real eigenvalues;
- the coefficient matrix should have a complete set of independent eigenvectors;

- the system of equations should include as limiting cases the single-phase flow of gas/vapor ($\alpha_g \rightarrow 1$), or liquid ($\alpha_l \rightarrow 1$), and homogeneous flow ($u_g = u_l$);
- the system of equations should be capable of representing existing experimental values for two-phase sound velocity end related critical flow conditions;
- the additional terms should not violate the second law of thermodynamics.

The "virtual" or "added" mass terms in the phasic momentum equations account for the effect of the local mass displacement in the case of a relative acceleration between the two phases. The existence of such a force was first deduced by Lamb [3] for frictionless (irrotational) flows around single spheres, which might be generalized to

$$F_i^{\mathrm{vm}} \sim f(\alpha_g)\varrho_m \left(\frac{d^l u_g}{dt} - \frac{d^g u_l}{dt} \right). \tag{5.7}$$

The final form of the "virtual mass" term, e.g., its dependence on the void fraction α_g or the formulation of the time and spatial derivatives, is still a matter of controversial discussions. In the present analysis, a specific form of the virtual mass force is introduced which represents a generalization of the "objective" form as proposed by Drew et al. [4],

$$F_i^{\mathrm{nv}} = \pm\alpha_g\alpha_l\varrho \left[c \left(\frac{d^l u_g}{dt} - \frac{d^g u_l}{dt} \right) + d(u_g - u_l) \left(\frac{\partial u_g}{\partial x} - \frac{\partial u_l}{\partial x} \right) \right]. \tag{5.8}$$

The expression still includes two open parameters. The factor c accounts for the actual spatial phase distribution of the two phases and, therefore, is expected to be flow regime dependent. The second parameter d, which was introduced by Drew in order to satisfy the postulate of objectivity, is expected to change the sign if the void fraction α_g varies from $\alpha_g = 0$ (pure liquid phase) to $\alpha_g = 1$ (pure gas phase). Both parameters c and d will be specified in the following section. It should be noted here that the introduction of only virtual mass forces in the two momentum equations does not result in a fully hyperbolic system of equations for all two-phase flow conditions.

A further term considered in the nonviscous part of the interfacial forces originates from the principal difference between the phasic pressure and the pressure at the interface as included in the general form of the momentum conservation equations (3.1) to (3.3) in Chapter 3. Although this pressure difference cannot be described explicitly in a single-pressure model, it can be shown that in the case when the pressure differences can be expressed by algebraic relationships, the two-pressure approach becomes equivalent to a single-pressure model where the pressure p represents an average pressure for the two-phase mixture. For these conditions the effect of the difference between phasic and interfacial pressure in the two momentum equations can be described by an expression of the form

$$F_i^{\Delta p} \sim \varrho_{\mathrm{cont}}(u_g - u_l)^2 \frac{\partial \alpha_g}{\partial x}. \tag{5.9}$$

In the following this is generalized to

$$F_i^{\Delta p} = \pm - \alpha_g\alpha_l e\varrho(u_g - u_l)^2 \frac{\partial \alpha_g}{\partial x} \tag{5.10}$$

with an open parameter, e, to be specified.

To date, nearly all attempts to obtain a complete formulation of the interfacial forces have been restricted to the assumption of incompressible phases. The effect of compressibility on the interfacial forces is therefore not well understood. In the present analysis the compressibility effects are considered by the terms

$$F^{\text{comp}} = \pm r_i (u_g - u_l) \frac{d^i \varrho_i}{dt}, \qquad i = g, l, \tag{5.11}$$

where r_g and r_l represent open parameters.

Combining the different contributions as introduced above, the following rather general form for the "non-viscous" contribution to the interfacial momentum coupling is obtained as

$$\left.\begin{aligned} F_g^{\text{nv}} =& -\underbrace{\alpha_g \alpha_l \varrho \left[c \left(\frac{d^l u_g}{dt} - \frac{d^g u_l}{dt} \right) - d(u_g - u_l) \left(\frac{\partial u_g}{\partial x} - \frac{\partial u_l}{\partial x} \right) \right]}_{F^{\text{vm}}} \\ & -\alpha_g \alpha_l e \varrho (u_g - u_l)^2 \frac{\partial \alpha_g}{\partial x} \\ & -\alpha_g \alpha_l \left[r_g (u_g - u_l) \frac{d^g \varrho_g}{dt} + r_l (u_g - u_l) \frac{d^l \varrho_l}{dt} \right] \\ F_l^{\text{nv}} =& -F_g^{\text{nv}} \end{aligned}\right\}, \tag{5.12}$$

where the substantive derivatives are defined as

$$\left.\begin{aligned} \frac{d^l u_g}{dt} &= \frac{\partial u_g}{\partial t} + u_l \frac{\partial u_g}{\partial x}, & \frac{d^g \varrho_g}{dt} &= \frac{\partial \varrho_g}{\partial t} + u_g \frac{\partial \varrho_g}{\partial x} \\ \frac{d^g u_l}{dt} &= \frac{\partial u_l}{\partial t} + u_g \frac{\partial u_l}{\partial x}, & \frac{d^l \varrho_l}{dt} &= \frac{\partial \varrho_l}{\partial t} + u_l \frac{\partial \varrho_l}{\partial x} \end{aligned}\right\}. \tag{5.13}$$

The open parameters c, d, e, r_g, and r_l in equation (5.12) are assumed to be algebraic functions of state and flow parameters. Introducing the nonviscous interfacial forces (5.12) the expanded forms of the one-dimensional balance equations become

$$\mathbf{A} \frac{\partial \mathbf{U}}{\partial t} + \mathbf{B} \frac{\partial \mathbf{U}}{\partial x} = \mathbf{C}, \tag{5.14}$$

with the state and source vectors $\mathbf{U}$ and $\mathbf{C}$ defined as

$$\mathbf{U} = \begin{bmatrix} p \\ u_g \\ u_l \\ \alpha \\ s_g \\ s_l \end{bmatrix}, \qquad \mathbf{C} = \begin{bmatrix} \sigma_g^M \\ \sigma_l^M \\ F_g^{\mathrm{v}} + \sigma_g^M (u^{\mathrm{ex}} - u_g) + F_g \\ F_l^{\mathrm{v}} + \sigma_l^M (u^{\mathrm{ex}} - u_l) + F_l \\ \sigma_g^S \\ \sigma_l^S \end{bmatrix} \tag{5.15}$$

with the total phasic entropy sources

$$\left.\begin{aligned} \sigma_g^S &= \frac{\sigma_g^M}{T_g}\left[(h^{\mathrm{ex}} - h_g) + \frac{1}{2}(u^{\mathrm{ex}} - u_g)^2\right] + \frac{F_g^{\mathrm{v}}\left(u^{\mathrm{int}} - u_g\right)}{T_g} + \frac{\sigma_g^Q}{T_g} + \frac{Q_g}{T_g} \\ \sigma_l^S &= \frac{\sigma_l^M}{T_l}\left[(h^{\mathrm{ex}} - h_l) + \frac{1}{2}(u^{\mathrm{ex}} - u_l)^2\right] + \frac{F_l^{\mathrm{v}}\left(u^{\mathrm{int}} - u_l\right)}{T_l} + \frac{\sigma_l^Q}{T_l} + \frac{Q_l}{T_l} \end{aligned}\right\} \tag{5.16}$$

and the matrices $\mathbf{A}$ and $\mathbf{B}$ for the time and space derives of the governing parameters.

The eigenvalues of the system of equations (5.14) are the roots of the characteristic equation

$$\det\left(\mathbf{B} - \lambda\,\mathbf{A}\right) = 0. \tag{5.17}$$

From equation (5.17) two eigenvalues can be obtained immediately as

$$\lambda_5 = u_g, \qquad \lambda_6 = u_l, \tag{5.18}$$

which represent the material transport velocities of the two phases.

By analogy to the simplified models discussed in the previous sections, the four remaining eigenvalues are expected to characterize the void/density waves (λ_1, λ_2) and pressure/density (sound) waves. Due to the assumption of equal local pressure values for the two phases, a unique value for the propagation of pressure waves (sound velocity a) is expected relative to an average mixture velocity u. This will result in two conjugate eigenvalues of the form

$$\lambda_{3,4} = u \pm a, \tag{5.19}$$

with the mixture velocity u and the mixture sound velocity a.

Rather comprehensive algebraic studies have been performed to verify the effect of the open parameters in the formulation of the nonviscous interfacial forces (5.12) on the eigenvalues and related eigenvectors. This required an enormous amount of algebraic manipulations which could be realized only by an extensive use of Computer Algebra Systems (CAS) such as Stephan Wolfram's "Mathematica, a System for Doing Mathematics by Computers" [6].

As an outcome of this effort, the most physically meaningful results with regard to the characteristic velocities were obtained for the following combination of parameters

$$\left.\begin{aligned} d &= -\frac{\alpha_g \varrho_l - \alpha_l \varrho_g}{\varrho} \\ e &= \frac{\varrho_l + \varrho_g}{\varrho} \\ r_g &= \alpha_g \frac{\varrho_l + \varrho_g}{\varrho_g} \\ r_l &= \alpha_l \frac{\varrho_l + \varrho_g}{\varrho_l} \end{aligned}\right\} . \tag{5.20}$$

The parameter c in the virtual mass term of (5.12) does not directly affect the existence of a hyperbolic system of equations. This parameter, in the following denoted as $k = c$, might be used to adjust the interfacial momentum coupling with respect to different flow regimes. For dispersed droplet or bubbly flow, values of $k = 0.5$ have been estimated, and for completely separated flows (e.g. stratified flow) it is expected that k approaches zero. In the case of churn-turbulent two-phase conditions with strong interfacial momentum coupling values of $k > 0.5$ might be more appropriate.

Introducing the parameters k, d, e, r_g, and r_l, the final form of the nonviscous interfacial forces becomes

$$\left.\begin{aligned} F_g^{\text{nv}} &= -\alpha_g \alpha_l \varrho \left[k \left(\frac{d^l u_g}{dt} - \frac{d^g u_l}{dt} \right) + \frac{\alpha_g \varrho_l - \alpha_l \varrho_g}{\varrho} (u_g - u_l) \left(\frac{\partial u_g}{\partial x} - \frac{\partial u_l}{\partial x} \right) \right] \\ &\quad - \alpha_g \alpha_l (\varrho_g + \varrho_l)(u_g - u_l)^2 \frac{\partial \alpha_g}{\partial x} \\ &\quad - \alpha_g \alpha_l (\varrho_g + \varrho_l)(u_g - u_l) \left[\frac{\alpha_g}{\varrho_g a_g^2} \frac{d^g p}{dt} + \frac{\alpha_l}{\varrho_l a_l} \frac{d^l p}{dt} \right] \\ &\quad - \alpha_g \alpha_l (\varrho_g + \varrho_l)(u_g - u_l) \left[\frac{\alpha_g}{\varrho_g a_g^2} \frac{d^g p}{dt} + \frac{\alpha_l}{\varrho_l a_l} \frac{d^l p}{dt} \right] \\ &\quad \alpha_g \alpha_l (\varrho_g + \varrho_l)(u_g - u_l) \left[\frac{\alpha_g}{\varrho_g} \left(\frac{\partial \varrho_g}{\partial s_g} \right) \frac{d^g s_g}{dt} + \frac{\alpha_l}{\varrho_l} \left(\frac{\partial \varrho_l}{\partial s_l} \right) \frac{d^l s_l}{dt} \right] \\ F_l^{\text{nv}} &= -F_g^{\text{nv}} . \end{aligned}\right\} . \tag{5.21}$$

This form of the nonviscous interfacial forces will be used through all the following model derivations. An extended version for two-dimensional flow conditions is presented in Section 5.2.

5.1.2 Final form of conservation equations

Introducing the nonviscous interfacial forces F_i^{nv} (5.21) the conservation equations for mass, momentum, and energy (5.1) to (5.4) can be written in a compact matrix form as

$$\mathbf{A}\frac{\partial \mathbf{U}}{\partial t} + \mathbf{B}\frac{\partial \mathbf{U}}{\partial x} = \mathbf{C}, \tag{5.22}$$

with the state and source term vectors already define by equation (5.15) and the coefficient matrices $\mathbf{A}$ and $\mathbf{B}$ for the time and space derivative terms as explicitly given in Tables B.27 and B.28 in Appendix B.

The system of governing equations (5.22) can be transposed into a more compact form by multiplying with $\mathbf{A}^{-1}$, which yields

$$\frac{\partial \mathbf{U}}{\partial t} + \mathbf{A}^{-1}\mathbf{B}\frac{\partial \mathbf{U}}{\partial x} = \mathbf{A}^{-1}\mathbf{C}, \tag{5.23}$$

$$\frac{\partial \mathbf{U}}{\partial t} + \mathbf{G}\frac{\partial \mathbf{U}}{\partial x} = \mathbf{D}. \tag{5.24}$$

The coefficient matrix $\mathbf{G} = \mathbf{A}^{-1}\mathbf{B}$ becomes

$$\mathbf{G} = \begin{bmatrix} u_g - \alpha_l \dfrac{\varrho_g}{\varrho_s}\dfrac{a_0^2 \Delta u}{a_l^2} & \alpha_g a_0^2 \dfrac{\varrho_l \varrho_g}{\varrho_s} & \alpha_l a_0^2 \dfrac{\varrho_l \varrho_g}{\varrho_s} & a_0^2 \Delta u \dfrac{\varrho_l \varrho_g}{\varrho_s} & 0 & 0 \\ \dfrac{\hat{\varrho}_l}{\hat{\varrho}^2} & \hat{\alpha}_g u_g + \hat{\alpha}_l u_l & -\hat{\alpha}_l \dfrac{\hat{\varrho}_l}{\hat{\varrho}_g}\Delta u & 0 & 0 & 0 \\ \dfrac{\hat{\varrho}_g}{\hat{\varrho}^2} & \hat{\alpha}_{gl}\dfrac{\hat{\varrho}_g}{\hat{\varrho}_l}\Delta u & \hat{\alpha}_g u_g + \hat{\alpha}_l u_l & 0 & 0 & 0 \\ \alpha_l \alpha_g \dfrac{\Delta u}{\varrho_s}\dfrac{a_0^2}{a_l^2 a_g^2} & \alpha_l \alpha_g \dfrac{\varrho_g}{\varrho_s}\dfrac{a_0^2}{a_l^2} & -\alpha_l \alpha_g \dfrac{\varrho_l}{\varrho_s}\dfrac{a_0^2}{a_g^2} & u_l + \alpha_l \dfrac{\varrho_g}{\varrho_s}\dfrac{a_0^2 \Delta u}{a_l^2} & 0 & 0 \\ 0 & 0 & 0 & 0 & u_g & 0 \\ 0 & 0 & 0 & 0 & 0 & u_l \end{bmatrix}, \tag{5.25}$$

with the additional abbreviations

$$\left.\begin{aligned} &\hat{\varrho}_g = \varrho_g + k\varrho, && \hat{\varrho}_l = \varrho_l + k\varrho, && \hat{\varrho} = \sqrt{\varrho_g \varrho_l + k\varrho^2}, \\ &\varrho_s = \alpha_g \varrho_l + \alpha_l \varrho_g, && \hat{\alpha}_g = \frac{\alpha_g \varrho_g \hat{\varrho}_l}{\hat{\varrho}^2}, && \hat{\alpha}_l = \frac{\alpha_l \varrho_l \hat{\varrho}_g}{\hat{\varrho}^2} \end{aligned}\right\} \tag{5.26}$$

and the condition

$$\hat{\alpha}_g + \hat{\alpha}_l = 1. \tag{5.27}$$

The expression a_0 in equation (5.25) is defined as

$$a_0 = \sqrt{\frac{\alpha_g \varrho_l + \alpha_l \varrho_g}{\dfrac{\alpha_g \varrho_l}{a_g^2} + \dfrac{\alpha_l \varrho_g}{a_l^2}}}, \tag{5.28}$$

which becomes identical with the sound velocity of the Wallis model for the specific case of equal phasic velocities $\Delta u = 0$. The new source vector $\mathbf{D} = \mathbf{A}^{-1}\mathbf{C}$, as introduced in equation (5.24), is given in Table B.30 in Appendix B.

5.1.3 Characteristic analysis – eigenvalues

For the system of equations

$$\frac{\partial \mathbf{U}}{\partial t} + \mathbf{G}\frac{\partial \mathbf{U}}{\partial x} = \mathbf{D}, \tag{5.29}$$

the corresponding characteristic function is defined as

$$f(\lambda) = \det\left(\mathbf{G} - \lambda\,\mathbf{I}\right), \tag{5.30}$$

or in a reduced form after removing the eigenvalues for the entropy waves

$$f^\star = \frac{f}{(u_g - \lambda)(u_l - \lambda)}. \tag{5.31}$$

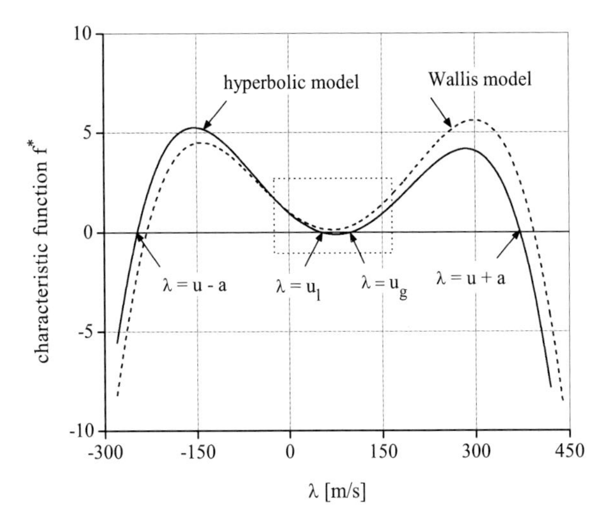

Fig. 5.1: Characteristic function for the hyperbolic and Wallis models, saturated *water/steam mixture* at $p = 10$ bar, void fraction $\alpha_g = 0.25$, phasic velocities: $u_g = 100$ m/s, $u_l = 50$ m/s

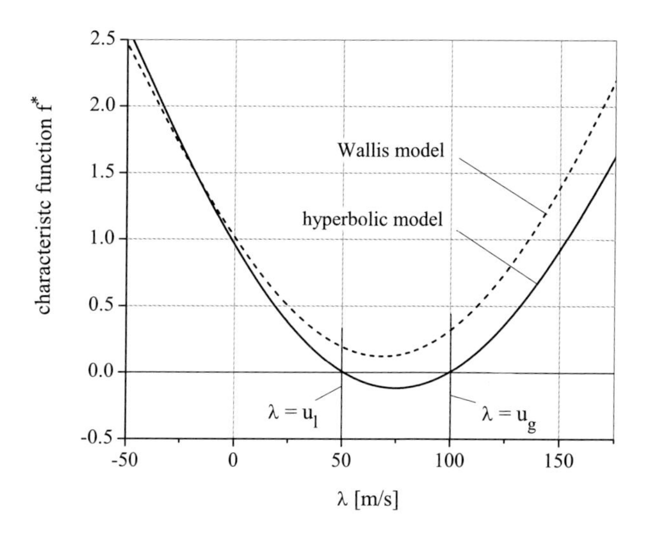

Fig. 5.2: Characteristic function for the hyperbolic and Wallis models, saturated *water/steam mixture* at $p = 10$ bar, void fraction $\alpha_g = 0.25$, phasic velocities: $u_g = 100$ m/s, $u_l = 50$ m/s

The characteristic function $f^{\star}(\lambda)$ for the present hyperbolic model and those obtained for the Wallis model are shown qualitatively in Figs. 5.1 and 5.2.

The figure indicates that the newly introduced nonviscous terms in the momentum equations, responsible for achieving a fully hyperbolic system of equations, represent only a moderate modification with respect to the original Wallis model. The figure also shows that, besides the already known two eigenvalues $\lambda_1 = u_g$ and $\lambda_2 = u_l$, there exist four further real solutions for the characteristic equations.

The eigenvalues of the coefficient matrix are determined as the roots of the characteristic equation (5.30)

$$\det\left(\,\mathbf{G} - \lambda \mathbf{I}\,\right) = 0, \tag{5.32}$$

which results in the following characteristic velocities (eigenvalues)

void waves:

$$\left.\begin{aligned} \lambda_1 &= u_g \\ \lambda_2 &= u_l \end{aligned}\right\} \tag{5.33}$$

pressure waves:

$$\left.\begin{aligned} \lambda_3 &= u + a \\ \lambda_4 &= u - a \end{aligned}\right\} \tag{5.34}$$

propagation of temperature/entropy fluctuations:

$$\left.\begin{aligned} \lambda_5 &= u_g \\ \lambda_6 &= u_l \end{aligned}\right\}. \tag{5.35}$$

The expressions for the two-phase mixture flow velocity u and the two-phase mixture sound velocity a have the following form:

$$u = \frac{(\alpha_g u_g + \alpha_l u_l) + k\left(\dfrac{\alpha_g \varrho_g u_g + \alpha_l \varrho_l u_l}{\alpha_g \varrho_g + \alpha_l \varrho_l}\right)}{1 + k\dfrac{\varrho^2}{\varrho_g \varrho_l}} \tag{5.36}$$

and

$$a^2 = \tilde{a}^2 - \Delta a^2, \tag{5.37}$$

with

$$\left.\begin{aligned} \tilde{a}^2 &= \frac{\alpha_g \varrho_l + \alpha_l \varrho_g}{\dfrac{\alpha_g \varrho_l}{a_g^2} + \dfrac{\alpha_l \varrho_g}{a_l^2}} \frac{1 + k\dfrac{\alpha_g \varrho_g + \alpha_l \varrho_l}{\alpha_g \varrho_l + \alpha_l \varrho_g}}{1 + k\dfrac{\varrho^2}{\varrho_g \varrho_l}}, \\ \Delta a^2 &= \alpha_g \alpha_l \varrho_g \varrho_l (u_g - u_l)^2 \frac{(\varrho_l + k\varrho)(\varrho_g + k\varrho)}{(\varrho_g \varrho_l + k\varrho^2)^2} \end{aligned}\right\}. \tag{5.38}$$

With the abbreviations as defined in equation (5.26) the mixture velocity (5.36) and the mixture sound velocity (5.37) can be written in a more compact form as

$$u = \hat{\alpha}_g u_g + \hat{\alpha}_l u_l \tag{5.39}$$

and

$$\tilde{a}^2 = \frac{\hat{\varrho}_s}{\varrho_s} \frac{\varrho_l \varrho_g}{\hat{\varrho}^2} a_0^2 - \hat{\alpha}_g \hat{\alpha}_l (u_g - u_l)^2,$$

with the basic sound velocity a_0 as already define by equation (5.28) and the newly introduced density

$$\hat{\varrho}_s = \varrho_s + k\varrho. \tag{5.40}$$

Calculated values for the sound velocity are shown in Fig. 5.3 for the special case of equal phase velocities ($u_g = u_0$). The figure indicates a strong effect of the virtual mass coefficient k where the two-phase sound velocity can be considerably smaller than the corresponding limiting values for pure liquid ($\alpha_g = 0$) or pure gas ($\alpha_g = 1$).

With the increase of the difference between the phasic velocities $\Delta u = (u_g - u_l)$, the interfacial momentum coupling becomes larger which results in a reduction of the two-phase

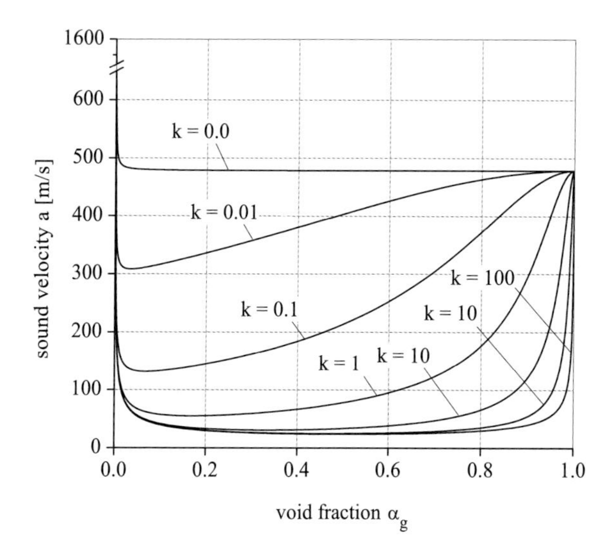

Fig. 5.3: Two-phase sound velocity as a function of the void fraction, effect of "virtual mass" coefficient, *saturated water/steam* at pressure $p = 1$ bar, equal phase velocities

sound velocity as shown in Fig. 5.4. However, the effect is rather small as long as the "slip" velocity is small compared with the sound velocity a_0. A noticeable reduction in sound velocity exists only for relatively large void fractions if the slip velocity becomes of the same order of magnitude as the sound velocity. Also for these conditions, the deviation from the strictly homogeneous case never exceeds a value of more than 10%.

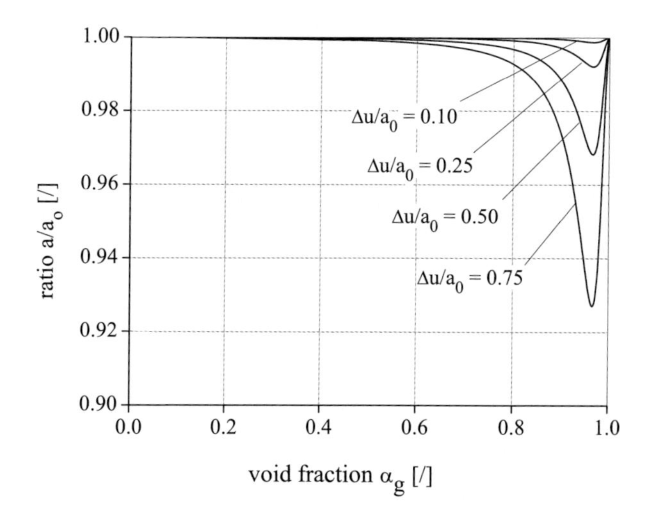

Fig. 5.4: Two-phase sound velocity as a function of the void fraction; effect of the slip velocity Δu, *saturated water/steam* at pressure $p = 1$ bar

The expression for the pressure wave propagation includes two limiting cases with respect to the "virtual mass" coefficient k.

(1) $k \to 0$:

$$\left.\begin{aligned} u &= u_1 = \alpha_g u_g + \alpha_l u_l \\ a^2 &= a_1^2 = \frac{\alpha_g \varrho_l + \alpha_l \varrho_g}{\dfrac{\alpha_g \varrho_l}{a_g^2} + \dfrac{\alpha_l \varrho_g}{a_l^2}} - \alpha_l \alpha_g (u_g - u_l)^2 \end{aligned}\right\} . \tag{5.41}$$

For the specific case of equal phase velocities ($u_g = u_l = u$), the sound velocity becomes

$$a_1^2 = \frac{\alpha_g \varrho_l + \alpha_l \varrho_g}{\dfrac{\alpha_g \varrho_l}{a_g^2} + \dfrac{\alpha_l \varrho_g}{a_l^2}} = a_0^2, \tag{5.42}$$

as also obtained for the Wallis model, equation (4.102) in Section 4.3.

(2) $k \to \infty$:

$$\left.\begin{aligned} u &= u_2 = \frac{\alpha_g \varrho_g u_g + \alpha_l \varrho_l u_l}{\alpha_g \varrho_g + \alpha_l \varrho_l} \\ a^2 &= a_2^2 = \frac{1}{\dfrac{\alpha_g \varrho}{\varrho_g a_g^2} + \dfrac{\alpha_l \varrho}{\varrho_l a_l^2}} - \frac{\alpha_l \alpha_g \varrho_l \varrho_g}{\varrho^2} (u_g - u_l)^2 \end{aligned}\right\} . \tag{5.43}$$

For equal phase velocities, the sound velocity becomes identical with the sound velocity for homogeneous nonequilibrium flow as given in Section 4.2 by equation (4.53)

$$a_2^2 = \frac{1}{\dfrac{\alpha_g \varrho}{\varrho_g a_g^2} + \dfrac{\alpha_l \varrho}{\varrho_l a_l^2}} = a_{\text{hom}}^2 . \tag{5.44}$$

The *first case* (1) describes flow conditions with spatially separated phases where the momentum coupling is reduced to the interfacial friction forces. For this condition, the characteristic velocity u_1 represents the average volumetric mixture velocity. The first term in the sound velocity a_1 is known as the "frozen" sound velocity for nonhomogeneous flow conditions with (instantaneous) equal flow velocities which can also be derived from the six-equation model if no derivative terms for the interfacial momentum coupling between the phases are considered (Wallis model).

If, as in the *second case* (2), the "virtual mass" force becomes dominating, the flow is strongly driven toward homogeneous conditions $u_g = u_l = u$. In this case, the characteristic flow velocity, u_2, is the average mass velocity of the two-phase mixture, and the first term in the expression for the mixture sound velocity, a_2, represents the "frozen" sound velocity for homogeneous two-phase flow defined as

$$\tilde{a}_2^2 = \left(\frac{\partial p}{\partial \varrho}\right)_{X, s_g, s_l = \text{const}} \tag{5.45}$$

with

$$\frac{1}{\varrho} = \frac{X}{\varrho_g} + \frac{1-X}{\varrho_l}. \tag{5.46}$$

Experimental data on sound velocities are rather scarce and mostly restricted to dispersed droplet or bubbly conditions. Figure 5.5 compares measured sound velocities for water/vapor mixtures from Nakoryakov et al. [5] with the values calculated using the present model. The calculated sound velocities are within the scatter of the measured data for a "virtual mass" coefficient of $k = 0.25$, which is not far from the value as determined for idealized dispersed flow conditions.

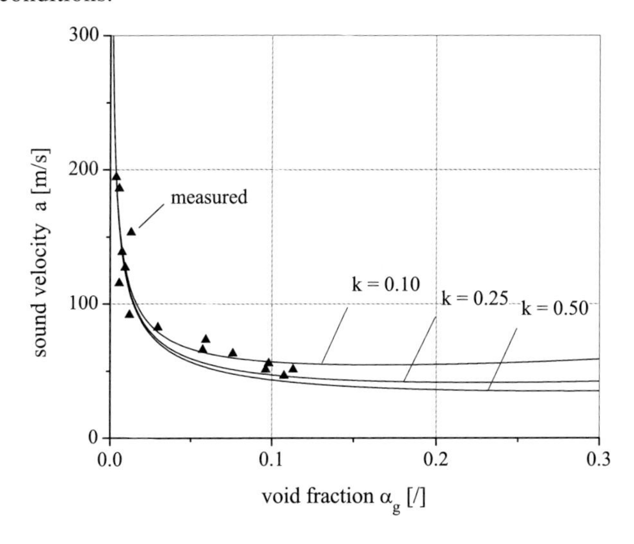

Fig. 5.5: Sound velocity in *water/steam media* as a function of the void fraction; saturated conditions at pressure $p = 1$ bar

5.1.4 Characteristic analysis – eigenvectors and splitting of coefficient matrix

For the six real eigenvalues, a complete set of independent eigenvectors can be derived which is the condition for the existence of a fully hyperbolic system of governing equations. These eigenvectors, which are determined up to an arbitrary factor, can be written as follows:

void wave: $\lambda_1 = u_g$

$$V_{R,1} = \left[\alpha_l \varrho_l (\Delta u)^2 \, , \, 0 \, , \, \Delta u \, , \, -\alpha_l \left(1 - \alpha_l \frac{(\Delta u)^2}{a_l^2}\right) \, , \, 0 \, , \, 0\right] \tag{5.47}$$

void wave: $\lambda_2 = u_l$

$$V_{R,2} = \left[\alpha_g \varrho_g (\Delta u)^2 \, , \, -\Delta u \, , \, 0 \, , \, \alpha_g \left(1 - \alpha_g \frac{(\Delta u)^2}{a_g^2}\right) \, , \, 0 \, , \, 0\right] \tag{5.48}$$

pressure wave: $\lambda_3 = u + a$

$$V_{R,3} = \left[\hat{\varrho}^2\, \tilde{a}^2\,,\; \hat{\varrho}_l \left(a - \frac{\alpha_l \varrho_l\, \hat{\varrho}_g}{\hat{\varrho}^2} \Delta u \right)\,,\; \hat{\varrho}_g \left(a + \frac{\alpha_g \varrho_g\, \hat{\varrho}_l}{\hat{\varrho}^2} \Delta u \right) \right.$$
$$\left. -\, \alpha_g \alpha_l \frac{\hat{\varrho}^2}{\hat{\varrho}_s} \left(\frac{\hat{\varrho}_g\, \tilde{a}^2}{\varrho_g a_g^2} - \frac{\hat{\varrho}_l\, \tilde{a}^2}{\varrho_l a_l^2} \right)\,,\; 0\,,\; 0 \right] \tag{5.49}$$

pressure wave: $\lambda_4 = u - a$:

$$V_{R,4} = \left[\hat{\varrho}^2\, \tilde{a}^2,\; \hat{\varrho}_l \left(-a - \frac{\alpha_l \varrho_l\, \hat{\varrho}_g}{\hat{\varrho}^2} \Delta u \right),\; \hat{\varrho}_g \left(-a + \frac{\alpha_g \varrho_g\, \hat{\varrho}_l}{\hat{\varrho}^2} \Delta u \right) \right.$$
$$\left. -\, \alpha_g \alpha_l \frac{\hat{\varrho}^2}{\hat{\varrho}_s} \left(\frac{\hat{\varrho}_g\, \tilde{a}^2}{\varrho_g a_g^2} - \frac{\hat{\varrho}_l\, \tilde{a}^2}{\varrho_l a_l^2} \right)\,,\; 0\,,\; 0 \right] \tag{5.50}$$

entropy wave: $\lambda_5 = u_g$

$$V_{R,5} = [\,0\,,\,0\,,\,0\,,\,0\,,\,1\,,\,0\,] \tag{5.51}$$

entropy wave: $\lambda_6 = u_l$

$$V_{R,6} = [\,0\,,\,0\,,\,0\,,\,0\,,\,0\,,\,1\,]\,. \tag{5.52}$$

The six right-hand side eigenvectors (5.47) to (5.50) can be combined in a matrix, $\mathbf{V}_R$, which is given in Table B.31 in Appendix B.4. Since all the eigenvectors are independent of each other, there exists a matrix $\mathbf{T}$ which allows a similarity transformation of the form

$$\mathbf{T}^{-1} \frac{\partial \mathbf{U}}{\partial t} + \mathbf{\Lambda}\, \mathbf{T}^{-1} \frac{\partial \mathbf{U}}{\partial x} = \mathbf{T}^{-1} \mathbf{D} = \mathbf{E}, \tag{5.53}$$

with the diagonal matrix of the eigenvalues

$$\mathbf{\Lambda} = \mathbf{T}^{-1} \mathbf{G}\, \mathbf{T} = \begin{bmatrix} u_g & 0 & 0 & 0 & 0 & 0 \\ 0 & u_l & 0 & 0 & 0 & 0 \\ 0 & 0 & u+a & 0 & 0 & 0 \\ 0 & 0 & 0 & u-a & 0 & 0 \\ 0 & 0 & 0 & 0 & u_g & 0 \\ 0 & 0 & 0 & 0 & 0 & u_l \end{bmatrix}. \tag{5.54}$$

The columns of the transformation matrix $\mathbf{T}$ are the right-hand side eigenvectors of $\mathbf{G}$,

$$\mathbf{T} = \mathbf{V}_R^T. \tag{5.55}$$

The inverse matrix $\mathbf{T}^{-1}$ is, apart from a common row factor, equivalent to the matrix of the left-hand side eigenvectors of the matrix $\mathbf{G}$ and, therefore, can also be written as

$$\mathbf{T}^{-1} = \mathbf{V}_L \tag{5.56}$$

with the matrix of left eigenvectors $\mathbf{V}_L$ as given by in Table B.32 in Appendix B.

From the characteristic equation (5.53), the following compatibility relations can be derived for nonhomogeneous two-phase flow:

void wave: $\dfrac{dx}{dt} = \lambda_1 = u_g$

$$\frac{\alpha_g \Delta u}{a_g^2}\frac{dp}{dt} + \alpha_g \varrho_g \frac{du_g}{dt} - \alpha_g \varrho_g \frac{\hat{\varrho}_l}{\hat{\varrho}_g}\frac{du_l}{dt} + \varrho_g\,\Delta u\,\frac{d\alpha_g}{dt} = E_1 \tag{5.57}$$

void wave: $\dfrac{dx}{dt} = \lambda_2 = u_l$

$$\frac{\alpha_l \Delta u}{a_l^2}\frac{dp}{dt} + \alpha_l \varrho_l \frac{\hat{\varrho}_g}{\hat{\varrho}_l}\frac{du_g}{dt} - \alpha_l \varrho_l \frac{du_l}{dt} - \varrho_l\,\Delta u\,\frac{d\alpha_g}{dt} = E_2 \tag{5.58}$$

pressure wave: $\dfrac{dx}{dt} = \lambda_3 = u + a$

$$(a\,\hat{\varrho}_s + \,\hat{\varrho}\Delta u \Delta a)\frac{dp}{dt} + \alpha_g\,\hat{\varrho}^2\left(a_2^2 + a\Delta u\frac{\varrho_g\,\hat{\varrho}_s}{\hat{\varrho}^2}\right)\frac{du_g}{dt}$$

$$+\alpha_l\,\hat{\varrho}^2\left(a_1^2 - a\Delta u\frac{\varrho_l\,\hat{\varrho}_s}{\hat{\varrho}^2}\right)\frac{du_l}{dt} + \,\hat{\varrho}^2\,\tilde{a}^2\Delta u\,\frac{d\alpha_g}{dt} = E_3 \tag{5.59}$$

pressure wave: $\dfrac{dx}{dt} = \lambda_4 = u - a$

$$(-a\,\hat{\varrho}_s + \,\hat{\varrho}\Delta u \Delta a)\frac{dp}{dt} + \alpha_g\,\hat{\varrho}^2\left(a_2^2 - a\Delta u\frac{\varrho_g\,\hat{\varrho}_s}{\hat{\varrho}^2}\right)\frac{du_g}{dt}$$

$$+\alpha_l\,\hat{\varrho}^2\left(a_1^2 + a\Delta u\frac{\varrho_l\,\hat{\varrho}_s}{\hat{\varrho}^2}\right)\frac{du_l}{dt} + \,\hat{\varrho}^2\,\tilde{a}^2\Delta u\,\frac{d\alpha_g}{dt} = E_4 \tag{5.60}$$

entropy wave: $\dfrac{dx}{dt} = \lambda_5 = u_g$

$$\frac{ds_g}{dt} = E_5 = \frac{\sigma_g^S}{\alpha_g \varrho_g} \tag{5.61}$$

entropy wave: $\dfrac{dx}{dt} = \lambda_6 = u_l$

$$\frac{ds_l}{dt} = E_6 = \frac{\sigma_l^S}{\alpha_l \varrho_l}. \tag{5.62}$$

In equations (5.59) and (5.60) additional abbreviations are introduced which are defined as

$$a_1^2 = a^2 - \frac{{\alpha_g}^2 \varrho_l \varrho_g \hat{\varrho}_l}{\hat{\varrho}^4} (\Delta u)^2 \tag{5.63}$$

and

$$a_2^2 = a^2 - \frac{{\alpha_l}^2 \varrho_l \varrho_g \hat{\varrho}_g^2}{\hat{\varrho}^4} (\Delta u)^2 . \tag{5.64}$$

The first four elements of the new source vector, E_1 to E_4, which are not yet defined, are given explicitly in Table B.33 in Appendix B.

Many numerical schemes dealing with hyperbolic flow equations need a splitting of the coefficient matrix $\mathbf{G}$ into elementary parts with respect to the individual eigenvalues λ_k. As was shown already for the case of single-phase gas or homogeneous two-phase flow, the split matrix $\mathbf{G}_k$ is obtained as

$$\mathbf{G}_k = \mathbf{T}\, \mathbf{\Lambda}_k \mathbf{T}^{-1} \qquad \text{with the condition} \qquad \mathbf{G} = \sum_{k=1}^{6} \mathbf{G}_k \tag{5.65}$$

where the diagonal matrix $\mathbf{\Lambda}_k$ includes only the kth eigenvalue.

The split matrices as introduced in Equation (5.65) can also be written in component notation as the product of the kth eigenvalue with the kth column vector of the matrix $\mathbf{T}$ and the kth row vector of the inverse matrix $\mathbf{T}^{-1}$

$$(G_k)_{i,j} = \lambda_k T_{i,k} T_{k,j}^{-1} . \tag{5.66}$$

The corresponding elementary split coefficient matrices $\mathbf{G}_k$ are given separately in Tables B.34 to B.31 in Appendix B.4.

As can be seen from Tables B.34 and B.35 the split matrices $\mathbf{G}_k$ for the two void waves, $\lambda_1 = u_g$ and $\lambda_1 = u_l$, exhibit singularities in the case of equal flow velocities $\Delta u \to 0$. Such homogeneous conditions might have been specified as initial conditions or might occur during a transient calculation when a new equilibrium steady state is reached. Since in the numerical methods the split matrices are sorted with regard to the sign of the eigenvalues, the problem can be avoided using the sum and difference of the split matrices depending whether co-current

$$\text{sign}(u_g) = \text{sign}(u_l): \qquad \mathbf{G}_{12}^{+} = \mathbf{G}_1 + \mathbf{G}_2 \tag{5.67}$$

or counter-current flow conditions

$$\text{sign}(u_g) \neq \text{sign}(u_l): \qquad \mathbf{G}_{12}^{-} = \mathbf{G}_1 - \mathbf{G}_2 \tag{5.68}$$

exist. It can be easily shown that for homogeneous conditions ($\lambda_1 = \lambda_2 = u$) the matrix remains well defined,

$$\mathbf{G}_{12}^{\text{hom}} = \lim_{\Delta u \to 0} (\mathbf{G}_1 + \mathbf{G}_2). \tag{5.69}$$

The resulting matrix $\mathbf{G}_{12}^{\text{hom}}$ is given in Table B.36 in Appendix B.4.

5.1.5 Homogeneous flow conditions as a limiting case

Under some circumstances it might be desirable to drive the flow toward homogeneous conditions, e.g., during the transition from two-phase to single-phase conditions when the liquid or gas phase disappears. Such a solution is provided as a limiting case for the parameter k introduced in the expression for the nonviscous forces as given by equation (5.21). For $k \to \infty$ and $\Delta u \to 0$ the coefficient matrix (5.25) simplifies to

$$\mathbf{G}^{\text{hom}} = \lim_{k \to \infty} \mathbf{G} = \begin{bmatrix} u & \alpha_g a_0^2 \dfrac{\varrho_l \varrho_g}{\varrho_s} & \alpha_l a_0^2 \dfrac{\varrho_l \varrho_g}{\varrho_s} & 0 & 0 & 0 \\ \dfrac{1}{\varrho} & u & 0 & 0 & 0 & 0 \\ \dfrac{1}{\varrho} & 0 & u & 0 & 0 & 0 \\ 0 & \alpha_l \alpha_g \dfrac{\varrho_g}{\varrho_s} \dfrac{a_0^2}{a_l^2} & -\alpha_l \alpha_g \dfrac{\varrho_l}{\varrho_s} \dfrac{a_0^2}{a_g^2} & u & 0 & 0 \\ 0 & 0 & 0 & 0 & u & 0 \\ 0 & 0 & 0 & 0 & 0 & u \end{bmatrix}. \tag{5.70}$$

With $u_g = u_l = u$ the second and the third column in (5.70) can be combined with results in a matrix with two identical lines for the momentum equation. If one of these lines is dropped a coefficient matrix is obtained which is identical with those derived for homogeneous thermal nonequilibrium conditions by (4.60) in Section 4.2,

$$\mathbf{G}^{\text{hom}} = \begin{bmatrix} u & \dfrac{\varrho_g \varrho_l a_g^2 a_l^2}{\alpha_g \varrho_l a_l^2 + \alpha_l \varrho_g a_g^2} & 0 & 0 & 0 \\ \dfrac{1}{\varrho} & u & 0 & 0 & 0 \\ 0 & \dfrac{\alpha_g \alpha_l (\varrho_l a_l^2 - \varrho_g a_g^2)}{\alpha_g \varrho_l a_l^2 + \alpha_l \varrho_g a_g^2} & u & 0 & 0 \\ 0 & 0 & 0 & u & 0 \\ 0 & 0 & 0 & 0 & u \end{bmatrix}. \tag{5.71}$$

The compatibility relations (5.57) to (5.62) include as a limiting case the specific form of homogeneous flow which can be obtained with $k \to \infty$ and setting $u_g = u_l = u$. For the first two eigenvalues, only one equation remains for the void waves *void wave:* $\dfrac{dx}{dt} = \lambda_{1,2} = u$

$$\alpha_l \alpha_g \left(\frac{1}{\varrho_g a_g^2} - \frac{1}{\varrho_g a_g^2} \right) \frac{dp}{dt} + \frac{d\alpha_g}{dt} = \frac{\varrho}{\varrho_l \varrho_g} \sigma_g^M - \frac{\alpha_l}{\varrho_g^2} \left(\frac{\partial \varrho_g}{\partial s_g} \right)_p \sigma_g^S - \frac{\alpha_g}{\varrho_l^2} \left(\frac{\partial \varrho_l}{\partial s_l} \right)_p \sigma_l^S, \tag{5.72}$$

The two pressure wave modes simplify to

pressure wave: $\dfrac{dx}{dt} = \lambda_3 = u + a$

$$\frac{dp}{dt} + \varrho\, a \frac{d\alpha_g}{dt} = \varrho\, a^2 \left(\frac{1}{\varrho_g} - \frac{1}{\varrho_l} \right) \sigma_g^M - \frac{\varrho\, a^2}{\varrho_g^2} \left(\frac{\partial \varrho_g}{\partial s_g} \right)_p \sigma_g^S - \frac{\varrho\, a^2}{\varrho_l^2} \left(\frac{\partial \varrho_l}{\partial s_l} \right)_p \sigma_l^S, \tag{5.73}$$

pressure wave: $\dfrac{dx}{dt} = \lambda_3 = u - a$

$$\frac{dp}{dt} - \varrho\, a \frac{d\alpha_g}{dt} = \varrho\, a^2 \left(\frac{1}{\varrho_g} - \frac{1}{\varrho_l} \right) \sigma_g^M - \frac{\varrho\, a^2}{\varrho_g^2} \left(\frac{\partial \varrho_g}{\partial s_g} \right)_p \sigma_g^S - \frac{\varrho\, a^2}{\varrho_l^2} \left(\frac{\partial \varrho_l}{\partial s_l} \right)_p \sigma_l^S, \tag{5.74}$$

propagating with the homogeneous equilibrium speed of sound a relative to the mixture velocity u.

For the remaining two eigenmodes which describe entropy (temperature) fluctuations propagating with the material transport velocity, the characteristic forms are unchanged,

$$\left.\begin{aligned} \frac{dx}{dt} &= \lambda_5 = u: \quad & \frac{ds_g}{dt} &= \sigma_g^S, \\ \frac{dx}{dt} &= \lambda_6 = u: \quad & \frac{ds_l}{dt} &= \sigma_l^S. \end{aligned}\right\} \tag{5.75}$$

Equations (5.72) to (5.75) are identical with the compatibility relations as were derived for homogeneous nonequilibrium flow in Section 4.2.

5.1.6 Use of conservative variables

The conservative form of the hyperbolic two-phase flow model can be directly obtained from the general balance equations (5.1) to (5.4) introducing the nonviscous contribution for the interfacial forces F_i^{nv} as defined by equation (5.21).

A more general way to derive the conservative flow equation as will be described in the following is through a similarity transformation of the "primitive" form,

$$\frac{\partial \mathbf{U}}{\partial t} + \mathbf{G} \frac{\partial \mathbf{U}}{\partial x} = \mathbf{D}. \tag{5.76}$$

Introducing the vectors of conserved variables $\mathbf{V}(\mathbf{U})$ and a corresponding flux vector $\mathbf{F}(\mathbf{U})$, equation (5.76) can be written as

$$\frac{\partial \mathbf{V}}{\partial t} + \mathbf{J}\mathbf{G}\mathbf{J}^{-1} \frac{\partial \mathbf{V}}{\partial x} = \mathbf{J}\mathbf{D}, \tag{5.77}$$

or

$$\frac{\partial \mathbf{V}}{\partial t} + \mathbf{H}\frac{\partial \mathbf{V}}{\partial x} = \mathbf{D}, \tag{5.78}$$

with the Jacobian matrix

$$\mathbf{J} = \frac{\partial \mathbf{V}}{\partial \mathbf{U}}, \tag{5.79}$$

and the new coefficient matrix $\mathbf{H}$ and the new state vector $\mathbf{D}$ defined as

$$\mathbf{H} = \mathbf{J}\,\mathbf{G}\,\mathbf{J}^{-1}, \qquad \mathbf{D} = \mathbf{J}\,\mathbf{D}. \tag{5.80}$$

In principle, any state vector can be used as long as the elements of the vector are linearly independent, which requires

$$\det(\mathbf{J}) \neq 0. \tag{5.81}$$

The eigenvalues of the new coefficient matrix $\mathbf{H}$ remain the same as for $\mathbf{G}$,

$$\lambda_1 = u_g, \qquad \lambda_2 = u_l, \qquad \lambda_{3,4} = u \pm a, \qquad \lambda_5 = u_g, \qquad \lambda_6 = u_l, \tag{5.82}$$

with the mixture velocity u and the mixture sound velocity a given by equations (5.36) to (5.38). The right- and left-hand side eigenvectors of the new coefficient matrix can be determined either directly from $\mathbf{H}$ or by the transformation rules

$$\left.\begin{aligned} \mathbf{V}_R &= \mathbf{V}_R^\star\,\mathbf{J}^T \\ \mathbf{V}_L &= \mathbf{V}_L^\star\,\mathbf{J}^{-1} \end{aligned}\right\} \tag{5.83}$$

where $\mathbf{V}_R^\star$ and $\mathbf{V}_L^\star$ represent the corresponding matrices of the right and left eigenvectors as were obtained for the coefficient matrix $\mathbf{G}$ (5.25) using the primitive form of balance equations.

The coefficient matrix $\mathbf{H}$ as introduced in equation (5.80) can be split into the elementary parts related to the individual eigenvalues

$$\mathbf{H}_k = \mathbf{T}\,\boldsymbol{\Lambda}_k\mathbf{T}^{-1}, \tag{5.84}$$

with the transformation matrix $\mathbf{T} = \mathbf{V}_R^T$ and its inverse $\mathbf{T}^{-1} = \mathbf{V}_L$, and the condition

$$\mathbf{H} = \sum_{k=1}^{6} \mathbf{H}_k. \tag{5.85}$$

Introducing the flux vector $\mathbf{F}$ into the equation, the conservative form of balance equations becomes

$$\frac{\partial \mathbf{V}}{\partial t} + \frac{\partial \mathbf{F}}{\partial x} + \mathbf{H}^{\mathrm{nc}}\frac{\partial \mathbf{V}}{\partial x} = \mathbf{E} \tag{5.86}$$

with the “nonconservative” contribution

$$\mathbf{H}^{\mathrm{nc}} = (\mathbf{J}\mathbf{G} - \mathbf{K})\,\mathbf{K}. \tag{5.87}$$

The Jacobian matrix in equation (5.87) represents the derivative of the flux vector $\mathbf{F}$ with respect to the vector of conserved variables $\mathbf{V}$,

$$\mathbf{K} = \frac{\partial \mathbf{F}}{\partial \mathbf{V}}. \tag{5.88}$$

The appearance of the "nonconservative" contribution is a specific peculiarity of the balance equation for nonhomogeneous two-phase flow, which results from the fact that some derivative terms in the separated momentum equations cannot be brought into a fully conservative form. Nevertheless, the general principle of conservation is maintained, since these terms cancel out for the sum of the phasic momentum equations.

So far no specific assumptions have been made for the form of the conserved and flux vectors. For simplicity entropy will be used here as a major thermodynamic variable with the state and flux vectors defined as

$$\mathbf{U} = \begin{bmatrix} p \\ u_g \\ u_l \\ \alpha_g \\ s_g \\ s_l \end{bmatrix}, \quad \mathbf{V}(\mathbf{U}) = \begin{bmatrix} \alpha_g \varrho_g \\ \alpha_l \varrho_l \\ \alpha_g \varrho_g u_g \\ \alpha_l \varrho_l u_l \\ \alpha_g \varrho_g s_g \\ \alpha_l \varrho_l s_l \end{bmatrix}, \quad \mathbf{F}(\mathbf{U}) = \begin{bmatrix} \alpha_g \varrho_g u_g \\ \alpha_l \varrho_l u_l \\ \alpha_g \varrho_g u_g^2 + \alpha_g p \\ \alpha_l \varrho_l u_l^2 + \alpha_l p \\ \alpha_g \varrho_g u_g s_g \\ \alpha_l \varrho_l u_l s_l \end{bmatrix}. \tag{5.89}$$

For this case the resulting Jacobian matrices $\mathbf{J}$ and $\mathbf{K}$ and the coefficient matrices $\mathbf{H}$, and $\mathbf{H}^{\mathrm{nc}}$ become rather compact as shown in Tables B.40 to B.45 in Appendix B.4. The derivation of the corresponding eigenvectors and split matrices is straightforward using the transformation rules given by equations (5.83) to (5.88).

For the correct prediction of flow discontinuities such as shock waves in gaseous media, it might be advisable to use the energy equations with the state and flux vectors as

$$\mathbf{U} = \begin{bmatrix} p \\ u_g \\ u_l \\ \alpha_g \\ e_g \\ e_l \end{bmatrix}, \quad \mathbf{V} = \begin{bmatrix} \alpha_g \rho_g \\ \alpha_l \rho_l \\ \alpha_g \rho_g u_g \\ \alpha_l \rho_l u_l \\ \alpha_g \rho_g \left(e_g + \frac{1}{2} u_g^2\right) \\ \alpha_l \rho_l \left(e_l + \frac{1}{2} u_l^2\right) \end{bmatrix}, \quad \mathbf{F} = \begin{bmatrix} \alpha_g \rho_g u_g \\ \alpha_l \rho_l u_l \\ \alpha_g (\rho_g u_g^2 + p) \\ \alpha_l (\rho_l u_l^2 + p) \\ \alpha_g \rho_g u_g (h_g + \frac{1}{2} u_g^2) \\ \alpha_l \rho_l u_l (h_l + \frac{1}{2} u_l^2) \end{bmatrix}. \tag{5.90}$$

For the characteristic analysis the same procedure can be applied as described above for using the "primitive" state vector .

5.1.7 Quasi-one-dimensional flow through channels of variable cross section

Up to this point, all the derivations have been reduced to strictly one-dimensional flow conditions. Many technical applications, however, deal with flow processes through pipes or channels of variable cross section as shown schematically in Fig. 5.6. These flow conditions can be approximated by a quasi one-dimensional approach, introducing average quantities over the pipe or channel flow cross section which leads to the "stream-tube" formulation of the flow.

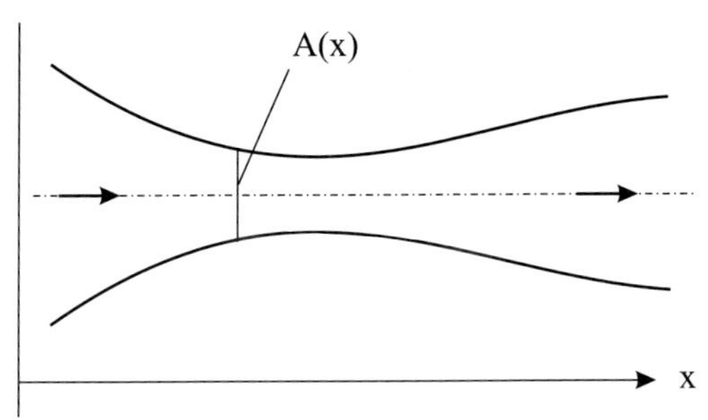

Fig. 5.6: Flow through a channel of variable cross section

For the quasi-one-dimensional flow through channels of variable cross section, the separated phasic balance equations for mass, momentum, energy, and entropy can be written as follows:

mass

$$\frac{\partial}{\partial t}(\alpha_g \varrho_g) + \frac{1}{A}\frac{\partial}{\partial x}(\alpha_g \varrho_g u_g A) = \sigma_g^M \tag{5.91}$$

$$\frac{\partial}{\partial t}(\alpha_l \varrho_l) + \frac{1}{A}\frac{\partial}{\partial x}(\alpha_l \varrho_l u_l A) = \sigma_l^M \tag{5.92}$$

momentum:

$$\frac{\partial}{\partial t}(\alpha_g \varrho_g u_g) + \frac{1}{A}\frac{\partial}{\partial x}(\alpha_g \varrho_g v_g^2 A) + \alpha_g \frac{\partial p}{\partial x} = F_g^{\mathrm{nv}} + F_g^{\mathrm{v}} + \sigma_g^M u^{\mathrm{ex}} \tag{5.93}$$

$$\frac{\partial}{\partial t}(\alpha_l \varrho_l u_l) + \frac{1}{A}\frac{\partial}{\partial x}(\alpha_l \varrho_l u_l^2 A) + \alpha_l \frac{\partial p}{\partial x} = F_l^{\mathrm{nv}} + F_l^{\mathrm{v}} + \sigma_l^M u^{\mathrm{ex}} \tag{5.94}$$

energy:

$$\frac{\partial}{\partial t}\left[\alpha_g\varrho_g\left(e_g+\frac{u_g^2}{2}\right)\right]+\frac{1}{A}\frac{\partial}{\partial x}\left[\alpha_g\varrho_g u_g A\left(e_g+\frac{p}{\varrho_g}+\frac{u_g^2}{2}\right)\right]+p\frac{\partial\alpha_g}{\partial t}$$
$$=\sigma_g^Q+Q_g+F_g^{\mathrm{nv}}u^{\mathrm{nv}}+F_g^{\mathrm{v}}u^{\mathrm{v}}+\sigma_g^M\left[h^{\mathrm{ex}}+\frac{(u^{\mathrm{ex}})^2}{2}\right] \tag{5.95}$$

$$\frac{\partial}{\partial t}\left[\alpha_l\varrho_l\left(u_l+e_l+\frac{u_l^2}{2}\right)\right]+\frac{1}{A}\frac{\partial}{\partial x}\left[\alpha_l\varrho_l u_l A\left(e_l+\frac{p}{\varrho_l}+\frac{u_l^2}{2}\right)\right]+p\frac{\partial\alpha_l}{\partial t}$$
$$=\sigma_l^Q+Q_l+F_l^{\mathrm{nv}}u^{\mathrm{nv}}+F_l^{\mathrm{v}}u^{\mathrm{v}}+\sigma_l^M\left[h^{\mathrm{ex}}+\frac{(u^{\mathrm{ex}})^2}{2}\right] \tag{5.96}$$

entropy:

$$\frac{\partial}{\partial t}(\alpha_g\varrho_g s_g)+\frac{1}{A}\frac{\partial}{\partial x}(\alpha_g\varrho_g u_g s_g A)=\frac{\sigma_g^M}{T_g}\left[s_gT_g+h^{\mathrm{ex}}-h_g+\frac{1}{2}(u^{\mathrm{ex}}-u_g)^2\right]$$
$$\frac{\sigma_g^Q}{T_g}+\frac{F_g^{\mathrm{v}}}{T_g}(u^{\mathrm{v}}-u_g)+\frac{Q_g}{T_g} \tag{5.97}$$

$$\frac{\partial}{\partial t}(\alpha_l\varrho_l s_l)+\frac{1}{A}\frac{\partial}{\partial x}(\alpha_l\varrho_l u_l s_l A)=\frac{\sigma_l^M}{T_l}\left[s_lT_l+h^{\mathrm{ex}}-h_l+\frac{1}{2}(u^{\mathrm{ex}}-u_l)^2\right]_l$$
$$+\frac{\sigma_l^Q}{T_l}+\frac{F_l^{\mathrm{v}}}{T_l}(u^{\mathrm{v}}-u)+\frac{Q_l}{T_l} \tag{5.98}$$

Expanding equations (5.91) to (5.98), the "primitive" form of the governing equations is obtained which will not be given here explicitly.

If the phasic entropy values are used in the state vector the resulting equations can be combined in the following vector form:

$$\mathbf{A}\frac{\partial\mathbf{U}}{\partial t}+\mathbf{B}\frac{\partial\mathbf{U}}{\partial x}=\mathbf{C}, \tag{5.99}$$

where the coefficient matrices $\mathbf{A}$ and $\mathbf{B}$ are identical with those of the strictly one-dimensional case. In the source term vector, however, new terms appear for the two mass conservation equations, having as a common factor the derivative of the cross section with respect to the axial direction, $\frac{1}{A}\frac{\partial A}{\partial x}$,

$$
\mathbf{U} = \begin{bmatrix} p \\ u_g \\ u_l \\ \alpha_g \\ s_g \\ s_l \end{bmatrix}, \qquad \mathbf{C} = \begin{bmatrix} \sigma_g^M - \alpha_g \varrho_g u_g \dfrac{1}{A}\dfrac{\partial A}{\partial x} \\ \sigma_l^M - \alpha_l \varrho_l u_l \dfrac{1}{A}\dfrac{\partial A}{\partial x} \\ F_g^{\mathrm{v}} + \sigma_g^M (u^{\mathrm{ex}} - u_g) + F_g \\ F_l^{\mathrm{v}} + \sigma_l^M (u^{\mathrm{ex}} - u_l) + F_l \\ \sigma_g^S \\ \sigma_l^S \end{bmatrix} \tag{5.100}
$$

with the total phasic entropy sources as defined in equation (5.15)

Multiplying equation (5.99) with $\mathbf{A}^{-1}$ yields the more compact form

$$
\frac{\partial \mathbf{U}}{\partial t} + \mathbf{G}\frac{\partial \mathbf{U}}{\partial x} = \mathbf{D}, \tag{5.101}
$$

where the coefficient matrix $\mathbf{G}$ is the same as for the strictly one-dimensional form as derived in Section 5.3. The new source term vector $\mathbf{D}$ is given for reference in Table B.30 in Appendix B.

Since the left-hand side of equation (5.99) is identical with that of the strictly one-dimensional case, all the results with respect to the characteristic analysis of nonhomogeneous two-phase flow derived so far also remain valid for the quasi-one-dimensional flow through channels of variable cross section. However, one specific feature of nonhomogeneous, nonequilibrium two-phase flow should be elaborated in more detail which concerns the locus of critical flow conditions in the case of stationary flow through convergent-divergent nozzles.

Assuming steady state flow conditions, the governing flow equations reduce to

$$
\mathbf{B}\frac{\partial \mathbf{U}}{\partial x} = \mathbf{C}, \tag{5.102}
$$

from which the following equation for the pressure gradient in the flow direction can be derived:

$$
\frac{\partial p}{\partial x} = \frac{\dfrac{\hat{\varrho}^2}{\hat{\varrho}_s}\tilde{u}^2 \left[\dfrac{1}{A}\dfrac{\partial A}{\partial x} + Z\right]}{1 - \dfrac{\tilde{u}^2}{\tilde{a}^2}}. \tag{5.103}
$$

The sound velocity, $\tilde{a}$, and the mixture flow velocity, $\tilde{u}$, are defined as

$$
\tilde{a}_{cr}^2 = \frac{\alpha_g \varrho_l + \alpha_l \varrho_g}{\dfrac{\alpha_g \varrho_l}{a_g^2} + \dfrac{\alpha_l \varrho_g}{a_l^2}} \; \frac{1 + k\dfrac{\alpha_g \varrho_g + \alpha_l \varrho l}{\alpha_g \varrho l + \alpha_l \varrho_g}}{1 + k\dfrac{\varrho^2}{\varrho_g \varrho_l}} \tag{5.104}
$$

and

$$\tilde{u}_{cr}^2 = \frac{(\alpha_g u_g^2 + \alpha_l u_l^2) + k\dfrac{\varrho^2}{\varrho_g \varrho_l}\dfrac{\alpha_g \varrho_g u_g^2 + \alpha_l \varrho_l u_l^2}{\alpha_g \varrho_g + \alpha_l \varrho_l}}{1 + k\dfrac{\varrho^2}{\varrho_g \varrho_l}}. \tag{5.105}$$

The function Z as introduced in the numerator of equation (5.103) depends exclusively on the algebraic interfacial coupling terms for mass, momentum, and energy and can be written in the following form:

$$\begin{aligned} Z = {} & \left[F_g^v + \sigma_g^M (u^{\text{ex}} - u_s)\right] \frac{\varrho_l u_l - \varrho_g u_g}{\hat{\varrho}^2 \, \tilde{u}^2 u_l u_g} \\ & - \left[\sigma_g^M (u_g - u_l)\right] \frac{\hat{\varrho}_s}{\hat{\varrho}^2 \, \tilde{u}^2} \\ & - \left[\sigma_g^M - \frac{1}{\varrho_g}\frac{\partial \varrho_g}{\partial s_g}\sigma_g^S\right] \frac{1}{\varrho_g u_g} \\ & - \left[\sigma_l^M - \frac{1}{\varrho_l}\frac{\partial \varrho_l}{\partial s_l}\sigma_l^S\right] \frac{1}{\varrho_l u_l}, \end{aligned} \tag{5.106}$$

with the "mirrored" velocity

$$u_s = \alpha_g u_l + \alpha_l u_g. \tag{5.107}$$

The general expression for the pressure gradient in the axial direction includes a number of limiting cases which are worth mentioning. If one of the phases disappears, for example the liquid phase, equation (5.103) simplifies to the well-known relation for compressible gas flow:

(1) *single-phase gas flow*: $\alpha_l \to 0, \quad \alpha_g \to 1$

$$\frac{\partial p}{\partial x} = \frac{\varrho_g u_g^2 \dfrac{1}{A}\dfrac{\partial A}{\partial x}}{1 - \dfrac{u_g^2}{a_g^2}}$$

as already given by equation (2.41) in Section 2.

(2) *two-phase homogeneous equilibrium flow*:

$$\frac{\partial p}{\partial x} = \frac{\varrho u^2 \dfrac{1}{A}\dfrac{\partial A}{\partial x}}{1 - \dfrac{u^2}{a^2}}, \tag{5.108}$$

where a represents the two-phase homogeneous equilibrium sound velocity as derived in Section 4.2 for two-component (gas/liquid) or one-component (vapor/liquid) conditions. Homogeneous equilibrium two-phase flow can be seen as a special case where the time constants for the interfacial coupling processes become very short compared with the characteristic time for the flow (e.g., for the flow through very long nozzles and ducts). Also the opposite conditions might be of interest where the interfacial exchange processes are sufficiently slow with respect to the fast character of the outer flow process (e.g., for the flow through very short nozzles or orifices).

(3) *"frozen" flow conditions*: $F_g^v \to 0, \quad \sigma_g^M \to 0, \quad \sigma_g^S \to 0, \quad \sigma_l^S \to 0$

$$\frac{\partial p}{\partial x} = \frac{\dfrac{\hat{\varrho}^2}{\hat{\varrho}_s}\tilde{u}^2\left[\dfrac{1}{A}\dfrac{\partial A}{\partial x}\right]}{1 - \dfrac{\tilde{u}^2}{\tilde{a}^2}}. \tag{5.109}$$

All the three cases described above show the same behavior for the adiabatic flow of compressible media through convergent–divergent nozzles. Assuming that the back pressure of the nozzle is sufficiently low, the flow will accelerate from subsonic to supersonic velocities with the sonic point, or critical flow conditions, exactly at the nozzle throat. At the nozzle throat, the flow is characterized by a saddle-point singularity for $u \to a$,

$$\frac{\partial p}{\partial x} = \frac{0}{0}. \tag{5.110}$$

However, the situation becomes completely different for the general case with finite interfacial transport processes as illustrated in Fig. 5.7.

Introducing an "effective" cross section, $A^\star$, as

$$\frac{1}{A^\star}\frac{\partial A^\star}{\partial x} = \frac{1}{A}\frac{\partial A}{\partial x} + Z \tag{5.111}$$

and using the function Z as already defined above, equation (5.106) for the pressure gradient can be written as

$$\frac{\partial p}{\partial x} = \frac{\dfrac{\hat{\varrho}^2}{\hat{\varrho}_s}\tilde{u}^2\left[\dfrac{1}{A^\star}\dfrac{\partial A^\star}{\partial x}\right]}{1 - \dfrac{\tilde{u}^2}{\tilde{a}^2}}. \tag{5.112}$$

As a result of the interfacial transfer for mass, momentum, and energy, the critical conditions, $u = a$, will no longer occur at the nozzle throat, but rather at a position further downstream in the divergent section where the "effective" cross-sectional area $A^\star$ reaches its minimum. The exact locus of this point is not *a priori* known and can only be determined by integration of the complete set of flow equations from the upstream reservoir to the critical cross section.

The occurrence of a saddle-point singularity for the two-phase nozzle flow as described above has an equivalence in the reactive gas flow situation if nonequilibrium effects are considered. A general method for the numerical integration through a saddle-point singularity has been developed by Emanuel [7] for one-dimensional nonequilibrium reactive gas flow through convergent–divergent nozzles which, as was shown by Städtke [8], can also be applied for nonequilibrium two-phase flow conditions.

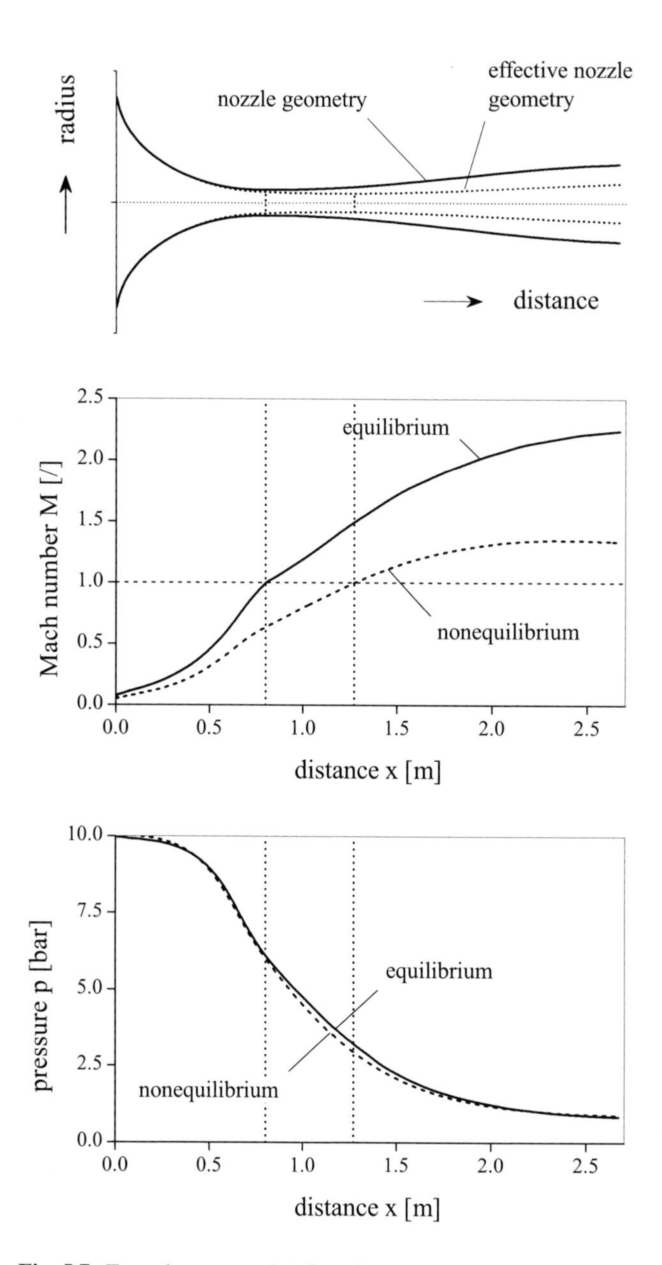

Fig. 5.7: Two-phase *water/air flow* through convergent–divergent nozzle; effect of nonequilibrium conditions on the critical flow

5.2 Two-dimensional two-phase flow conditions

5.2.1 Basic flow equations for two-dimensional flow

The extension of the nonviscous interfacial forces as defined by equation (5.12) is not unique; however, a rather straightforward form can be obtained as follows:

$$\left.\begin{aligned}
\vec{F}_g^{\mathrm{nv}} &= -\alpha_g \alpha_l \rho k \left(\frac{d^l \vec{u}_g}{dt} - \frac{d^g \vec{u}_l}{dt} \right) \\
&\quad + \alpha_g \alpha_l \left(\alpha_g \rho_l - \alpha_l \rho_g \right) \left[\left(\vec{u}_g - \vec{u}_l \right) \cdot \nabla \right] \left(\vec{u}_g - \vec{u}_l \right) \\
&\quad - \alpha_g \alpha_l \left(\rho_g + \rho_l \right) \left(\vec{u}_g - \vec{u}_l \right) \nabla \cdot \alpha_g \\
&\quad - \alpha_g \alpha_l \left(\rho_g + \rho_l \right) \left(\vec{u}_g - \vec{u}_l \right) \frac{\alpha_g}{\rho_g} \frac{d\rho_g}{dt} + \frac{\alpha_l}{\rho_l} \frac{d\rho_l}{dt} \\
\vec{F}_l^{\mathrm{nv}} &= -\vec{F}_g^{\mathrm{nv}}
\end{aligned}\right\}, \tag{5.113}$$

with the total derivatives

$$\frac{d^l \vec{u}_g}{dt} = \frac{\partial \vec{u}_g}{\partial t} + \left(\vec{u}_l \cdot \nabla \right) \vec{u}_g, \qquad \text{and} \qquad \frac{d^g \vec{u}_l}{dt} = \frac{\partial \vec{u}_l}{\partial t} + \left(\vec{u}_g \cdot \nabla \right) \vec{u}_l. \tag{5.114}$$

Introducing the expression for the nonviscous contribution to the interfacial forces into the general balance equations (3.16) to (3.20) in Chapter 3, a system of partial differential equations is obtained which can be combined in a compact vector form for two-dimensional flow conditions,

$$\mathbf{A} \frac{\partial \mathbf{U}}{\partial t} + \mathbf{B}_x \frac{\partial \mathbf{U}}{\partial \mathbf{x}} + \mathbf{B}_y \frac{\partial \mathbf{U}}{\partial \mathbf{y}} = \mathbf{C}, \tag{5.115}$$

with the vector of “primitive” variables $\mathbf{U}$, the source term vector $\mathbf{C}$, and the coefficient matrices $\mathbf{A}$, $\mathbf{B}_x$, and $\mathbf{B}_y$ related to the time and space derivatives in the x- and y-directions. For the state and source term vectors two alternatives exist depending on whether the entropy equations or the full energy equations are used.

For the characteristic analysis of the flow equations the use of phasic entropies in the state vector is preferred since this leads to an immediate separation of the “entropy waves”. The corresponding state and source term vectors then become

$$\mathbf{U} = \begin{bmatrix} p \\ \vec{u}_g \\ \vec{u}_l \\ \alpha_g \\ s_g \\ s_l \end{bmatrix}, \qquad \mathbf{C} = \begin{bmatrix} \sigma_g^M \\ \sigma_l^M \\ \vec{F}_g^{\mathrm{v}} + \sigma_g^M \vec{u}^{\mathrm{ex}} + \vec{F}_g \\ \vec{F}_l^{\mathrm{v}} + \sigma_l^M \vec{u}^{\mathrm{ex}} + \vec{F}_l \\ \sigma_g^S \\ \sigma_l^S \end{bmatrix}. \tag{5.116}$$

Multiplying equation (5.115) by $\mathbf{A}^{-1}$ yields

$$\frac{\partial \mathbf{U}}{\partial t} + \mathbf{G}_x \frac{\partial \mathbf{U}}{\partial \mathbf{x}} + \mathbf{G}_y \frac{\partial \mathbf{U}}{\partial \mathbf{y}} = \mathbf{D}, \tag{5.117}$$

with the new coefficient matrices $\mathbf{G}_x$ and $\mathbf{G}_y$ for the x- and y-directions and the new source term vector defined as

$$\mathbf{G}_x = \mathbf{A}^{-1}\mathbf{B}_x, \qquad \mathbf{G}_y = \mathbf{A}^{-1}\mathbf{B}_y \qquad \text{and} \qquad \mathbf{D} = \mathbf{A}^{-1}\mathbf{C}. \tag{5.118}$$

For the projection of the governing equations in an arbitrary direction in the flow field given by the unit vector $\vec{n}$ (Fig. 5.8) one obtains

$$\frac{\partial \mathbf{U}}{\partial t} + \mathbf{G}_n \frac{\partial \mathbf{U}}{\partial n} = \mathbf{D}, \tag{5.119}$$

with the corresponding coefficient matrix $\mathbf{G}_n = n_x\mathbf{G}_x + n_y\mathbf{G}_y$. The coefficient matrices $\mathbf{G}_x$, $\mathbf{G}_y$, and $\mathbf{G}_n$ are given in Tables B.48 to B.50 in Appendix B.4. As can be seen, from the matrices $\mathbf{G}_x$ and $\mathbf{G}_y$ there appear some transverse coupling terms between the momentum in the x- and y-directions which are proportional to the difference in phasic velocities Δu_x and Δu_y.

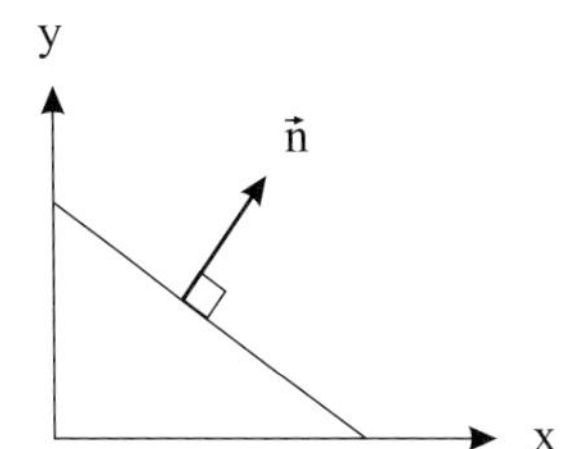

Fig. 5.8: Direction of wave propagation

5.2.2 Eigenvalues and split matrices

As for the one-dimensional case the eigenvalues of the coefficient matrix $\mathbf{G}_n$ are determined as the roots of the characteristic function,

$$f(\lambda) = \det(\mathbf{G}_n - \lambda \mathbf{I}) = 0, \tag{5.120}$$

which yields for the different wave modes

void waves:

$$\left.\begin{aligned} \lambda_1 &= \vec{u}_g \cdot \vec{n} = u_{g,n} \\ \lambda_2 &= \vec{u}_l \cdot \vec{n} = u_{l,n} \end{aligned}\right\} \tag{5.121}$$

pressure waves:

$$\left.\begin{aligned} \lambda_3 &= \vec{u} \cdot \vec{n} + a = u_n + a \\ \lambda_4 &= \vec{u} \cdot \vec{n} - a = u_n - a \end{aligned}\right\} \tag{5.122}$$

shear waves:

$$\left.\begin{aligned} \lambda_5 &= \vec{u}_g \cdot \vec{n} = u_{g,n} \\ \lambda_6 &= \vec{u}_l \cdot \vec{n} = u_{l,n} \end{aligned}\right\} \tag{5.123}$$

temperature/entropy waves:

$$\left.\begin{aligned} \lambda_7 &= \vec{u}_g \cdot \vec{n} = u_{g,n} \\ \lambda_8 &= \vec{u}_l \cdot \vec{n} = u_{l,n} \end{aligned}\right\}. \tag{5.124}$$

The mixture flow velocity $\vec{u}$ and the mixture sound velocity a as introduced for the pressure waves in equation (5.122) are defined as

$$\vec{u} = \frac{\varrho_g \varrho_l \left(\alpha_g \vec{u}_g + \alpha_l \vec{u}_l\right) + k\varrho \left(\alpha_g \varrho_g \vec{u}_g + \alpha_l \varrho_l \vec{u}_l\right)}{\varrho_g \varrho_l + k\varrho^2} \tag{5.125}$$

and

$$a^2 = \tilde{a}^2 - \Delta a^2, \tag{5.126}$$

with

$$\left.\begin{aligned} \tilde{a}^2 &= \frac{\alpha_g \varrho_l + \alpha_l \varrho_g}{\dfrac{\alpha_g \varrho_l}{a_g^2} + \dfrac{\alpha_l \varrho_g}{a_l^2}} \, \frac{1 + k \dfrac{\alpha_g \varrho_g + \alpha_l \varrho_l}{\alpha_g \varrho_l + \alpha_l \varrho_g}}{1 + k \dfrac{\varrho^2}{\varrho_g \varrho_l}} \\ \Delta a^2 &= \alpha_g \alpha_l \varrho_g \varrho_l \frac{(\varrho_l + k\varrho)(\varrho_g + k\varrho)}{(\varrho_g \varrho_l + k\varrho^2)^2} (\vec{u}_g - \vec{u}_l)^2 \end{aligned}\right\}. \tag{5.127}$$

Apart from the *shear waves* $\lambda_{5,6}$ (5.123) all other characteristic velocities represent straightforward extensions of the expressions derived for the one-dimensional flow conditions by equations (5.36) to (5.38).

As in the one-dimensional case, the eigenvalues corresponding to the entropy waves can be immediately deduced from the coefficient matrix $\mathbf{G}_n$ as

$$\left.\begin{aligned} \lambda_7 &= u_{g,n}: & V_{R,7} &= [0,0,0,0,0,0,0,1] \\ \lambda_8 &= u_{g,n}: & V_{R,7} &= [0,0,0,0,0,0,1,0] \end{aligned}\right\}. \tag{5.128}$$

For the remaining eigenmodes a specific problem arises from the fact that multiple eigenvectors $u_{g,n}$ and $u_{l,n}$ appear for the void and shear waves, which does not allow us to define a complete set of independent eigenvectors. Nevertheless, since for the numerical methods the split matrices are sorted with regard to eigenvalues, this problem can be easily solved by introducing a small auxiliary value ϵ for some elements in the coefficient matrix $\mathbf{G}_n$,

$$\left.\begin{aligned} G_{4,4} &= u_{g,n} + \epsilon \\ G_{5,5} &= u_{l,n} - \epsilon \end{aligned}\right\}, \tag{5.129}$$

as shown in Table B.50 of Appendix B. The eigenvalues for the shear waves then change to

$$\left.\begin{aligned} \lambda_7 &= \vec{u}_g \cdot \vec{n} = u_{g,n} + \epsilon \\ \lambda_8 &= \vec{u}_l \cdot \vec{n} = u_{l,n} - \epsilon \end{aligned}\right\}, \tag{5.130}$$

different from the corresponding values $\lambda_{1,2}$ for the void waves. Under these conditions, a complete independent set of right eigenvectors $\mathbf{V}_R$ and left eigenvectors $\mathbf{V}_L$ can be derived and the split matrices become as for the one-dimensional case

$$\mathbf{G}_{n,k} = \mathbf{T}\,\mathbf{\Lambda}_k\mathbf{T}^{-1}, \tag{5.131}$$

with the transformation matrix $\mathbf{T} = \mathbf{V}_R^T$ and $\mathbf{T}^{-1} = \mathbf{V}_L$ and the diagonal matrix $\mathbf{\Lambda}_k$ including only the kth eigenvalue. The common split matrix for the void and shear waves for the gas and liquid phases can then be obtained as the limiting values

$$\begin{aligned} \lambda_{1,5} &= u_{g,n}: & \mathbf{G}_{n,k} &= \lim_{\epsilon\to 0}[\mathbf{G}_{n,1} + \mathbf{G}_{n,5}] \\ \lambda_{2,6} &= u_{l,n}: & \mathbf{G}_{n,k} &= \lim_{\epsilon\to 0}[\mathbf{G}_{n,2} + \mathbf{G}_{n,6}]. \end{aligned} \tag{5.132}$$

All the other split matrices for pressure and entropy waves remain as originated from equation (5.125). The procedure for obtaining the split matrices does not violate the condition

$$\mathbf{G}_n = \sum_{k=1}^{6} \mathbf{G}_{n,k}. \tag{5.133}$$

5.2.3 Conservative form of flow equations

The conservative form of the balance equations can be easily obtained by a similarity transformation as already shown for the one-dimensional case, resulting in the vector of conserved variables $\mathbf{V}$,

$$\frac{\partial \mathbf{V}}{\partial t} + \mathbf{H}\,\nabla\mathbf{V} = \mathbf{E}, \tag{5.134}$$

with the state vectors of "primitive" and conserved state parameters $\mathbf{U}$, $\mathbf{V}$, and the related flux vector $\mathbf{F}$

$$\mathbf{U} = \begin{bmatrix} p \\ \vec{u}_g \\ \vec{u}_l \\ \alpha_g \\ s \\ s_l \end{bmatrix}, \quad \mathbf{V} = \begin{bmatrix} \alpha_g\rho_g \\ \alpha_l\rho_l \\ \alpha_g\rho_g\vec{u}_g \\ \alpha_l\rho_l\vec{u}_l \\ \alpha_g\rho_g s_g \\ \alpha_l\rho_l s_l \end{bmatrix}, \quad \mathbf{F} = \begin{bmatrix} \alpha_g\rho_g\vec{u}_g \\ \alpha_l\rho_l\vec{u}_l \\ \alpha_g(\rho_g\vec{u}_g\,\vec{u}_g + p\mathbf{I}) \\ \alpha_l(\rho_l\vec{u}_l\,\vec{u}_l + p\mathbf{I}) \\ \alpha_g\rho_g\vec{u}_g s_g \\ \alpha_l\rho_l\vec{u}_l s_l \end{bmatrix}. \tag{5.135}$$

The new coefficient matrix $\mathbf{H}$, and the related new source vector are defined as

$$\mathbf{H} = \mathbf{J}\,\mathbf{G}\,\mathbf{J}^{-1}, \tag{5.136}$$

and the related new source vector defined as

$$\mathbf{E} = \mathbf{J}\,\mathbf{D}. \tag{5.137}$$

The Jacobian matrix in equation (5.136) describes the derivative of the conserved state vector with respect to the vector of primitive state parameters,

$$\mathbf{J} = \frac{\partial \mathbf{V}}{\partial \mathbf{U}}. \tag{5.138}$$

The eigenvalues of the new coefficient matrix $\mathbf{H}$ remain the same as for $\mathbf{G}$. The final conservative form of the flow equations is obtained introducing the flux vector $\mathbf{F}$ resulting in

$$\frac{\partial \mathbf{V}}{\partial t} + \nabla\cdot\mathbf{F} + \mathbf{H}^{\mathrm{nc}}\nabla\mathbf{F} = \mathbf{E}, \tag{5.139}$$

where the nonconservative part $\mathbf{H}^{\mathrm{nc}}$ is defined as

$$\mathbf{H}^{\mathrm{nc}} = (\mathbf{J}\,\mathbf{G} - \mathbf{K})\;\mathbf{K}^{-1} \tag{5.140}$$

and the Jacobian matrix

$$\mathbf{K} = \frac{\partial \mathbf{F}}{\partial \mathbf{U}}.$$

If instead of the entropy relations the full energy equations are used, the internal phasic energies appear as major dependent thermodynamic variables and the primitive state parameters $\mathbf{U}$, the vector of conserved state parameters $\mathbf{V}$, and the corresponding flux vector $\mathbf{F}$ become

$$\mathbf{U} = \begin{bmatrix} p \\ \vec{u}_g \\ \vec{u}_l \\ \alpha_g \\ e_g \\ e_l \end{bmatrix}, \quad \mathbf{V} = \begin{bmatrix} \alpha_g \rho_g \\ \alpha_l \rho_l \\ \alpha_g \rho_g \vec{u}_g \\ \alpha_l \rho_l \vec{u}_l \\ \alpha_g \rho_g \left(e_g + \frac{1}{2} u_g^2\right) \\ \alpha_l \rho_l \left(e_l + \frac{1}{2} u_l^2\right) \end{bmatrix}, \quad \mathbf{F} = \begin{bmatrix} \alpha_g \rho_g \vec{u}_g \\ \alpha_l \rho_l \vec{u}_l \\ \alpha_g (\rho_g \vec{u}_g \, \vec{u}_g + p\mathbf{I}) \\ \alpha_l (\rho_l \vec{u}_l \, \vec{u}_l + p\mathbf{I}) \\ \alpha_g \rho_g \vec{u}_g (h_g + \frac{1}{2} u_g^2) \\ \alpha_l \rho_l \vec{u}_l (h_l + \frac{1}{2} u_l^2) \end{bmatrix}. \qquad (5.141)$$

For both cases a complete algebraic evaluation of the eigenspectrum and a splitting of the coefficient matrices can be obtained. However, since the corresponding vectors and matrices become more voluminous, they will not be explicitly given here.

5.3 Final remarks to the hyperbolic two-phase flow model

The hyperbolic two-fluid model as described above includes several unique features which make it very attractive as a modeling basis for the numerical simulation of complex two-phase flows governed by compressibility effects and wave propagation processes. The advantages of the model includes in particular the following:

1. Although based on a rather general formulation of the single-pressure two-fluid model, it allows a complete and consistent algebraic evaluation of the full eigenspectrum of the governing equations resulting in explicit formulation of all eigenvalues and related right and left eigenvectors.

2. The hyperbolicity of the model is maintained over the full spectrum of volumetric concentration of gas or liquid ($0 \leq \alpha_g \leq 1$). The "virtual mass" coefficient k remains an open flow parameter which might be used to represent different flow regimes such as dispersed bubbly or droplet flow regimes or to force homogeneous flow conditions ($u_g = u_l$).

3. The flow of single-phase gas or liquid is included as natural limiting values for $\alpha_g \rightarrow 1$ or $\alpha_g \rightarrow 0$, respectively.

4. The model allows the use of any (thermodynamically consistent) set of state equations for the gas/vapor and liquid phases.

5. Thanks to the strictly hyperbolic character of the flow equations, an extension to differently defined state and flux vectors can be easily derived just by a similarity transformation for the coefficient matrix and related source term vectors. The only necessary requirement for such transformation is that state and flux vectors are uniquely defined as indicated by a complete and independent set of row/column vectors for the Jacobian matrices involved.

6. The model provides all necessary information in the algebraic form for the implementation of "high resolution" numerical schemes which make explicit use of the hyperbolic nature of the flow equations. This allows the development of rather compact and economic numerical algorithms as will be shown in Chapter 7.

The model has been extensively tested during the development of the Advanced Two-phase Flow Module (ATFM) which covered a wide spectrum of flow conditions and phenomena. From the large number of test cases, no deficiencies or shortcomings of the model have been found which could be related to the specific formulation of the nonviscous interfacial forces. The major limitations of the model are related to the assumption of a single (local) pressure value of the two phases which might become crucial for two-phase flow conditions governed by surface tension effects such as collapsing of (small) vapor bubbles and cavitation phenomena.

References

[1] Städtke, G. Franchello and B. Worth, *Towards a High-Resolution Numerical Simulation of Transient Two-Phase Flow*, Third International Conference on Multi-Phase Flow, ICMF'98, Lyon, France, 1998.

[2] H. Städtke, B. Worth, and G. Franchello, *On the Hyperbolic Nature of Two-phase Flow Equations: Characteristic Analysis and Related Numerical Methods*, in Godunov Methods, Theory and Application, Kluwer, 841–862, 2001.

[3] H. Lamb *Hydrodynamics*, 6th ed., Cambridge University Press, Cambridge 1932.

[4] D. Drew, I. Cheng, and R.T. Drew *The Analysis of Virtual Mass Effects in Two-phase Flow*, International Journal of Multiphase Flow, 5, 233–242, 1979.

[5] V.E. Nakoryakov et al., *The Wave Dynamics of Vapor -Liquid Medium*, International Journal of Multiphase Flow 14, 6, 655–677, 1988.

[6] Stephan Wolfram, *Mathematica: A System for Doing Mathematics by Computer*, Addison-Wesley, Reading. MA, 1991.

[7] G. Emanuel, *A General Method for the Numerical Integration through a Saddle-Point Singularity with Application to One-Dimensional Nozzle Flow*, Arnold Engineering Development Centre, Technical Documentary Report AEDC-TDR-29, January 1964.

[8] H. Städtke, *Gasdynamische Aspekte der Nichtgleichgewichts-Zweiphasenströmung durch Düsen*, Ph.D. thesis, Technical University Berlin, 1975.

6 Dispersion of Sound Waves

Within the framework of the characteristic analysis as given in Chapter 5 only the time and space derivatives of flow parameters enter and, therefore, the obtained "sound velocity" does not include any effect of the algebraic source terms describing mass, momentum, and energy transfer processes at the interface. A more physical insight into sound wave propagation phenomena can be obtained from the "acoustic approximation" where the effect of the algebraic source terms are taken into account.

6.1 Acoustic approximation of flow equations

For the acoustic approximation, the basic flow equations of the hyperbolic two-phase flow model, as given in Chapter 5,

$$\mathbf{A}\frac{\partial \mathbf{U}}{\partial t} + \mathbf{B}\frac{\partial \mathbf{U}}{\partial x} = \mathbf{C} \tag{6.1}$$

are linearized, assuming small time- and space-dependent disturbances superimposed on a steady state flow,

$$\mathbf{U}\left(x,t\right) = \mathbf{U}_0\left(x\right) + \mathbf{U}'\left(x,t\right) \quad \text{with} \quad \left|\mathbf{U}'\right| \ll \left|\mathbf{U}_0\right|. \tag{6.2}$$

Neglecting higher order terms, this transforms the flow equations (6.1) into a system of linear equations for the disturbances $\mathbf{U}'$

$$\mathbf{A}_0\frac{\partial \mathbf{U}'}{\partial t} + \mathbf{B}_0\frac{\partial \mathbf{U}'}{\partial x} - \left(\frac{\partial \mathbf{C}}{\partial \mathbf{U}}\right)_0 \mathbf{U}' = 0, \tag{6.3}$$

where for the basic steady flow, thermal and mechanical equilibrium conditions are assumed for simplicity, e.g., $T_{g,0} = T_{l,0} = T_0$, $u_g = u_l = u_0$.

Any monochromatic wave can be expressed in a general form as

$$\mathbf{U}' = \delta\mathbf{U} \, \exp\left[i\omega\left(t - \frac{K\,x}{\omega}\right)\right], \tag{6.4}$$

with the complex wave number $K = K_r + i\,K_i$, the circular frequency ω, and the amplitude $\delta\mathbf{U}$. The wave propagates with the velocity $a = \omega/K_r$. The imaginary part of the wave number K_i is a measure for the wave attenuation.

Gasdynamic Aspects of Two-Phase Flow. Herbert Städtke

ISBN: 3-527-40578-X

Introducing the formulation (6.4) into the linearized flow equations (6.3) a system of linear equations is obtained for the amplitudes $\delta\mathbf{U}$ of the different wave modes,

$$\delta\mathbf{U}\left[\mathbf{A}_0-\mathbf{B}_0\left(\frac{K}{\omega}\right)+\left(\frac{\partial\mathbf{C}}{\partial\mathbf{U}}\right)_0\frac{i}{\omega}\right]=0. \tag{6.5}$$

A non-trivial solution ($\delta\mathbf{U}\neq\mathbf{0}$) of equation (6.5) exists only in the case where the coefficient determinant vanishes, which leads to the dispersion relation

$$f=\det\left[\mathbf{A}_0-\mathbf{B}_0\left(\frac{K}{\omega}\right)+\left(\frac{\partial\mathbf{C}}{\partial\mathbf{U}}\right)_0\frac{i}{\omega}\right]=0 \tag{6.6}$$

or

$$f=\det\left[\mathbf{A}_0-\mathbf{B}_0\left(\frac{K}{\omega}\right)+\mathbf{J}_0\frac{i}{\omega}\right]=0, \tag{6.7}$$

with the Jacobian matrix

$$\mathbf{J}_0=\left(\frac{\partial\mathbf{C}}{\partial\mathbf{U}}\right)_0. \tag{6.8}$$

Equation (6.7) describes the functional dependence of the wave velocity a and attenuation d on the frequency ω, as will be shown in the following for dispersed gas/liquid flow conditions.

6.2 Dispersion analysis of gas–particle flows

Assuming that mono-dispersed spherical droplets of constant radius r_p are homogeneously dispersed in the carrier gas, the particle volume fraction can be expressed as

$$\alpha_p=n_p\frac{4}{3}\pi r_p^3 \tag{6.9}$$

or evaluated for the particle number density

$$n_p=\frac{3}{4}\frac{\alpha_p}{\pi r_p^3}. \tag{6.10}$$

The friction force on a single droplet can be written as

$$F_p^{\mathrm{v}}=\pi r_p^2 C^D\frac{1}{2}\varrho_g\left|u_g-u_l\right|\left(u_g-u_l\right) \tag{6.11}$$

with the "drag" coefficient $C^D=f\left(Re\right)$, and the Reynolds number

$$Re=\frac{\varrho_g 2r_p\left|u_g-u_l\right|}{\mu_g}. \tag{6.12}$$

The interfacial friction force then becomes $F_l^{\mathrm{v}}=n_pF_p^{\mathrm{v}}$, or more specific

$$F_l^{\mathrm{v}}=-F_g^{\mathrm{v}}=n_pF_p^{\mathrm{v}}=\frac{3}{8}C^D\frac{\alpha_p}{r_p}\varrho_g\left|u_g-u_l\right|\left(u_g-u_l\right). \tag{6.13}$$

With the restriction to low Reynolds numbers $Re \leq 1$, the Stokes law $C^D = 24/Re$ can be applied resulting in the interfacial friction forces:

$$F_l^{\text{v}} = -F_g^{\text{v}} = \frac{9}{2} \frac{\mu_g \alpha_p}{r_p^2} (u_g - u_l) . \tag{6.14}$$

For the heat transfer between gas and droplets, a similar relation can be derived for the interfacial heat source,

$$\sigma_l^Q = -\sigma_g^Q = n_p 4\pi r_p^2 \mathcal{H}^q (T_g - T_l) \tag{6.15}$$

or

$$\sigma_l^Q = -\sigma_g^Q = \frac{3}{2} \frac{\alpha_p}{r_p^2} Nu \, \lambda_g (T_g - T_l) , \tag{6.16}$$

where the heat transfer coefficient $\mathcal{H}^q$ is expressed by the Nusselt numbers as

$$\mathcal{H}^q = Nu \, \lambda_g / 2 \, r_p . \tag{6.17}$$

For low Reynolds numbers, the Nusselt number becomes $Nu = 2$ and the heat source term simplifies to

$$\sigma_l^Q = -\sigma_g^Q = 3 \frac{\alpha_p}{r_p^2} \lambda_g (T_g - T_l) . \tag{6.18}$$

The equations for the interfacial friction and heat transfer can be written in a more compact form as

$$\left. \begin{aligned} F_l^{\text{v}} &= -F_g^{\text{v}} = \frac{\varrho (u_g - u_l)}{\tau_i} \\ \sigma_l^Q &= -\sigma_g^Q = \frac{\varrho_g \bar{C}^p (T_g - T_l)}{\tau_h} \end{aligned} \right\} \tag{6.19}$$

with the characteristic time values for the interfacial friction and heat transfer

$$\left. \begin{aligned} \tau_i &= \frac{2}{9} \frac{\varrho r_p^2}{\mu_g \alpha_p} \\ \tau_h &= \frac{1}{3} \frac{\bar{C}^p \, \varrho r_p^2}{\lambda_g \, \alpha_p} \end{aligned} \right\} . \tag{6.20}$$

For the definition of the characteristic time values, the mixture density $\varrho = \alpha_g \varrho_g + \alpha_l \varrho_l$ and mixture specific heat for constant pressure $\bar{C}^p = X_g C_g^p + X_l C_l^p$ are used as reference values.

With equations (6.18) and (6.19) the source term and state vectors become

$$\mathbf{C} = \begin{bmatrix} 0 \\ 0 \\ -\dfrac{\varrho\,(u_g - u_l)}{\tau_i} \\ +\dfrac{\varrho\,(u_g - u_l)}{\tau_i} \\ -\dfrac{\varrho\,\bar{C}^p(T_g - T_l)}{\tau_i} \\ +\dfrac{\varrho\,\bar{C}^p(T_g - T_l)}{\tau_i} \end{bmatrix} \qquad \mathbf{U} = \begin{bmatrix} p \\ u_g \\ u_l \\ \alpha_g \\ T_g \\ T_l \end{bmatrix}, \tag{6.21}$$

which results in the Jacobian matrix

$$\mathbf{J} = \frac{\partial \mathbf{C}}{\partial \mathbf{U}} = \begin{bmatrix} 0 & 0 & 0 & 0 & 0 & 0 \\ 0 & 0 & 0 & 0 & 0 & 0 \\ 0 & -\dfrac{\varrho}{\tau_i} & +\dfrac{\varrho}{\tau_i} & 0 & 0 & 0 \\ 0 & +\dfrac{\varrho}{\tau_i} & -\dfrac{\varrho}{\tau_i} & 0 & 0 & 0 \\ 0 & 0 & 0 & 0 & -\dfrac{\varrho\,\bar{C}^p}{\tau_h} & +\dfrac{\varrho\,\bar{C}^p}{\tau_h} \\ 0 & 0 & 0 & 0 & +\dfrac{\varrho\,\bar{C}^p}{\tau_h} & -\dfrac{\varrho\,\bar{C}^p}{\tau_h} \end{bmatrix}. \tag{6.22}$$

Introducing the coefficient matrices $\mathbf{A}_0$, $\mathbf{B}_0$ and the Jacobian $\mathbf{J}$ into the characteristic equation (6.7), one obtains the following relation for the complex wave number K:

$$\left(\frac{K}{\omega}\right)^2 = \frac{1}{a_{\text{equ}}^2} \frac{1 + i\,\alpha_l\,\alpha_g \dfrac{\hat{\varrho}^2}{\varrho^2}\,(\tau_i\omega)}{1 + i\,\alpha_l\,\alpha_g \dfrac{\hat{\varrho}_s}{\varrho}\,(\tau_i\omega)} \, \frac{1 + i\,\alpha_l\,\alpha_g \dfrac{C_l^p C_g^p}{(\bar{C}^p)^2}\,\dfrac{\varrho_l\,\varrho_g\,\hat{\varrho}_s}{\varrho\,\hat{\varrho}^2}\left(\dfrac{a_{\text{equ}}}{a_{\text{fr}}}\right)^2 \dfrac{\tau_h}{\tau_i}\,(\tau_i\omega)}{1 + i\,\alpha_l\,\alpha_g \dfrac{C_l^p C_g^p}{(\bar{C}^p)^2}\,\dfrac{\varrho_l\,\varrho_g}{\varrho^2}\,\dfrac{\tau_h}{\tau_i}\,(\tau_i\omega)}, \tag{6.23}$$

with the abbreviation for the homogeneous and frozen sound velocities a_{equ} and a_{fr} respectively. Apart from the ratio of time constants for the interfacial heat transfer and interfacial friction τ_h/τ_i, equation (6.23) does not include any further assumption on the actual values for the interfacial coupling terms. From the dispersion relation (6.23) the following limiting values can be obtained for the sound velocity $a = \omega/K_r$ for $\tau_i\omega \to \infty$ and $\tau_i\omega \to 0$:

1. For high frequencies $\tau_i\omega \to \infty$ the "frozen" speed of sound is obtained as

$$a = a_{\text{fr}} = \sqrt{\frac{\hat{\varrho}_s}{\hat{\varrho}^2} \frac{1}{\dfrac{\alpha_g}{\varrho_g a_g^2} + \dfrac{\alpha_l}{\varrho_l a_l^2}}} \tag{6.24}$$

with the abbreviations for the densities $\hat{\varrho}^2 = \varrho_l\, \varrho_g + k\varrho^2$ and $\hat{\varrho}_s = \alpha_l\, \varrho + \alpha_g \varrho_l + k\varrho$. The "frozen" sound speed is (apart from the assumption of instantaneous equal phase velocities) identical with the sound velocity as appears in the eigenvalues of the hyperbolic model presented in Chapter 5. If the virtual mass effect is neglected ($k = 0$) the expression for the frozen sound velocity simplifies to

$$a_{\text{fr}} = \sqrt{\frac{\alpha_g \varrho_l + \alpha_l\, \varrho_g}{\dfrac{\alpha_g \varrho_l}{a_g^2} + \dfrac{\alpha_l \varrho_g}{a_l^2}}}. \tag{6.25}$$

2. For low frequencies ($\tau_i\omega \to 0$) the sound velocity approaches the limit

$$a = a_{\text{equ}} = \frac{1}{\varrho} \sqrt{\frac{1}{\left[\dfrac{X_g\gamma_g}{\varrho_g} + \dfrac{X_l\gamma_l}{\varrho_l}\right] - \dfrac{T}{\bar{C}p}\left[\dfrac{X_g\beta_g}{\varrho_g} + \dfrac{X_l\beta_l}{\varrho_l}\right]^2}} \tag{6.26}$$

as was derived in Section 4.2 for the homogeneous equilibrium flow for two-component two-phase flow. Applying simplified state equations for liquid and gas and replacing the mass fractions X_i by the volume fractions α_i, the homogeneous equilibrium sound velocity becomes

$$a_{\text{equ}} = \sqrt{\frac{1}{\dfrac{\alpha_g \varrho}{\varrho_g \bar{a}_g^2} + \dfrac{\alpha_l \varrho}{\varrho_l \bar{a}_l^2}}}, \tag{6.27}$$

with

$$\bar{a}_g^2 = \bar{\varkappa}_g \frac{p}{\varrho_g} \quad \text{and} \quad \bar{a}_g^2 = \frac{1}{\varrho_l \gamma_l}. \tag{6.28}$$

The sound velocity $a = \omega/K_r$ and the dimensionless damping coefficient $d = K_i\, a_{\text{fr}}/\omega$ as functions of the dimensionless frequency are given in Figs. 6.1 and 6.2 for a two-component water/air mixture at atmospheric pressure. The figures show a continuous transition region and an asymptotic approach to the "frozen" and equilibrium condition for high and low frequencies, respectively. The damping coefficient d of the sound wave exhibits a maximum value within the transition region and vanishes when the upper and low values of the sound velocity are reached. As long as the characteristic time for the interfacial heat transfer is of the same order as the corresponding value for the interfacial friction ($\tau_h \approx \tau_i$) the dispersion curves show only one point of inflection for the sound velocity and one maximum for the .

For relatively low characteristic time values for the heat transfer ($\tau_h \ll \tau_i$) thermal equilibrium conditions are reached already at very high frequencies where the interfacial friction is still in the "frozen" condition. This condition was predicted for $\tau_h/\tau_i < 0.1$, as indicated in Figs. 6.1 and 6.2, together with the occurrence of a second point of inflection for the sound velocity and a second maximum for the attenuation coefficient..

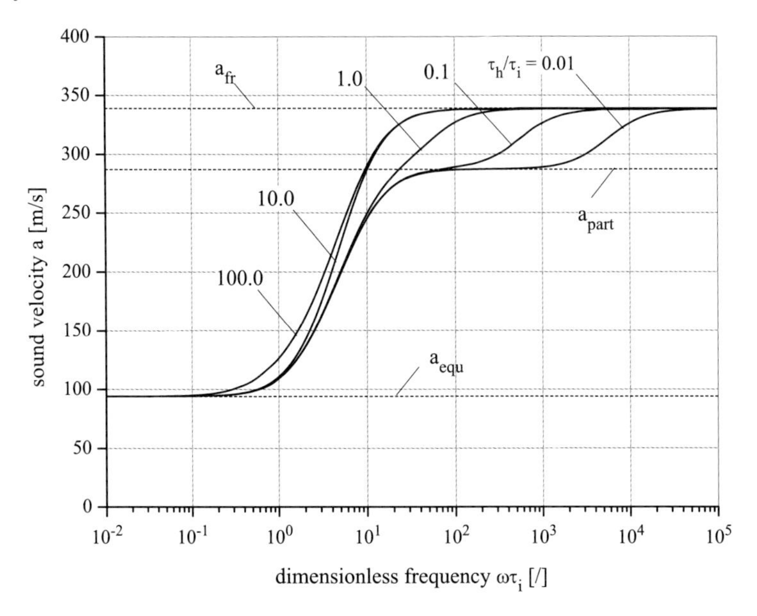

Fig. 6.1: Sound velocity as a function of the dimensionless sound frequency, *water/air mixtures* at atmospheric conditions, dispersed droplet flow, virtual mass coefficient $k = 0$

The value of the sound velocity for the partial equilibrium conditions, a_{part}, can be obtained from (6.23),

$$\left(\frac{K_{\text{part}}}{\omega}\right)^2 = \lim_{\tau_h/\tau_{i.}\to 0}\left(\frac{K}{\omega}\right)^2 \quad \text{and} \quad \frac{1}{a_{\text{part}}^2} = \lim_{\omega\tau_{i.}\to\infty}\left(\frac{K_{\text{part}}}{\omega}\right)^2, \tag{6.29}$$

which results in

$$a_{\text{part}} = a_{\text{equ}}\sqrt{\frac{\varrho\hat{\varrho}_s}{\hat{\varrho}^2}}, \tag{6.30}$$

with the equilibrium sound velocity as given by equation (6.26) or (6.27), respectively.

The calculated values for the equilibrium, partial equilibrium, and frozen sound velocities as a function of the void fraction are given in Figs. 6.3 and 6.4 for a water/air mixture at atmospheric pressure. As one can see, the condition $a_{\text{fr}} \geq a_{\text{part}} \geq a_{\text{equ}}$ remains valid over the whole range of void fraction from pure liquid ($\alpha_g = 0$) to pure gas ($\alpha_g = 1$) independent of the chosen value for the "virtual mass" coefficient.

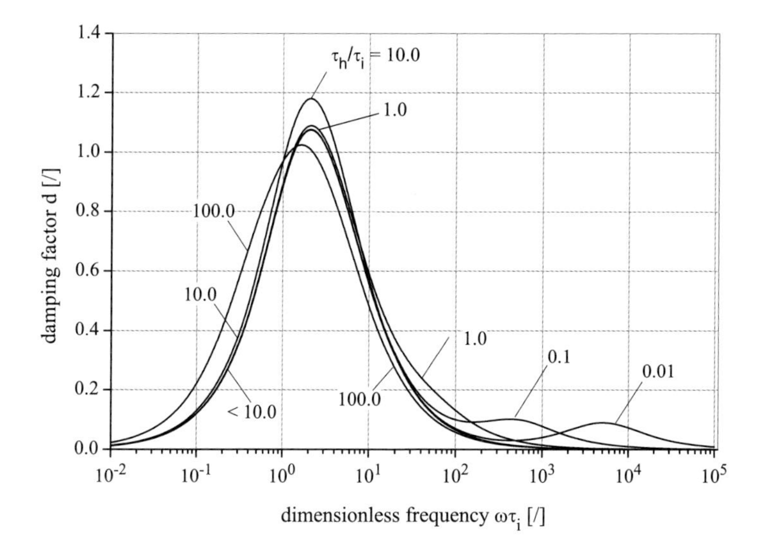

Fig. 6.2: Attenuation of sound waves as a function of the dimensionless sound frequency, *water/air mixtures* at atmospheric conditions, dispersed droplet flow, virtual mass coefficient $k = 0.5$

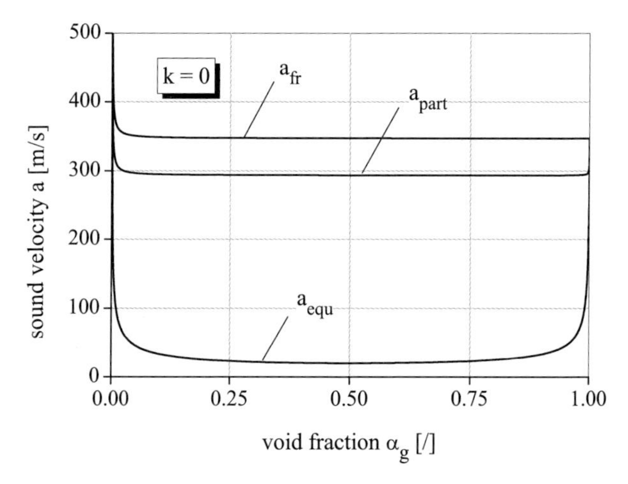

Fig. 6.3: Equilibrium, partial equilibrium, and frozen sound velocity for two-phase *water/air mixtures* at atmospheric conditions; virtual mass coefficient $k = 0$

Figures 6.3 and 6.4 clearly indicate the strong effect of the virtual mass coefficient k on the frozen sound velocity as already explained in Chapter 5. Nevertheless, for the present case of dispersed droplet flow a value of $0.25 \leq k \leq 0.5$ might be appropriate as indicated in Fig. 5.5.

The dispersion analysis as outlined above for the rather simple case of mono-dispersed droplet flows in gas/liquid media can be easily extended to more complex flow regimes or to one-component liquid/steam mixtures as described for example by Ardron and Duffey in [2].

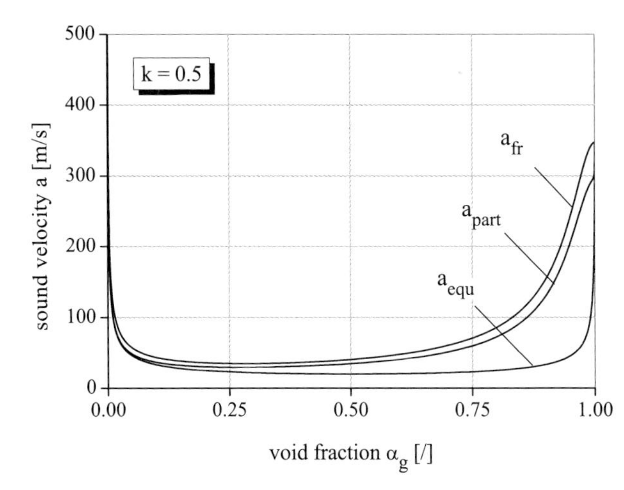

Fig. 6.4: Equilibrium, partial equilibrium, and frozen sound velocity for two-phase *water/air mixtures* at atmospheric conditions; virtual mass coefficient $k = 0.5$

Nevertheless some qualitative conclusions from the present analysis might be drawn which are of importance for the interpretation of critical flow or shock waves in two-phase flow.

- Due to the presence of nonequilibrium effects the sound velocity in two-phase media is no longer a single value determined by simple thermodynamic state properties as in single-phase gas media. Instead, it becomes a function of the sound frequency with an upper (a^{fr}) or lower (a^{eq}) limiting value for very high or very low frequency values depending on whether either "frozen" or equilibrium conditions are reached.

- The sound velocity as it appears in the characteristic analysis of the governing two-phase flow equations is independent of the algebraic interfacial coupling terms and as such it represents the upper "frozen" limit a^{fr} for the sound propagation as obtained from the dispersion analysis.

- In steady state flow situations critical flow conditions are reached where the fastest wave speed becomes stationary which is also obtained by single-pressure two-fluid model for the condition $u = a^{\text{fr}}$, with the mixture flow velocity u and the mixture sound velocity $a = a^{\text{fr}}$ as resulting from the characteristic analysis of the governing equations.

- The presence (or absence) of critical flow conditions is not *a priori* an indication whether or not a maximum value for the mass flow (choking) in a nozzle or pipe is reached. As will be shown in a number of test cases presented in Chapter 9, a maximum discharge mass flow might occur much earlier under subsonic conditions ($u < a^{\text{fr}}$) due to the governing effects of interfacial heat and mass transfer processes.

References

[1] G.B. Whitham, *Linear and Nonlinear Waves*, Wiley Interscience, New York, 1973.

[2] K.H. Ardron and R.B. Duffey, *Acoustic Wave propagation in a Flowing Liquid-Vapor Mixture*, International Journal of Multiphase Flow, 4, 303–322, 1978.

7 Numerical Methods for Hyperbolic Two-Phase Flow System Equations

7.1 Mathematical nature of two-phase flow equations

The general balance equations of the two-fluid model as described in Chapter 3 can be written in a compact vector form as

$$\frac{\partial \mathbf{U}}{\partial t} + \mathbf{G}^{\mathrm{adv}} \nabla \mathbf{U} + \mathbf{G}^{\mathrm{dif}} \nabla^2 \mathbf{U} = \mathbf{D} \tag{7.1}$$

with the vector of "primitive" state parameters $\mathbf{U}$, the advection or Euler part of the flow equations, $\mathbf{G}^{\mathrm{adv}}$, the diffusive part, $\mathbf{G}^{\mathrm{dif}}$, including the molecular or turbulent viscosity and heat conduction effects, and an algebraic source term vector $\mathbf{D}$.

For the numerical integration of equation (7.1), an operator splitting technique might be appropriate which allows us to apply the most suitable method for the different major physical processes involved:

1. The hyperbolic advection or transport part of the equations

   $$\frac{\partial \mathbf{U}}{\partial t} + \mathbf{G}^{\mathrm{adv}} \nabla \mathbf{U} = \mathbf{0} \tag{7.2}$$

   is characterized by the propagation of information with finite velocities resulting in bounded regions of dependence and influence. As a result, discontinuous solutions might exist in the flow field representing pressure waves (e.g., shock waves) or void waves with abrupt changes in flow velocity, density, or volumetric concentration of the gas or liquid. The most appropriate numerical schemes for such processes are based on techniques which make explicit use of the hyperbolic nature of the flow equations, in the following called "hyperbolic methods".

2. The parabolic diffusive part of the flow equations,

   $$\frac{\partial \mathbf{U}}{\partial t} + \mathbf{G}^{\mathrm{dif}} \nabla^2 \mathbf{U} = \mathbf{0}, \tag{7.3}$$

   is less challenging mathematically and a central differencing (or finite-volume equivalent) scheme might be adequate for the numerical solution.

Gasdynamic Aspects of Two-Phase Flow. Herbert Städtke

ISBN: 3-527-40578-X

3. The effect of the algebraic source term represents a system of "stiff" ordinary differential equations

$$\frac{\partial \mathbf{U}}{\partial t} = \mathbf{C}, \tag{7.4}$$

where the characteristic times for various interfacial transport processes can vary over several orders of magnitude. A fully implicit time integration scheme is therefore essential for robustness and computational efficiency.

7.2 Overview on hyperbolic numerical methods

In the following only essentially non-oscillatory high-resolution numerical methods will be considered as were originally developed for single phase gas flows. This includes the methods based on the *Approximate Riemann Solver*, the *Flux Vector Splitting* technique, and the *Split Coefficient Matrix* method. The former two belong to the Godunov-type methods based on a finite volume discretization assuming a constant (or linear) parameter distribution within a computational cell. Both techniques can be easily extended to unstructured grids. The Split Coefficient Method instead represents a finite-difference techniques where the new-time values of all grid points are calculated by solving the linearized characteristic form of the governing equations along characteristic lines within the framework of a regular Cartesian grid. Common to all these methods is the concept of "upwinding" which combines the preservation of wave propagation processes along the characteristic directions with the conservation property for mass, momentum, and energy, for the solution of the advection problem.

All these methods require a fully hyperbolic system of flow equations of the form

$$\frac{\partial \mathbf{U}}{\partial t} + \mathbf{G}\,\nabla \mathbf{U} = \mathbf{0} \tag{7.5}$$

where the coefficient matrix $\mathbf{G}$ is characterized by exclusively real eigenvalues and a complete set of independent eigenvectors.

Some of the methods are based on the conservative form of the balance equation which can be derived from the "primitive" form of balance equations (7.5) by a similarity transformation as

$$\frac{\partial \mathbf{V}}{\partial t} + \mathbf{H}\,\nabla\,\mathbf{V} = \mathbf{0} \tag{7.6}$$

with the state vector of conserved quantities $\mathbf{V}$ and the new coefficient matrix $\mathbf{H}$ defined as

$$\mathbf{H} = \mathbf{J}\,\mathbf{G}\,\mathbf{J}^{-1} \tag{7.7}$$

using the Jacobian matrix

$$\mathbf{J} = \frac{\partial \mathbf{V}}{\partial \mathbf{U}}. \tag{7.8}$$

Introducing the flux vector $\mathbf{F}$ into equation (7.6) results in the quasi-conservative form of the balance equation

$$\frac{\partial \mathbf{V}}{\partial t} + \mathbf{H}^{\mathrm{nc}} \nabla \mathbf{U} + \nabla \mathbf{F} = \mathbf{0}, \tag{7.9}$$

with the "nonconservative" contribution resulting from some time and space derivative terms in the separated momentum equations as was explained in Chapter 5. The presence of this non-conservative contribution does not affect the use of the numerical integration as long as the method is based on the "primitive" form of the balance equations. However, this term might become crucial for all Godunov-type finite volume numerical schemes such as Approximate Riemann solvers or Flux Vector Splitting techniques.

7.3 The Split Coefficient Matrix method

The Split Coefficient Matrix (SCM) method of Chakravarthy [1] represents an extension of the CIR method of Courant Isaacson and Rees [2]. The method might be seen as a finite difference analogy to the Method of Characteristics applied on a Cartesian grid. The technique, originally developed for gasdynamic problems, has been later applied by Romstedt [3] for homogeneous two-phase flows.

The CIR and SCM methods are usually based on the "primitive" form of the conservation equations for mass, momentum, and energy which can be written for the one-dimensional case as

$$\frac{\partial \mathbf{U}}{\partial t} + \mathbf{G} \frac{\partial \mathbf{U}}{\partial x} = \mathbf{0}, \tag{7.10}$$

with the state vector $\mathbf{U}$, the coefficient matrix $\mathbf{G}$, and the vector of source terms $\mathbf{D}$.

Assuming that the system of equations (7.10) is hyperbolic, the coefficient matrix $\mathbf{G}$ can be diagonalized as

$$\mathbf{\Lambda} = \mathbf{T} \mathbf{G} \mathbf{T}^{-1} \quad \text{or} \quad \mathbf{G} = \mathbf{T}^{-1} \mathbf{\Lambda} \mathbf{T}, \tag{7.11}$$

where the diagonal matrix $\mathbf{\Lambda}$ contains all (real) eigenvalues and the columns of the transformation matrix $\mathbf{T}$ are the right eigenvectors of $\mathbf{G}$. Multiplying equation (7.10) with the inverse of the transformation matrix $\mathbf{T}^{-1}$, the characteristic form of the governing equations is obtained as

$$\mathbf{T}_0^{-1} \frac{\partial \mathbf{U}}{\partial t} + \mathbf{\Lambda}\, \mathbf{T}^{-1} \frac{\partial \mathbf{U}}{\partial x} = \mathbf{0}. \tag{7.12}$$

For small time steps (7.12) might be linearized resulting in

$$\mathbf{T}^{-1} \frac{\partial \mathbf{U}}{\partial t} + \mathbf{\Lambda}_0 \mathbf{T}_0^{-1} \frac{\partial \mathbf{U}}{\partial x} = \mathbf{0} \tag{7.13}$$

and respectively

$$\frac{\partial \mathbf{W}}{\partial t} + \mathbf{\Lambda}_0 \frac{\partial \mathbf{W}}{\partial x} = \mathbf{0}, \tag{7.14}$$

with Riemann invariants

$$\delta\mathbf{W} = \mathbf{T}_0^{-1}\delta\mathbf{U}. \tag{7.15}$$

Equation (7.14) represents a system of coupled differential equations for the change of the Riemann invariants W_k within the characteristic directions,

$$\frac{\partial W_k}{\partial t} + \lambda_{k,0}\frac{\partial W_k}{\partial x} = \mathbf{0}. \tag{7.16}$$

From equation (7.16) it follows that the Riemann invariants W_k remain constant along the characteristic lines, apart from (usually small) contributions coming from the source term vector; hence, the Riemann invariants $W_{k,i}^{n+1}$ can be calculated as

$$W_{k,i}^{n+1} = W_{k,\xi}^{n} + E_{k,i}^{n}\Delta t, \tag{7.17}$$

where $W_{k,i}^{n+1}$ represents the corresponding value at the intersection of the characteristic curve with the previous time level ξ_k .

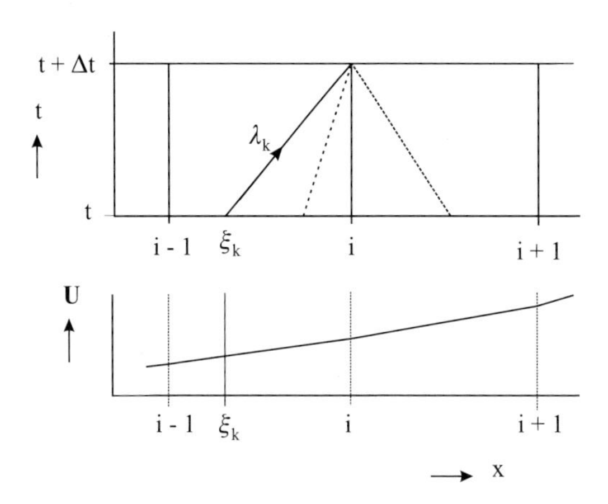

Fig. 7.1: First-order upwind scheme in a fixed space-time grid

Since the characteristic curves become straight lines within the time interval Δt as shown in Fig. 7.1, characteristic variables at the intersection can be calculated by a linear interpolation using the corresponding values at the adjacent grid points,

$$\left.\begin{aligned} W_{k,i}^{n+1} &= W_{k,i}^{n} - (\xi_k - x_i)\,\frac{W_{k,i}^{n} - W_{k,i-1}^{n}}{x_i - x_{i-1}} \quad &\text{for} \quad \lambda_k > 0 \\ W_{k,i}^{n+1} &= W_{k,i}^{n} + (\xi_k - x_i)\,\frac{W_{k,i+1}^{n} - W_{k,i}^{n}}{x_{i+1} - x_i} \quad &\text{for} \quad \lambda_k < 0 \end{aligned}\right\} \tag{7.18}$$

with the spatial difference $\xi_k - x_i = \lambda_k \Delta t$ the kth wave is traveling in the time interval Δt, and the finite difference formulation for the "left" and "right" spatial gradients of the Riemann invariances.

Introducing equations (7.18) into equation (7.16) the finite difference equation for the new time Riemann invariants becomes

$$\left.\begin{aligned} W_{k,i}^{n+1} &= W_{k,i}^{n} - \frac{\lambda_k \Delta t}{x_i - x_{i-1}}\left(W_{k,i}^{n} - W_{k,i-1}^{n}\right) + E_{k,i}^{n}\Delta t \quad \text{for } \lambda_k > 0 \\ W_{k,i}^{n+1} &= W_{k,i}^{n} - \frac{\lambda_k \Delta t}{x_{i+1} - x_i}\left(W_{k,i+1}^{n} - W_{k,i}^{n}\right) + E_{k,i}^{n}\Delta t \quad \text{for } \lambda_k < 0 \end{aligned}\right\} \tag{7.19}$$

or in the vector form

$$\mathbf{W}_i^{n+1} = \mathbf{W}_i^n - \mathbf{\Lambda}^+ \frac{\mathbf{W}_i^n - \mathbf{W}_{i-1}^n}{x_i - x_{i-1}}\Delta t - \mathbf{\Lambda}^- \frac{\mathbf{W}_{i+1}^n - \mathbf{W}_i^n}{x_{i+1} - x_i}\Delta t, \tag{7.20}$$

with the diagonal matrices $\mathbf{\Lambda}^+$and $\mathbf{\Lambda}^-$ containing only the positive or negative eigenvalues, respectively

$$\left.\begin{aligned} \Lambda_{k,k}^{+} &= \lambda_k \quad \text{for} \quad \lambda_k \geq 0 \quad \text{and} \quad \Lambda_{k,k}^{+} = 0 \quad \text{otherwise} \\ \Lambda_{k,k}^{-} &= \lambda_k \quad \text{for} \quad \lambda_k < 0 \quad \text{and} \quad \Lambda_{k,k}^{-} = 0 \quad \text{otherwise} \end{aligned}\right\}.$$

With the linear relationship between the vectors for the state and characteristic variables given by equation (7.15), the equation for the new time equation "primitive" state vector $\mathbf{U}$ is obtained as

$$\mathbf{U}_i^{n+1} = \mathbf{U}_i^n - \mathbf{G}_0^+ \frac{\left(\mathbf{U}_i^n - \mathbf{U}_{i-1}^n\right)}{x_i - x_{i-1}}\Delta t + \mathbf{G}_0^- \frac{\left(\mathbf{U}_{i+1}^n - \mathbf{U}_i^n\right)}{x_{i+1} - x_i}\Delta t \tag{7.21}$$

or

$$\mathbf{U}_i^{n+1} = \mathbf{U}_i^n - \mathbf{G}_0^+ \left(\nabla^+ \mathrm{U}\right)\Delta t + \mathbf{G}_0^- \left(\nabla^- \mathrm{U}\right)\Delta t + \mathbf{E}_i^n \Delta t \tag{7.22}$$

with the split matrices

$$\left.\begin{aligned} \mathbf{G}_0^+ &= \mathbf{T}_0\, \mathbf{\Lambda}_0^+\, \mathbf{T}_0^{-1} = \sum_{k,\lambda_k \geq 0} \mathbf{G}_k \\ \mathbf{G}_0^- &= \mathbf{T}_0\, \mathbf{\Lambda}_0^-\, \mathbf{T}_0^{-1} = \sum_{k,\lambda_k < 0} \mathbf{G}_k \end{aligned}\right\} \tag{7.23}$$

and the finite difference operator

$$\left.\begin{aligned} \nabla^+ \mathbf{U} &= \frac{\left(\mathbf{U}_i^n - \mathbf{U}_{i-1}^n\right)}{x_i - x_{i-1}} \\ \nabla^- \mathbf{U} &= \frac{\left(\mathbf{U}_{i+1}^n - \mathbf{U}_i^n\right)}{x_{i+1} - x_i} \end{aligned}\right\}. \tag{7.24}$$

The Split Coefficient Matrix method is sometimes preferred since it can be easily implemented on structured Cartesian grids. The spatial resolution of the methods can be increased using higher order differential operators in equation (7.19). For the time integration in principle any ODE solver can be used. As in all characteristic upwind schemes the Split Coefficient Matrix method requires a full hyperbolic system of equations. A further peculiarity of the method is that it does not *a priori* conserve mass, momentum, and energy, which can be a drawback in particular for the calculation of long-lasting slow transients.

7.4 Godunov methods and Approximate Riemann solver

7.4.1 General Godunov approach

The method of Godunov already published in 1959 [4] was developed for gasdynamic application; however, it can in principle be applied for any hyperbolic flow problem described by a system of nonlinear hyperbolic conservation laws of the form

$$\frac{\partial \mathbf{V}}{\partial t} + \nabla \mathbf{F} = \mathbf{0} \tag{7.25}$$

with the vector of conserved state variables $\mathbf{V}$ and the flux vector $\mathbf{F}$ and the source term vector $\mathbf{D}$. Equation (7.25) can also be written in the expanded form as

$$\frac{\partial \mathbf{V}}{\partial t} + \mathbf{H}\,\nabla \mathbf{V} = \mathbf{0}, \tag{7.26}$$

where the coefficient matrix $\mathbf{H}$ is identical with the Jacobian matrix

$$\mathbf{H} = \frac{\partial \mathbf{F}}{\partial \mathbf{V}}. \tag{7.27}$$

In the following the method will be explained for the one-dimensional case where equation (7.25) simplifies to

$$\frac{\partial \mathbf{V}}{\partial t} + \frac{\partial \mathbf{F}}{\partial x} = \mathbf{0} \tag{7.28}$$

and

$$\frac{\partial \mathbf{V}}{\partial t} + \mathbf{G}\frac{\partial \mathbf{V}}{\partial x} = \mathbf{0}. \tag{7.29}$$

Integrating the governing parameters over the cell

$$\mathbf{V}_i^{n+1} = \int_{x_{i-1/2}}^{x_{i+1/2}} \mathbf{V}^{n+1} dx \tag{7.30}$$

results in a piecewise constant distribution of flow parameters with a discontinuous change of parameters at the cell interfaces as shown in Fig. 7.2.

The discontinuities at the cell boundaries are then treated as a sequence of Riemann problems for the calculation of the new time parameter distributions in the adjacent cells as indicated in Fig. 7.2. At the end of the time step, all parameters are averaged according to equation (7.30) in order to provide the initial conditions for the new time step.

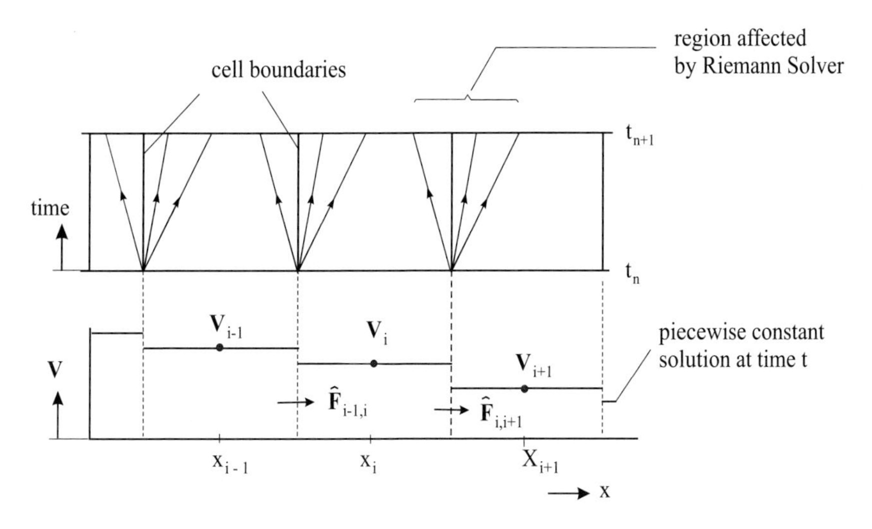

Fig. 7.2: Godunov scheme for the one-dimensional flow condition

To avoid wave interference, the method is bounded by the Courant number criteria,

$$\left| \frac{\Delta t}{\Delta x \, \lambda_k} \right| \leq 1 \qquad \text{for all } k. \tag{7.31}$$

In most practical applications, the method described above is further simplified where the Riemann solver is limited to the prediction of the numerical flux at the cell interfaces,

$$\hat{\mathbf{F}}_{i,1+1} = f(\mathbf{V}_i, \mathbf{V}_{i+1}), \tag{7.32}$$

which is then used for updating the new averaged conservative parameters from the space and time discretization form of the flow equations (7.28) as

$$\mathbf{V}_i^{n+1} = \Delta t_i^n + \frac{\Delta t}{\Delta x}(\hat{\mathbf{F}}_{i,i+1} - \hat{\mathbf{F}}_{i-1,i}). \tag{7.33}$$

As mentioned above the Godunov method requires the solution of the Riemann problem for each cell boundary and each time step. Although this can theoretically be done (at least for single-phase flow of gas), however, it needs an iterative procedure for solving the associated non-linear equations which is not only computationally rather costly but might become extremely difficult for more complex state equations or two-phase flow conditions. Therefore, the exact solution is in most cases replaced by an approximated Riemann solver providing the necessary information for the calculation of the "Godunov fluxes" at the cell interfaces as a function of the state properties on the left- and right-hand side cells $\hat{\mathbf{F}} = f(\mathbf{V}_l, \mathbf{V}r)$. A rather comprehensive review of the proposed Riemann solvers and their specific merits and limitations is provided by Toro in [5]. Many details on numerical methods for hyperbolic flow equations can also be found in the book of LeVeque [7].

7.4.2 The linearized Riemann solver

Within the framework of the Godunov scheme the discontinuities at the cell interfaces are treated as Riemann problems which are described by the one-dimensional form of the hyperbolic system of conservation equations,

$$\frac{\partial \mathbf{U}}{\partial t} + \mathbf{G}(\mathbf{U})\frac{\partial \mathbf{U}}{\partial x} = 0, \tag{7.34}$$

and the initial data as illustrated in Fig. 7.3

$$\mathbf{U}(x, t_0) = \mathbf{U}_0(x) = \begin{cases} \mathbf{U}_L & \text{for } x < 0 \\ \\ \mathbf{U}_R & \text{for } x < 0 \end{cases} . \tag{7.35}$$

The linearization of equation (7.34) leads to a constant coefficient matrix $\mathbf{G}_0(\bar{\mathbf{U}})$,

$$\frac{\partial \mathbf{U}}{\partial t} + \mathbf{G}_0\frac{\partial \mathbf{U}}{\partial x} = \mathbf{0}, \tag{7.36}$$

based on an average state vector $\bar{\mathbf{U}}(\mathbf{U}_L, \mathbf{U}_{R)}$ as a function of the corresponding states at the left and right side of the discontinuity. In the most simplified way, the arithmetic mean value might be used

$$\bar{\mathbf{U}} = \frac{1}{2}(\mathbf{U}_L + \mathbf{U}_{R)}. \tag{7.37}$$

Since the system of equations is hyperbolic, it can be transformed into the characteristic form

$$\mathbf{T}_0^{-1}\frac{\partial \mathbf{U}}{\partial t} + \mathbf{\Lambda}_0\mathbf{T}_0^{-1}\frac{\partial \mathbf{U}}{\partial x} = \mathbf{0} \tag{7.38}$$

or

$$\frac{\partial \mathbf{W}}{\partial t} + \mathbf{\Lambda}_0\frac{\partial \mathbf{W}}{\partial x} = \mathbf{0} \tag{7.39}$$

with the diagonal matrix $\mathbf{\Lambda}_0$ containing all eigenvalues of the matrix $\mathbf{G}_0$ and the characteristic variables

$$\mathbf{W} = \mathbf{T}_0^{-1}\mathbf{U}. \tag{7.40}$$

As already explained for the derivation of the SCM method in Section 6.2, the transformation matrix $\mathbf{T}$ is the transpose of the matrix of “right” eigenvectors of $\mathbf{G}_0$. Equation (7.39) describes a system of decoupled waves propagating with the constant velocities λ_k,

$$\frac{\partial W_k}{\partial t} + \lambda_{k,0}\frac{\partial W_k}{\partial x} = 0, \tag{7.41}$$

with the solution

$$W_k\,(x,t) = W_{k,0}\,(x - \lambda_{k,0}\,t)\,. \tag{7.42}$$

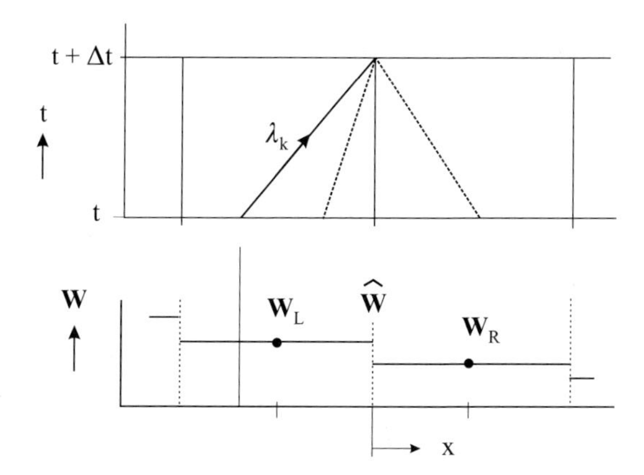

Fig. 7.3: Initial conditions for a Riemann problem

For a space and time discretized system with piecewise constant parameter distributions as shown in Fig. 7.3, the new time characteristic parameters at the cell interface become

$$\left.\begin{aligned} \hat{W}_k^{n+1} &= W_{k,L}^n \quad \text{for } \lambda_k > 0 \\ \hat{W}_k^{n+1} &= W_{k,R}^n \quad \text{for } \lambda_k < 0 \end{aligned}\right\}. \tag{7.43}$$

Due to the linearization the different wave modes are completely decoupled and, therefore, can be superimposed leading to the new time vector of characteristic variables,

$$\hat{\mathbf{W}} = \sum_k \hat{W}_k^{n+1} = \sum_{k,\lambda_k \geq 0} W_{k,L} + \sum_{k,\lambda_k < 0} W_{k,R} \tag{7.44}$$

or

$$\hat{\mathbf{W}} = \mathbf{I}^+ \mathbf{W}_L + \mathbf{I}^- \mathbf{W}_R \tag{7.45}$$

with the diagonal matrices having the elements

$$\left.\begin{aligned} I_{k,k}^+ &= 1 \quad \text{for} \quad \lambda_k \geq 0 \quad \text{and} \quad I_{k,k}^+ = 0 \quad \text{otherwise} \\ I_{k,k}^- &= 1 \quad \text{for} \quad \lambda_k < 0 \quad \text{and} \quad I_{k,k}^- = 0 \quad \text{otherwise} \end{aligned}\right\} \tag{7.46}$$

With relation (7.36), equation (7.45) can be transformed back into the space of the state vector $\mathbf{U}$

$$\mathbf{T}_0 \hat{\mathbf{W}} = (\mathbf{T}_0 \mathbf{I}^+ \mathbf{T}_0^{-1}) \mathbf{T}_0 \mathbf{W}_L + (\mathbf{T}_0 \mathbf{I}^- \mathbf{T}_0^{-1}) \mathbf{T}_0 \mathbf{W}_r \tag{7.47}$$

or

$$\hat{\mathbf{U}} = \tilde{\mathbf{G}}^+ \mathbf{U}_L + \tilde{\mathbf{G}}^- \mathbf{U}_R. \tag{7.48}$$

The reduced split matrices $\tilde{\mathbf{G}}_0^+$ and $\tilde{\mathbf{G}}_0^-$ are ordered with respect to the sign of the corresponding eigenvalues

$$\left.\begin{aligned} \tilde{\mathbf{G}}_0^+ &= \mathbf{T}_0\,\mathbf{I}_0^+\,\mathbf{T}_0^{-1} = \sum_{k,\lambda_k\geq 0} \frac{1}{\lambda_k}\mathbf{G}_k \\ \tilde{\mathbf{G}}_0^- &= \mathbf{T}_0\,\mathbf{I}_0^-\,\mathbf{T}_0^{-1} = \sum_{k,\lambda_k< 0} \frac{1}{\lambda_k}\mathbf{G}_k \end{aligned}\right\}. \tag{7.49}$$

With

$$\tilde{\mathbf{G}}_0^+ + \tilde{\mathbf{G}}_0^- = \mathbf{T}_0\,(\mathbf{I}_0^+ + \mathbf{I}_0^-)\,\mathbf{T}_0^{-1} = \mathbf{T}_0\,\mathbf{I}_0\,\mathbf{T}_0^{-1} = \mathbf{I} \tag{7.50}$$

equation (7.48) can also be written as

$$\left.\begin{aligned} \hat{\mathbf{U}} &= \mathbf{U}_R + \tilde{\mathbf{G}}^+(\mathbf{U}_L - \mathbf{U}_R) \\ \hat{\mathbf{U}} &= \mathbf{U}_L - \tilde{\mathbf{G}}^-(\mathbf{U}_L - \mathbf{U}_R) \end{aligned}\right\}. \tag{7.51}$$

For the specific cases where all eigenvalues are positive or negative, respectively, the intercell state vector becomes equal to the upstream value

$$\left.\begin{aligned} \hat{\mathbf{U}} &= \mathbf{U}_L \quad \text{if} \quad \lambda_k > 0 \quad \text{for all } k \\ \hat{\mathbf{U}} &= \mathbf{U}_L \quad \text{if} \quad \lambda_k < 0 \quad \text{for all } k \end{aligned}\right\}. \tag{7.52}$$

From the new state vector $\hat{\mathbf{U}}$ the Godunov fluxes $\hat{\mathbf{F}}(\hat{\mathbf{U}})$ can be calculated, which then are used to update the conservative state parameters in the computation cells as given by equation (7.33). The linearized Riemann solver, as descried above for the vector of primitive state parameters, can be based on any type of independent state vector as long as the governing system of equation remains hyperbolic.

7.4.3 The Roe solver

One of the most popular Riemann solvers is that of Roe (1981) [6] which is based on the conservative form of the flow equations

$$\frac{\partial \mathbf{V}}{\partial t} + \mathbf{H}\frac{\partial \mathbf{V}}{\partial x} = 0, \tag{7.53}$$

with the coefficient matrix

$$\mathbf{H} = \frac{\partial \mathbf{F}}{\partial \mathbf{V}}. \tag{7.54}$$

As described above the exact Riemann problem is replaced by an approximate (linearized) problem

$$\frac{\partial \mathbf{V}}{\partial t} + \mathbf{H}^{\mathrm{Roe}}\frac{\partial \mathbf{V}}{\partial x} = 0, \tag{7.55}$$

where the Jacobian matrix $\bar{\mathbf{H}}^{\text{Roe}}(\mathbf{V}_R, \mathbf{V}_L)$ introduced here is assumed to satisfy the following conditions:

(a) the matrix $\bar{\mathbf{H}}^{\text{Roe}}$ has only real eigenvalues and is diagonalizable,

(b) for equal states in adjacent cells, the "Roe matrix" becomes identical with the original coefficient matrix

$$\mathbf{H}^{\text{Roe}} = \mathbf{H} \qquad \text{for} \qquad \mathbf{V}_L = \mathbf{V}_R = \bar{\mathbf{V}}, \tag{7.56}$$

(c) the matrix $\bar{\mathbf{H}}^{\text{Roe}}$ satisfies the Rankine Hugoniot condition

$$\mathbf{F}_L - \mathbf{F}_R = \mathbf{H}^{\text{Roe}}\,(\mathbf{V}_L - \mathbf{V}_R)\,. \tag{7.57}$$

The first two conditions are rather trivial and are valid for all approximate Riemann solvers. The third condition (c) results from the need to handle flow discontinuities related to the occurrence of shock waves in single-phase gasdynamic problems.

The construction of the Roe average matrix $\mathbf{H}^{\text{Roe}}$ is not unique; nevertheless, as shown by Roe for the specific case of single-phase gas flow, the conditions (a) to (c) are satisfied for the following averaged state parameters for density, flow velocity, and enthalpy

$$\left.\begin{aligned} \bar{\varrho} &= \sqrt{\varrho_L + \varrho_R} \\ \bar{u} &= \frac{\sqrt{\varrho_L}u_L + \sqrt{\varrho_L}u_L}{\sqrt{\varrho_L + \varrho_R}} \\ \bar{h} &= \frac{\sqrt{\varrho_L}h_L + \sqrt{\varrho_L}h_L}{\sqrt{\varrho_L + \varrho_R}} \end{aligned}\right\}. \tag{7.58}$$

Following the same procedure as explained in the previous section the new time value for the conservative state vector at the cell-to-cell interface can be obtained according to equation (7.52) as discussed by Tuomi and Kumbaro [11]

$$\left.\begin{aligned} \hat{\mathbf{V}} &= \mathbf{V}_R + \tilde{\mathbf{H}}^{+}(\mathbf{V}_L - \mathbf{V}_R) \\ \hat{\mathbf{V}} &= \mathbf{V}_L - \tilde{\mathbf{H}}^{-}(\mathbf{V}_L - \mathbf{V}_R) \end{aligned}\right\}. \tag{7.59}$$

Combining equation (7.59) with condition (c) for the Roe matrix as given by equation (7.57) results in the Godunov fluxes $\hat{\mathbf{F}}$ at the cell boundaries,

$$\left.\begin{aligned} \hat{\mathbf{F}} &= \mathbf{F}_R + \mathbf{H}^{+}(\mathbf{V}_L - \mathbf{V}_R) \\ \hat{\mathbf{F}} &= \mathbf{F}_L - \mathbf{H}^{-}(\mathbf{V}_L - \mathbf{V}_R) \end{aligned}\right\}. \tag{7.60}$$

One should note the difference between the split matrices as used for the state vectors in equation (7.59) and for the flux vectors in equation (7.60) which are related by

$$\mathbf{H}_k = \lambda_k \tilde{\mathbf{H}}_k. \tag{7.61}$$

The update for the conservative state vector at each cell is then obtained from the finite volume formulation as was already given by equation (7.33).

The Roe solver as described above is strictly linked to the existence of a fully conservative form of the flow equations as is the case for the flow of single-phase gas or for homogeneous two-phase flow. However, this is not the case for the more general two-fluid representation of two-phase flow where some differential coupling terms in the separated momentum equations cannot be brought into a conservative form. Nevertheless, under certain hypotheses a Roe-type numerical scheme can be constructed as shown by Tuomi and Kumbaro in [11]. The validity of the additional modeling assumptions might need some further assessment and justification.

7.5 Flux Vector Splitting method

The Flux Vector Splitting scheme (FVS) belongs to the finite volume Godunov class of methods which are based on the conservative form of the flow equations

$$\frac{\partial \mathbf{V}}{\partial t} + \nabla \mathbf{F} = 0, \tag{7.62}$$

with the vectors of conserved quantities $\mathbf{V}$ and corresponding flux vector $\mathbf{F}$. For an explicit finite volume discretization of equation (7.62) the new time value of the conserved parameter $\mathbf{V}$ in the computational cell i can be expressed as

$$\mathbf{V}_i^{n+1} = \mathbf{V}_i^n - \frac{\Delta t}{\Delta x}(\hat{\mathbf{F}}_{i-1,i}^n - \hat{\mathbf{F}}_{i,i+1}^n) \tag{7.63}$$

with the numerical fluxes $\hat{\mathbf{F}}_{i-1,i}^n$ and $\hat{\mathbf{F}}_{i,i+1}^n$ at the cell interfaces on the left and right side as schematically shown in Fig. 7.4.

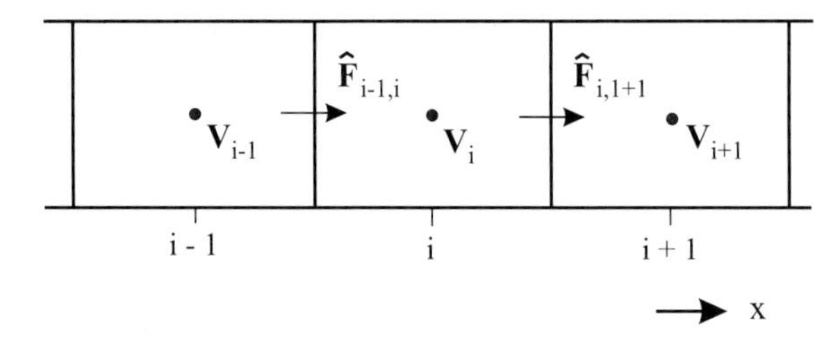

Fig. 7.4: Finite volume space discretization

There are several versions of the FVS scheme which differ mainly in the way how the numerical fluxes are calculated.

Introducing the Jacobian matrix

$$\mathbf{H} = \frac{\partial \mathbf{F}}{\partial \mathbf{V}}.$$

the flow equation (7.62) can also be written in an expanded form as

$$\frac{\partial \mathbf{V}}{\partial t} + \mathbf{H} \nabla \mathbf{V} = \mathbf{0}.$$

The original approach of Steger and Warming [8] makes use of the homogeneity property of the Euler equations

$$\mathbf{F} = \mathbf{H}\mathbf{V}, \tag{7.64}$$

which is strictly valid only for gas flows under the assumption of ideal state equations. Since the governing equations are hyperbolic, the coefficient matrix $\mathbf{H}$ can be split into elementary parts related to the eigenvalues,

$$\mathbf{H} = \sum_k \mathbf{H}_k, \tag{7.65}$$

or sorted with respect to the positive or negative sign (direction) of the eigenvalues,

$$\mathbf{H} = \sum_{k,\lambda_k \geq 0} \mathbf{H}_k + \sum_{k,\lambda_k < 0} \mathbf{H}_k = \mathbf{H}^+ + \mathbf{H}^-. \tag{7.66}$$

According to the homogeneity property (7.64) the fluxes for the computational cells can then be split as

$$\mathbf{F}_i(\mathbf{V}_i) = \mathbf{H}_i^+ \mathbf{V}_i + \mathbf{H}_i^- \mathbf{V}_i = \mathbf{F}_i^+ + \mathbf{F}_i^-. \tag{7.67}$$

The numerical flux at the cell interfaces is then calculated according to the contributions coming from the right (positive) and left (negative) running wave modes as

$$\hat{\mathbf{F}}_{i,i+i}(\mathbf{V}_i, \mathbf{V}_{i+1}) = \mathbf{F}_i^+(\mathbf{V}_i) + \mathbf{F}_i^-(\mathbf{V}_{i+1}). \tag{7.68}$$

An alternative way to calculate the numerical fluxes at the cell boundaries as proposed by Städtke et al. [9] and [10] does not require the homogeneity property for the flux splitting. The method is based on the solutions of a linearized, quasi-one-dimensional Riemann problem using the fluxes as major dependent parameters. The equation for fluxes is obtained from the corresponding equation for primitive state parameters (7.10) by a similarity transformation, resulting in

$$\frac{\partial \mathbf{F}}{\partial t} + \mathbf{R}\frac{\partial \mathbf{F}}{\partial x} = \mathbf{0} \tag{7.69}$$

with the new coefficient matrix $\mathbf{R}$

$$\mathbf{R} = \mathbf{K}\,\mathbf{G}\,\mathbf{K}^{-1} \tag{7.70}$$

and the Jacobian matrix $\mathbf{K}$

$$\mathbf{K} = \frac{\partial \mathbf{F}}{\partial \mathbf{U}}. \tag{7.71}$$

The eigenvalues of the "flux" matrix $\mathbf{R}$ are the same as those obtained for $\mathbf{G}$; the new eigenvectors can be calculated from the corresponding eigenvectors of the $\mathbf{G}$ matrix as

$$\mathbf{V}_R^* = \mathbf{V}_R\,\mathbf{K}^T \qquad \mathbf{V}_L^* = \mathbf{V}_L\,\mathbf{K}^{-1}. \tag{7.72}$$

As already described for the equations using the primitive state vector, the coefficient matrix $\mathbf{R}_n$ can be split into elementary parts with respect to the individual eigenvalues,

$$\mathbf{R} = \sum_k \mathbf{R}_k, \tag{7.73}$$

where the split matrices for the "primitive" state variables are defined as

$$\mathbf{R}_k = \mathbf{K}\,\mathbf{G}_k\,\mathbf{K}^{-1}. \tag{7.74}$$

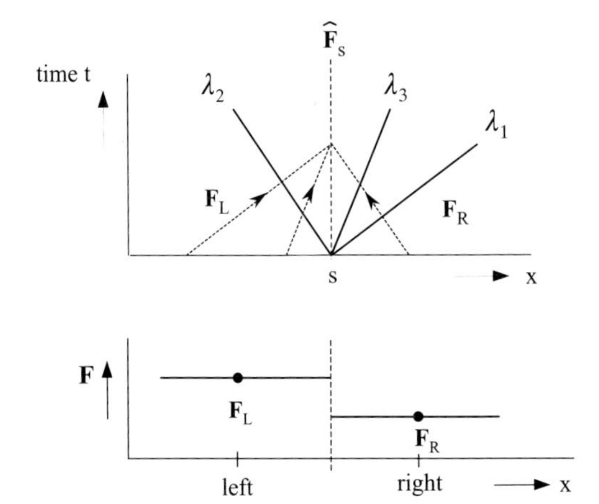

Fig. 7.5: Linearized Riemann problem for fluxes

The "Godunov fluxes" $\hat{\mathbf{F}}_s$ at the cell interface are calculated from the corresponding fluxes in the "left" and "right" cells as (see Fig. 7.5)

$$\hat{\mathbf{F}} = \sum_{k,\lambda_k \geq 0} \tilde{\mathbf{R}}_k \mathbf{F}_L + \sum_{k,\lambda_k \leq 0} \tilde{\mathbf{R}}_k \mathbf{F}_R \tag{7.75}$$

with the "weighting" factors as the sum of the split coefficient matrices for fluxes ordered with respect to the sign of the corresponding eigenvalues λ_k,

$$\tilde{\mathbf{R}}_k = \frac{1}{\lambda_k}\mathbf{R}_k, \tag{7.76}$$

with the condition

$$\sum_k \tilde{\mathbf{R}}_k = \mathbf{I}. \tag{7.77}$$

The Flux Vector Splitting technique can be easily applied for unstructured grids and second-order accuracy in space as will be described in Chapter 8.

References

[1] S.R. Chakravarthy, D.A. Andersen, and M.D. Salas, *The Split Coefficient Matrix Method for Hyperbolic Systems of Gasdynamic Equations*, AIAA 18th Science Meeting, Paper 80-0268, Pasadena, CA, 1980.

[2] R. Courant, E. Isaacson, and M. Rees, *On the Solution of Nonlinear Hyperbolic Differential Equations by Finite Differences*, Communication in Pure and Applied Mathematics, 5, 243–255, 1955.

[3] P. Romstedt, *A Split-Matrix Method for the Numerical Solution of Two-Phase Flow Equations*, Int. Top. Meeting on Advances in Reactor Physics, Mathematics and Computation, Paris, France, 1987.

[4] S.K. Godunov, *A Finite Difference Method for the Numerical Computation of Discontinuous Solutions of the Equation of Fluid Dynamics*, Mat. Sbornik, 47, 271–290, 1959.

[5] F. Toro, *Riemann Solvers and Numerical Methods for Fluid Dynamics*, Springer Berlin/Heidelberg, 1999.

[6] P. Roe, *Approximate Riemann Solvers, Parameter Vectors, and Difference Schemes*, Journal of Computational Physics, 43, 357–372, 1981.

[7] R.J. LeVeque, *Numerical Methods for Conservation Laws*, Birkhäuser Verlag, Basel/ Boston/ Berlin, 1990.

[8] J.L. Steger and R.R .Warming, *Flux Vector Splitting of the Inviscid Gasdynamic Equations with Application to Finite Difference Methods*, Journal of Computational Physics, 40, 263–293, 1981.

[9] H. Städtke, G. Franchello and B. Worth, *Towards a High-Resolution Numerical Simulation of Transient Two-Phase Flow*, Third Int. Conference on Multi-Phase Flow, ICMF'98, Lyon, France, 1998.

[10] H. Städtke, B. Worth and G. Franchello, *On the Hyperbolic Nature of Two-phase Flow Equations: Characteristic Analysis and Related Numerical Methods*, in Godunov Methods, Theory and Application, Kluwer Dortrecht, 841–862, 2001.

[11] I. Toumi and A. Kumbaro, *An Approximate Linearized Riemann Solver for a Two-Fluid Model,* Journal of Computational Physics, 124, 286–300, 1996.

[12] B. Van Leer, *Towards the Ultimate Conservative Difference Scheme: A Second Order Sequel to Godunov's Method*, Journal of Computational Physics, 32, 101–136, 1979.

8 Remarks on the Advanced Two-Phase Flow Module

The Advanced Two-phase Flow Module (ATFM) has been developed at the European Commission's Joint Research Centre Ispra (JRC Ispra) with the specific aim to study new modeling and numerical concepts for the numerical simulation of transient two-phase flow. The code package includes:

1. a numerical solver for transient one- and two-dimensional two-phase flow,
2. a visualization package allowing a detailed online display of predicted results,
3. the Vector Processing Language (VPL) for data handling and graphic output, and
4. a database for storage of state and transport properties as well as predicted results.

The code offers various options for modeling and numerical details like selection of state property routines, description of algebraic source terms for mass momentum and energy, first and second order spatial accuracy or degree of implicitness for time integration and related automatic time step control. Additional time-dependent procedures can be added for the definition of boundary conditions or complex output parameters. The code also includes a restart capability at user defined time frequencies.

In the following the basic modeling and related numerical solution strategies as implemented in the ATFM code will be summarized. Some more information to the ATFM code can be found in [2].

8.1 Basic modeling approach

8.1.1 Balance equations of two-fluid model

The ATFM code is based on the hyperbolic single-pressure two-fluid model as described in detail in Chapter 5. In order to guaranty the conservation of mass, momentum, and energy the conservative form of the balance equations is applied for the numerical integration which can be written as:

mass:

$$\frac{\partial}{\partial t}\left(\alpha_i \varrho_i\right) + \nabla \cdot \left(\alpha_i \varrho_i\, \vec{u}_i\right) = \sigma_i^M \qquad \text{with} \qquad \sum_{i=g,l} \sigma_i^M = 0 \tag{8.1}$$

Gasdynamic Aspects of Two-Phase Flow. Herbert Städtke

ISBN: 3-527-40578-X

momentum:

$$\frac{\partial}{\partial t}(\alpha_i \varrho_i \vec{u}_i) + \nabla \cdot (\alpha_i \varrho_i \vec{u}_i \vec{u}_i) + \alpha_i \nabla p - \nabla \cdot (\alpha_i \bar{\mathbf{T}}_i) = \sigma_i^M \vec{u}_i^{\mathrm{ex}} + \vec{F}_i^{\mathrm{int}} + \vec{F}_i$$

$$= \vec{\sigma}_i^J + \vec{F}_i \qquad \text{with} \qquad \sum_{i=g,l} \vec{\sigma}_i^J = 0 \tag{8.2}$$

energy:

$$\frac{\partial}{\partial t}\left[\alpha_i \varrho_i \left(e_i + \frac{v_i^2}{2}\right)\right] + \nabla \cdot \left[\alpha_i \varrho_i \vec{u}_i \left(h_i + \frac{v_i^2}{2}\right)\right] + \nabla \cdot (\alpha_i \vec{q}_i) - \nabla (\alpha_i \bar{\mathbf{T}}_i \cdot \vec{u}_i)$$

$$= \sigma_i^M \left(h + \frac{v_i^2}{2}\right) + \sigma_i^Q + \vec{F}_i^{\mathrm{int}} \cdot \vec{u}_i^{\mathrm{int}} + Q_i + \vec{F}_i \cdot \vec{u}_i$$

$$= \sigma_i^E + Q_i + \vec{F}_i \cdot \vec{u}_i \qquad \text{with} \qquad \sum_{i=g,l} \sigma_i^E = 0 \tag{8.3}$$

entropy:

$$\frac{\partial}{\partial t}(\alpha_i \varrho_i s_i) + \nabla \cdot (\alpha_i \varrho_i \vec{u}_i s_i) = \sigma_i^{S,\mathrm{int}} + \frac{Q_i}{T_i}$$

with

$$\sigma_i^{S,\mathrm{int}} = \sigma_i^M s_i + \frac{\vec{F}_i^{\mathrm{int}}}{T_i}(\vec{u}^{\mathrm{int}} - \vec{u}_i) + \frac{\sigma_i^M}{T_i}\left[h^{\mathrm{ex}} - h_i + \frac{1}{2}(\vec{u}^{\mathrm{ex}} - \vec{u}_i)^2\right]. \tag{8.4}$$

and the conditions of compliance with the second law of thermodynamics

$$\sum_{i=g,l} \sigma_i^{S,\mathrm{int}} \geq 0.$$

Whether the full energy equation (8.3) is used or the entropy balance (8.4) is an option for the code user depending on the problem being considered.

The interfacial forces, introduced in equations (8.2) and (8.3), have been split into two parts: the interfacial friction force $\vec{F}_i^{\mathrm{v}}$ and the nonviscous forces $\vec{F}_i^{\mathrm{nv}}$

$$\vec{F}_i^{\mathrm{int}} = \vec{F}_i^{\mathrm{v}} + \vec{F}_i^{\mathrm{nv}}. \tag{8.5}$$

The hyperbolicity of the system of balance equations is achieved with the following form of the nonviscous contribution to the interfacial forces $\vec{F}_i^{\mathrm{nv}}$,

$$\left.\begin{aligned} \vec{F}_i^{\mathrm{nv}} = &-\alpha_g \alpha_l \rho k \left(\frac{d^l \vec{u}_g}{dt} - \frac{d_l^g \vec{u}}{dt}\right) \\ &+\alpha_g \alpha_l (\alpha_g \hat{\rho}_l - \alpha_l \hat{\rho}_g)(\vec{u}_g - \vec{u}_l) \nabla \cdot (\vec{u}_g - \vec{u}_l) \\ &-\alpha_g \alpha_l (\hat{\rho}_g + \hat{\rho}_l)(\vec{u}_g - \vec{u}_l) \nabla \cdot \alpha_g \\ &-\alpha_g \alpha_l (\hat{\rho}_g + \hat{\rho}_l)(\vec{u}_g - \vec{u}_l)\left(\frac{\alpha_g}{\rho_g}\frac{d\rho_g}{dt} + \frac{\alpha_l}{\rho_l}\frac{d\rho_l}{dt}\right) \end{aligned}\right\}. \tag{8.6}$$

The abbreviations introduced in equation (8.6) are the total derivatives

$$\left.\begin{aligned} \frac{d^g\vec{u}_g}{dt} &= \frac{\partial \vec{u}_g}{\partial t} + (\vec{u}_g \cdot \nabla)\,\vec{u}_g, & \frac{d^g\varrho_g}{dt} &= \frac{\partial \varrho_g}{\partial t} + \vec{u}_g \cdot \nabla \varrho_g \\ \frac{d^l\vec{u}_l}{dt} &= \frac{\partial \vec{u}_l}{\partial t} + (\vec{u}_l \cdot \nabla)\,\vec{u}_l, & \frac{d^l\varrho_l}{dt} &= \frac{\partial \varrho_l}{\partial t} + \vec{u}_l \cdot \nabla \varrho_l \end{aligned}\right\} \tag{8.7}$$

and the "densities"

$$\hat{\rho}_g = \varrho_g + k\rho, \quad \text{and} \quad \hat{\rho}_l = \varrho_l + k\rho. \tag{8.8}$$

8.1.2 Flow topology and interfacial area

Bubbly as well as droplet flow regimes are considered depending on whether liquid or gas/vapor is the continuous "carrier" fluid. As indicated schematically in Fig. 8.1, both phases might be present simultaneously in a spatial or time dependent transition process, characterized by the volumetric "weighting" function with the condition

$$X_b + X_d = 1. \tag{8.9}$$

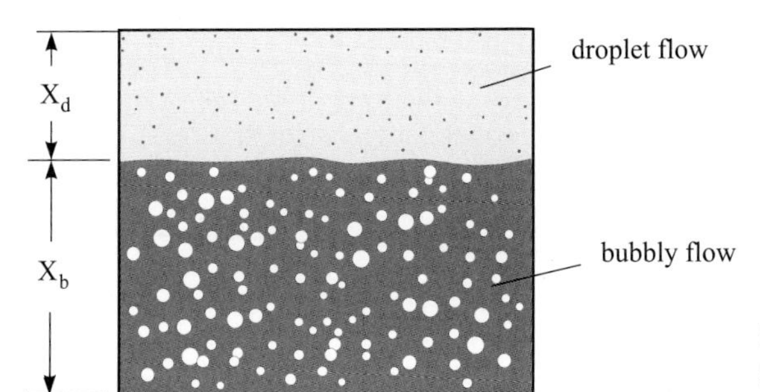

Fig. 8.1: Flow topology (schematically)

Introducing the local bubble and droplet volumetric fraction α_b^* and α_d^*, respectively, the following relations for the gas/vapor and liquid volume fractions can be obtained:

$$\left.\begin{aligned} \alpha_g &= \alpha_b^* X_b + (1 - \alpha_d^*)\,X_d \\ \alpha_l &= \alpha_d^* X_d + (1 - \alpha_b^*)\,X_b \end{aligned}\right\}. \tag{8.10}$$

Assuming also mono-dispersed bubbly and/or droplet flow, the contributions of both flow regimes to the interfacial area per (total) volume are given by

$$\left.\begin{aligned} a_b^{\text{int}} &= \tfrac{3}{2} C_b^{\text{int}}\,\alpha_b^*\,X_b = \tfrac{3}{2} C_b \alpha_b \\ a_d^{\text{int}} &= \tfrac{3}{2} C_d^{\text{int}}\,\alpha_d^*\,X_d = \tfrac{3}{2} C_d\,\alpha_d \end{aligned}\right\} \tag{8.11}$$

with the bubble and droplet volumetric fractions related to the total volume

$$\alpha_b = \alpha_b^*\,X_b, \qquad \alpha_d = \alpha_d^*\,X_d. \tag{8.12}$$

Note that the sum of α_b and α_d is generally different from 1

$$\alpha_b + \alpha_d \neq 1. \tag{8.13}$$

The curvature of the interface for bubbly and droplet flow depending on the average particle radius is given by

$$C_b^{\text{int}} = \frac{2}{r_b}, \qquad C_d = \frac{2}{r_d}.$$

The total interfacial area concentration is the sum of bubbly and droplet flow regimes as

$$a^{\text{int}} = \frac{3}{2}\left(C_b^{\text{int}}\,\alpha_b + C_d^{\text{int}}\,\alpha_d\right). \tag{8.14}$$

Equation (8.14) includes the two limiting cases:

1. For low void fraction values $\alpha_g < \alpha_{g,\text{cr}}$ a bubbly flow regime is expected and with $X_b = 1$, and $\alpha_b = \alpha_g$ the interfacial area concentration becomes

$$a^{\text{int}} = \tfrac{3}{2} C_b^{\text{int}} \alpha_g. \tag{8.15}$$

2. For high void fraction values $\alpha_g > (1 - \alpha_{l,\text{cr}})$ a disperse droplet flow regimeis expected and with $X_d = 1$, and $\alpha_d = 1 - \alpha_g$ the interfacial area concentration is

$$a^{\text{int}} = \tfrac{3}{2} C_d^{\text{int}} \left(1 - \alpha_g\right). \tag{8.16}$$

In the intermediate region, $\alpha_{b,\text{cr}} < \alpha_g < (1 - \alpha_{d,\text{cr}})$, a smooth transition is assumed where the gas and liquid volume concentration in the bubbly and droplet subregions are approaching to a prescribed maximum values $\alpha^*_{b,\max}$ and $\alpha^*_{d,\max}$ for increased gas of liquid volume fraction. This is schematically shown in Fig. 8.2 for $\alpha_{b,\text{cr}} = \alpha_{d,cr} = 0.15$ and $\alpha_{b,\max} = \alpha_{d,\max} = 0.35$.

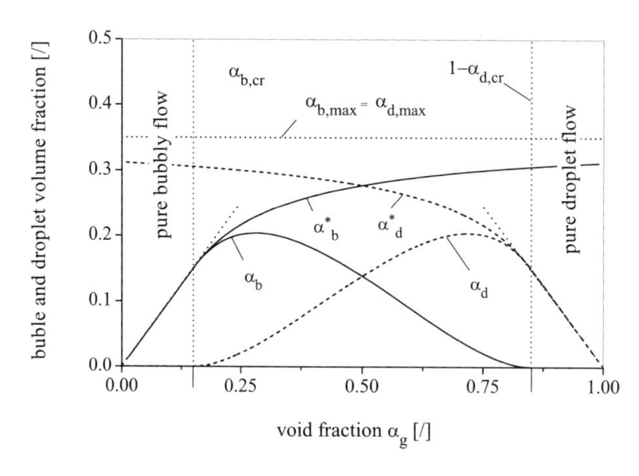

Fig. 8.2: Flow regime transition

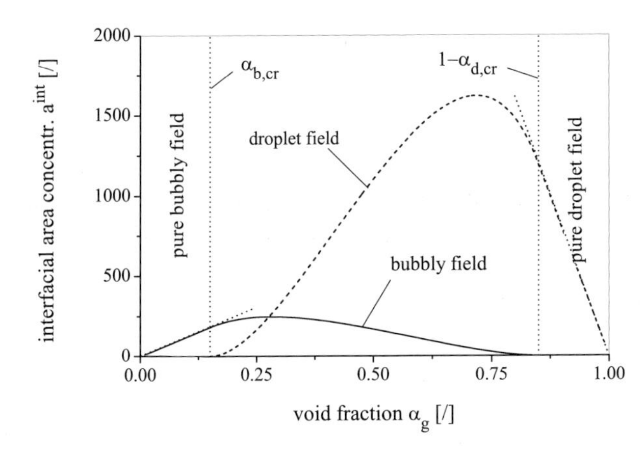

Fig. 8.3: Interfacial area concentration for mono-dispersed bubbly and droplet fields

The corresponding values for the interfacial area for the bubbly and droplet fields are given in Fig. 8.3 assuming mono-dispersed conditions with the constant bubble and droplet diameters of $d_{\text{bub}} = 4.0$ mm and $d_{\text{drop}} = 0.8$ mm, respectively.

For the modeling of the bubble and droplet diameter and related curvature, empirical correlations are used based on stability criteria with regard to particle breakup and coalescence.

8.1.3 Algebraic source terms

For the *interfacial friction forces*, a general resistance law is applied which can be formulated with respect to the square of the "slip" velocity between the gas and liquid phases, and the interfacial area concentration as

$$F_{g,l}^{v} = F_{g,l}^{D} = -\frac{1}{8}\left(C_b^D a_b^{\text{int}} \rho_l + C_d^D a_d^{\text{int}} \rho_g\right) |\vec{u}_g - \vec{u}_l| \left(\vec{u}_g - \vec{u}_l\right). \tag{8.17}$$

The drag coefficients for bubbly and droplet flow, C_b^D and C_d^D, are assumed to be a function of the Reynolds number.

The source terms for *interfacial heat and mass transfer* between the phases are determined by the sum and the difference of heat fluxes from the bulk of the corresponding phases to the interface, resulting in

$$\sigma_{g,l}^{M} = \pm\frac{1}{\Delta h^s}\left[\left(a_b^{\text{int}}\mathcal{H}_{b,g}^q + a_d^{\text{int}}\mathcal{H}_{d,g}^q\right)\Delta T_g + \left(a_b^{\text{int}}\mathcal{H}_{b,l}^q + a_d^{\text{int}}\mathcal{H}_{d,l}^q\right)\Delta T_l\right] \tag{8.18}$$

$$\sigma_{g,l}^{Q} = \pm\left[a_b^{\text{int}}\left(\mathcal{H}_{b,g}^q + \mathcal{H}_{b,l}^q\right) + a_d^{\text{int}}\left(\mathcal{H}_{d,g}^q + \mathcal{H}_{d,l}^q\right)\right]\left(T_g - T_l\right), \tag{8.19}$$

with semi-empirical heat transfer coefficients for the bubble and droplet fields, $\mathcal{H}_{b,i}^q$ and $\mathcal{H}_{d,i}^q$ on the gas ($i = g$) and the liquid ($i = l$) side of the interface. For the driving temperature differences in the the expression for the evaporation rate (8.18), only the metastable contributions are taken into account

$$\Delta T_g = \min\left(T_g - T^{\text{sat}}, 0\right), \qquad \Delta T_l = \max\left(T_l - T^{\text{sat}}, 0\right). \tag{8.20}$$

8.1.4 State properties

For the calculation of state properties the code user has a choice of three different property packages including:

1. The Global Equation of State (GOST) package which has been developed at the Joint Research Centre at Ispra (JRC) for the prediction of water/steam properties at saturated conditions. The package is based on a "global" canonical equation of state for the free enthalpy as function of temperature and specific volume $f(T, v)$ (Helmholtz function). From this function and its algebraic derivatives with respect to temperature and specific volume, a consistent set of thermodynamic variables for liquid, vapor, and supercritical states are derived without any further assumptions or numerical manipulations. The method has been proven to reproduce the International Skeleton Table for water and steam [1] over a wide range of pressure and temperature and prescribed tolerance limits. In case of metastable conditions (subcooled vapor or superheated liquid), the properties are extrapolated from the values at saturated conditions and simplified state laws.

2. The SPWAT package for the calculation of water/steam state properties at saturated conditions. From a completely defined reference point (T_0, p_0), all state variables are extrapolated using simplified state equations for the gas phase

$$v_g(T,p) = v_{g,0}\frac{T/T_0}{p/p_0}\left[1 - \gamma_{g,0}p_0 \ln\left(\frac{p}{p_0}\right) + (\beta_{g,0}T_0 - 1)\ln\left(\frac{T}{T_0}\right)\right], \quad (8.21)$$

   and the liquid phase

$$v_l(p,T) = v_{l,0}\left[1 - \gamma_{l,0}(p - p_0) + \beta_{l,0}(T - T_0)\right]. \quad (8.22)$$

   From equations (8.21), (8.22) and corresponding derivatives with respect to specific volume v_i and temperature T_i all thermodynamic variables for the two phases can be easily calculated.

3. The MIXTURE package providing the calculation of state and transport properties for gas mixtures including steam and noncondensibles. For simplicity Dalton's law is applied for the gas mixture. As for the SPWAT package, the user has to specify all relevant state parameters for a reference point (p_0, T_0) from which all state variables are then extrapolated using simplified state equations for water and gas.

8.2 Numerical method

8.2.1 Conservative form of flow equations

The numerical approach in the ATFM code represents a multi-dimensional extension of the one-dimensional Flux Vector Splitting method as described in Section 4. The method is based on the conservative form of the basic equations which, arranged in a compact matrix form, can be written as

$$\frac{\partial \mathbf{V}}{\partial t} + \nabla\,\mathbf{F} + \mathbf{H}^{\mathrm{nc}}\nabla\,\mathbf{U} = \mathbf{E} \quad (8.23)$$

where the "non-conservative" part $\mathbf{H}^{\mathrm{nc}}$ is defined as

$$\mathbf{H}^{\mathrm{nc}} = (\mathbf{J}\,\mathbf{G} - \mathbf{K})\,\mathbf{K}^{-1} = \mathbf{X}\mathbf{K}^{-1}, \tag{8.24}$$

with the Jacobian matrices

$$\mathbf{J} = \frac{\partial \mathbf{V}}{\partial \mathbf{U}} \qquad \text{and} \qquad \mathbf{K} = \frac{\partial \mathbf{F}}{\partial \mathbf{U}}. \tag{8.25}$$

The code user has the choice of two different sets of the state vector $\mathbf{V}$ and the corresponding flux vector $\mathbf{F}$ depending on whether the entropy balance equations or the full energy equations are used. These vectors are:

using entropy equation:

$$\mathbf{U} = \begin{bmatrix} p \\ \vec{u}_g \\ \vec{u}_l \\ \alpha_g \\ s_g \\ s_l \end{bmatrix}, \quad \mathbf{V} = \begin{bmatrix} \alpha_g \rho_g \\ \alpha_l \rho_l \\ \alpha_g \rho_g \vec{u}_g \\ \alpha_l \rho_l \vec{u}_l \\ \alpha_g \rho_g s_g \\ \alpha_l \rho_l s_l \end{bmatrix}, \quad \mathbf{F} = \begin{bmatrix} \alpha_g \rho_g \vec{u}_g \\ \alpha_l \rho_l \vec{u}_l \\ \alpha_g (\rho_g \vec{u}_g\, \vec{u}_g + p\mathbf{I}) \\ \alpha_l (\rho_l \vec{u}_l\, \vec{u}_l + p\mathbf{I}) \\ \alpha_g \rho_g \vec{u}_g s_g \\ \alpha_l \rho_l \vec{u}_l s_l \end{bmatrix} \tag{8.26}$$

using energy equation:

$$\mathbf{U} = \begin{bmatrix} p \\ \vec{u}_g \\ \vec{u}_l \\ \alpha_g \\ e_g \\ e_l \end{bmatrix}, \quad \mathbf{V} = \begin{bmatrix} \alpha_g \rho_g \\ \alpha_l \rho_l \\ \alpha_g \rho_g \vec{u}_g \\ \alpha_l \rho_l \vec{u}_l \\ \alpha_g \rho_g (e_g + \frac{1}{2} u_g^2) \\ \alpha_l \rho_l (e_l + \frac{1}{2} u_l^2) \end{bmatrix}, \quad \mathbf{F} = \begin{bmatrix} \alpha_g \rho_g \vec{u}_g \\ \alpha_l \rho_l \vec{u}_l \\ \alpha_g (\rho_g \vec{u}_g\, \vec{u}_g + p\mathbf{I}) \\ \alpha_l (\rho_l \vec{u}_l\, \vec{u}_l + p\mathbf{I}) \\ \alpha_g \rho_g \vec{u}_g (h_g + \frac{1}{2} u_g^2) \\ \alpha_l \rho_l \vec{u}_l (h_l + \frac{1}{2} u_l^2) \end{bmatrix} \tag{8.27}$$

For many of the test cases described in Chapter 9, there is practically no difference between the two choices, particularly for slow or subsonic conditions. The use of the entropy in the state vector might be preferred due to the reduced complexity of the coefficient matrices and eigenvectors. However, there are other applications where the use of the energy equation becomes crucial, for example in the prediction of shock waves in single phase gas or high void fraction, and dispersed two-phase flow conditions.

8.2.2 Finite volume discretization

For the numerical solution scheme, the governing equations (8.23) are transformed into a finite volume approximation for arbitrary polygon-shaped computational cells i with the volume V_i, the boundary segment area A_s, and the perimeter index s (Fig. 8.4), as given by

$$\mathbf{V}_i^{n+1} = \mathbf{V}_i^n - \frac{\Delta t}{V_i}\sum_s A_s \left(\hat{\mathbf{F}}_s\right)_i^{n+1} - \frac{\Delta t}{V_i}\sum_s A_s \left(\mathbf{H}_s^{\mathrm{nc}}\right)_i^{n+1} \left(\hat{\mathbf{F}}_s\right)_i^{n+1} + \mathbf{D}_i^{n+1}\Delta t, \tag{8.28}$$

where the intercell fluxes $\hat{\mathbf{F}}_s$, the nonconservative part of the coefficient matrix $\mathbf{H}_s^{\mathrm{nc}}$, and the source term vector $\mathbf{D}_i$ are evaluated implicitly at the new time level.

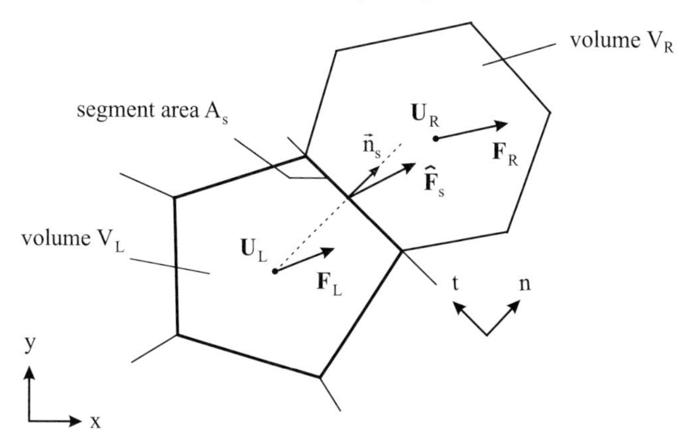

Fig. 8.4: Finite volume discretization

The numerical fluxes at the cell boundaries $\hat{\mathbf{F}}_s$ can be calculated from a series of linearized, quasi one-dimensional Riemann problems normal to the specific surface areas of the computational cell boundary segments.

With a projection of the primitive form of the governing equations

$$\frac{\partial \mathbf{U}}{\partial t} + \mathbf{G}\nabla\mathbf{U} = \mathbf{D} \tag{8.29}$$

normal to the cell section $\vec{n}_n$ (Fig. 8.4) one obtains

$$\frac{\partial \mathbf{U}}{\partial t} + \mathbf{G}_n \frac{\partial \mathbf{U}}{\partial n} = \mathbf{D}. \tag{8.30}$$

For this purpose, the governing equations (8.29) are transformed using fluxes as major dependent variables.

$$\frac{\partial \mathbf{F}}{\partial t} + \mathbf{R}_n \frac{\partial \mathbf{F}}{\partial n} = \mathbf{K}\,\mathbf{C}_n. \tag{8.31}$$

The new coefficient matrix $\mathbf{R}$ is defined as

$$\mathbf{R}_n = \mathbf{K}\,\mathbf{G}_n\,\mathbf{K}^{-1}, \tag{8.32}$$

with the Jacobian matrix $\mathbf{K}$

$$\mathbf{K} = \frac{\partial \mathbf{F}}{\partial \mathbf{U}}. \tag{8.33}$$

The "Godunov" fluxes at the cell interface are calculated from the corresponding fluxes in the "left" and "right" cell as

$$\hat{\mathbf{F}}_s(\mathbf{F}_l, \mathbf{R}_k) = \sum_{k,\lambda_k \geq 0} (\mathbf{R}_k)_s\,\mathbf{F}_l + \sum_{k,\lambda_k \leq 0} (\mathbf{R}_k)_s\,\mathbf{F}_r$$

with the "weighting" factors as the sums of the split coefficient matrices for fluxes ordered with respect to the sign of the corresponding eigenvalues λ_k

$$\tilde{\mathbf{R}}_k = \frac{1}{\lambda_k}\mathbf{R}_k, \tag{8.34}$$

with the condition

$$\sum_k \tilde{\mathbf{R}}_k = \mathbf{I}. \tag{8.35}$$

8.2.3 Second-order accuracy

A near second-order accuracy is obtained by a linear reconstruction of the solution in all computational cells following the Monotonic Upwind Scheme for Conservation Laws (MUSCL) approach as indicated schematically in Fig. 8.5.

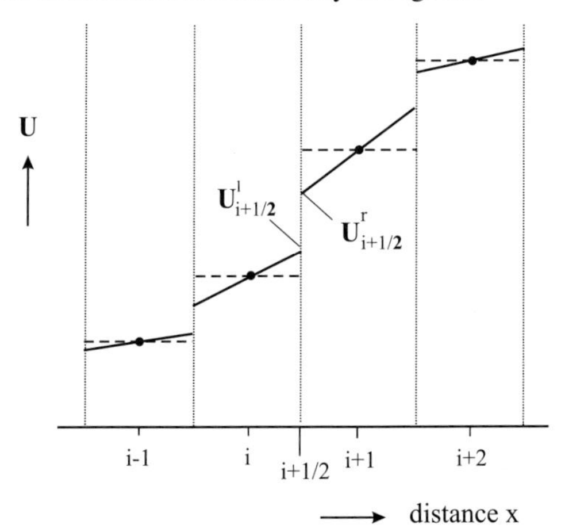

Fig. 8.5: MUSCL approach: linear reconstruction of solution

New values for the "primitive" parameters at the left and right side of the cell interfaces are calculated by a linear extrapolation from the adjacent cells as

$$\left.\begin{aligned} \mathbf{U}^l_{i+1/2} &= \mathbf{U}_i + \sigma_i \frac{(\Delta x)_i}{2} \\ \mathbf{U}^r_{i+1/2} &= \mathbf{U}_{i+1} - \sigma_{i+1} \frac{(\Delta x)_{i+1}}{2} \end{aligned}\right\}. \tag{8.36}$$

Slope limiter functions are then applied in order to maintain a monotonic behavior of the solution

$$\sigma_i = f_{\lim}\left[\left(\frac{\partial \mathbf{U}}{\partial x}\right)_{i-1/2}, \left(\frac{\partial \mathbf{U}}{\partial x}\right)_{i+1/2}\right]. \tag{8.37}$$

The limiter function used in the following numerical examples combine some properties of both the "minmod" and "superbee" limiters.

8.2.4 Implicit time integration

As indicated in equation (8.28), the intercell fluxes, the nonconservative part of the coefficient matrix and the source term vector are evaluated implicitly at the new time level. For the implicit time integration, the conservative variables, the source terms, and the fluxes are evaluated by a first-order Taylor expansion up to the new time-step value

$$\left.\begin{aligned} \mathbf{V}_i^{n+1} &= \mathbf{V}_i^n + \left(\frac{\partial \mathbf{V}}{\partial \mathbf{W}}\right)_i^n \left(\mathbf{W}_i^{n+1} - \mathbf{W}_i^n\right) \\ \mathbf{D}_i^{n+1} &= \mathbf{D}_i^n + \left(\frac{\partial \mathbf{D}}{\partial \mathbf{W}}\right)_i^n \left(\mathbf{W}_i^{n+1} - \mathbf{W}_i^n\right) \\ \mathbf{F}_i^{n+1} &= \mathbf{F}_i^n + \left(\frac{\partial \mathbf{F}}{\partial \mathbf{W}}\right)_i^n \left(\mathbf{W}_i^{n+1} - \mathbf{W}_i^n\right) \end{aligned}\right\} \tag{8.38}$$

with the base vector $\mathbf{W}$

$$\left.\begin{aligned} \mathbf{W}_1 &= \{p, \alpha_g, \alpha_g\rho_g, \vec{u}_g, \alpha_l\rho_l, \vec{u}_l, \alpha_g\rho_g, s_g, \alpha_l\rho_l, s_l\}^T \\ \mathbf{W}_2 &= \{p, \alpha_g, \alpha_g\rho_g, \vec{u}_g, \alpha_l\rho_l, \vec{u}_l, \alpha_g\rho_g, e_g, \alpha_l\rho_l, e_l\}^T \end{aligned}\right\} \tag{8.39}$$

depending on whether the entropy balance or the energy conservation equation is used.

The elements of the vectors $\mathbf{W}_1, \mathbf{W}_2$ have been chosen in a way to avoid singularities in the resulting sparse system matrix. The Jacobian matrices for the derivation of the source term and flux vectors are evaluated algebraically in order to save computational time. The final solution for the new-time conservative parameters is done by a Newton–Raphson iteration using a sparse matrix solver with complete lower and upper preconditioning. The primitives are calculated after each iteration

$$\mathbf{U}_i^{n+1} = \mathbf{U}_i^n + \left(\frac{\partial \mathbf{U}}{\partial \mathbf{W}}\right)_i^n \left(\mathbf{W}_i^{n+1} - \mathbf{W}_i^n\right). \tag{8.40}$$

Experience shows that convergence is achieved within less than two to three iteration steps. More details on the FVS method can be found in [3]. The automatic time step control is based on the following criteria:

1. check whether the Courant number is above a user-defined maximum value
2. error check: evaluate the difference between mass in each computational cell with the anticipated value based on a Taylor expansion from previous time value,
3. check whether the increments of quantities selected from the primitive state vector are above user-defined threshold values.

In the case that one of these conditions is not satisfied, the iteration procedure is repeated with half of the time step size. The predictive capability of the ATFM code has been demonstrated by a large number of numerical and physical benchmark test cases. Some selected results are given by Städtke et al. in [3] and [5].

References

[1] *Properties of Water and Steam in SI-Units*, 2nd Revised and Updated Printing, Springer, Berlin, 1979.

[2] Städtke, G. Franchello and B. Worth, *Towards a High-Resolution Numerical Simulation of Transient Two-Phase Flow*, Third International Conference on Multi-Phase Flow, ICMF'98, Lyon, France, 1998.

[3] H. Städtke, B. Worth, and G. Franchello, *On the Hyperbolic Nature of Two-phase Flow Equations: Characteristic Analysis and Related Numerical Methods*, in Godunov Methods, Theory and Application, Kluwer, 841–862, 2001.

[4] B. van Leer, *Towards the Ultimate Conservative Difference Scheme, Second Order Sequel to Godunov's Method*, Journal of Computational Physics. 32, 101–136, 1979.

[5] H. Städtke, G. Franchello, and B. Worth, *Assessment of the JRC Advanced Two-Phase Flow Module Using ASTAR Benchmark Test Cases*, ASTAR International Workshop on Advanced Numerical Methods for Multidimensional Simulation of Two-Phase Flow, September 15–16, 2003, GRS Garching, Germany.

9 Numerical Results and Applications

In this Chapter numerical results for various two-phase flow phenomena are presented which are governed by wave propagation phenomena. This includes the propagation of volumetric phase concentration (void waves), pressure waves, flow discontinuities and shock waves, fast depressurization processes, and related critical flow phenomena. All numerical results shown are obtained with the Advanced Two-Phase Flow Module (ATFM), as described in Chapter 8.

The examples presented fall into two different categories: (1) base-line test cases using simplified boundary conditions and constitutive modeling to provide insight into physical phenomena and to test the accuracy and robustness of the numerical method applied, and (2) engineering type applications using physically more realistic state and interfacial transport models. Where available the numerical results are compared with analytical solutions or with experimental data.

9.1 Phase separation and void waves

This is an isothermal transient test case to investigate gravity-induced phase separation and related counter-current flow conditions. It tests the ability of the methods to predict counter-current flow conditions as exist in many reactor safety-related transients. Initial conditions represent a vertical pipe of height $h = 2.0$ m filled with a homogeneous two-phase mixture with a void fraction of $\alpha_g = 0.5$. The specific challenge is the prediction of two steep void waves travelling simultaneously from the top and bottom ends into the pipe, which, when meeting at the middle section, results in the formation of a sharp interface (liquid level) after phase separation is complete. As schematically shown in Fig. 9.1, three different regions can be distinguished: single-phase gas and liquid conditions at the top and bottom part of the pipe, respectively, and a quasi-stationary two-phase flow region at the middle section.

9.1.1 Analytical model

In the undisturbed middle section of the pipe, quasi-steady-state flow conditions exist where the volumetric upward and downward fluxes of the two phases compensate each other. Neglecting the momentum flux terms and virtual mass forces, compared to the gravity and interfacial friction forces, the phasic momentum equations simplify to

$$\alpha_g \frac{\partial p}{\partial y} = F_g^{\mathrm{v}} - \alpha_g \varrho_g \, g, \tag{9.1}$$

Gasdynamic Aspects of Two-Phase Flow. Herbert Städtke

ISBN: 3-527-40578-X

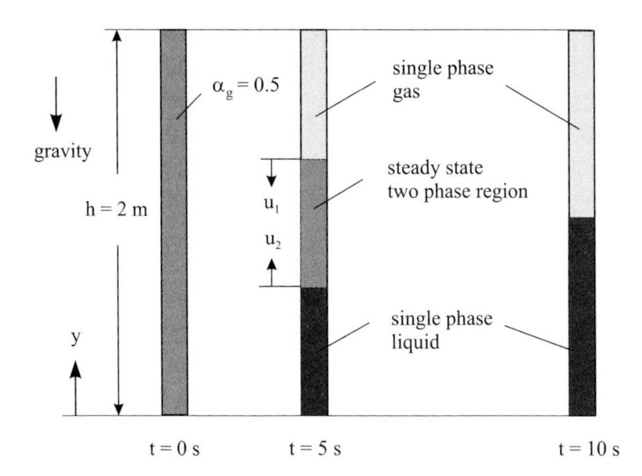

Fig. 9.1: Phase separation in a vertical pipe due to gravity (schematically)

$$\alpha_l \frac{\partial p}{\partial y} = F_l^{\mathrm{v}} - \alpha_l \varrho_l \, g. \tag{9.2}$$

With the simplified expression for the interfacial drag as specified for the ASTAR benchmark cases [2], the interfacial friction force becomes

$$F_g^{\mathrm{v}} = \frac{1}{8} C_D \, a^{\mathrm{int}} \varrho_m \left(u_g - u_l\right)^2, \tag{9.3}$$

with the interfacial area per unit volume

$$a^{\mathrm{int}} = \frac{3\alpha_g \alpha_l}{r_p} \tag{9.4}$$

and a unique "particle" radius r_p. The sum and the difference of (9.1) and (9.2) result in the pressure gradient due to the buoyancy forces

$$\frac{\partial p}{\partial x} = -\varrho_m \, g \quad \text{with} \quad \varrho_m = \alpha_g \varrho_g + \alpha_l \varrho_l \tag{9.5}$$

and the "slip" velocity in the middle two-phase flow region as

$$\Delta u = u_g - u_l = \sqrt{\frac{8\left(\varrho_l - \varrho_g\right) r_p g}{3 C_D \, \varrho_m}}. \tag{9.6}$$

Introducing the density values and the specified values for the "drag" coefficient $C_D = 0.44$ and particle radius $r_p = 0.5 \times 10^{-3}$ m, the slip velocity becomes $\Delta u = 0.24$ m/s. Due to the compensating volumetric fluxes across the void wave,

$$\bar{u} = \alpha_g u_g + \alpha_l u_l = 0 \tag{9.7}$$

the propagation of the void waves can be predicted using the slip velocity as given by equation (9.6) as

$$\left.\begin{aligned} u_1 &= u_g = \alpha_l \Delta u \\ u_2 &= u_l = -\alpha_g \Delta u \end{aligned}\right\} . \tag{9.8}$$

For an initial void fraction of $\alpha_g = 0.5$, one obtains for the upward void front $u_1 = 0.12$ m/s, and for the downward void front $u_2 = -0.12$ m/s.

9.1.2 Numerical results

The prediction was performed for a water/air mixture in a closed pipe with the initial conditions of a constant void fraction of $\alpha_{g,0} = 0.5$ and a pressure of $p_0 = 1$ bar. For the one-dimensional case a second-order Flux Vector Splitting scheme was used with 500 computational cells in the y-direction. Heat transfer between the phases as well as viscous and wall friction effects are neglected.

Calculated results for void fraction, gas and liquid velocities, and pressure distributions for various consecutive time values are shown in Fig. 9.2. As can be seen the numerical simulation largely confirms the analytical results including the positions and amplitudes of the two void waves and the (nearly) discontinuous changes of void fraction and phasic velocities across the wave fronts. The discontinuities in the pressure gradient coincides with the passage of the void waves marking the change of the gravity head between the two-phase regions and the and single-phase regions of pure liquid and pure gas. A new steady state is reached at $t = 8.3$ s when the two waves have merged at the middle of the pipe. The prediction does not show any anomaly or numerically induced instability during the transition from two-phase to pure single-phase gas or liquid conditions. The void wave itself is represented by two to three grid points.

During the initial phase of the transient some weak pressure wave propagation and reflection appeared in the prediction which are related to the somehow artificial (mechanical disequilibrium) initial conditions. However, these waves vanish when the nearly steady pressure gradient is reached depending on the density values in the gas, two-phase and liquid region. At the end of the transient, the pressure in the gas space has returned to the initial pressure value as expected.

The dependence of numerical results in case of a progressively increased number of grid points n is given in Fig. 9.3 for the void fraction distribution at a fixed time of $t = 4$ s. Figure 9.3 clearly indicates the continuous convergence toward the discontinuous analytical solution which is practically reached for $n = 500$. The position of the void wave remains unaffected by the change of the grid resolution.

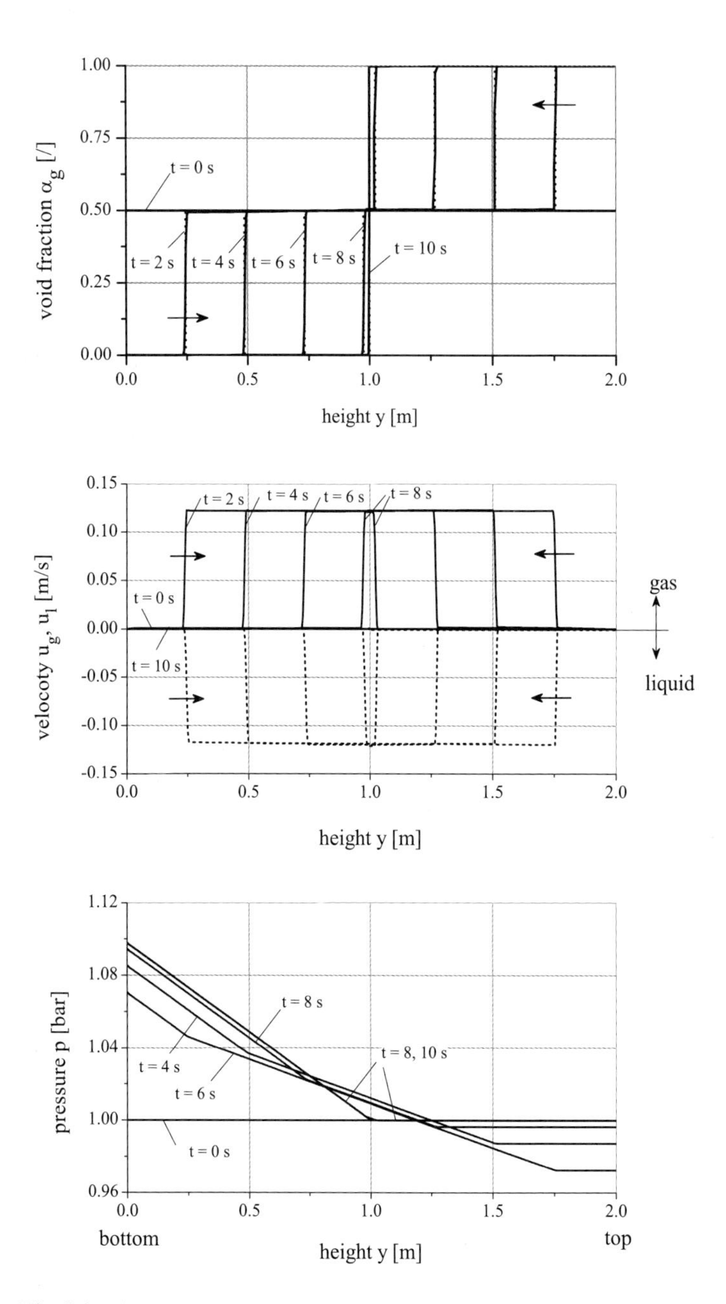

Fig. 9.2: Phase separation in a verical pipe due to gravity; void fraction, phasic velocities, and pressure distributions along pipe length at various time values

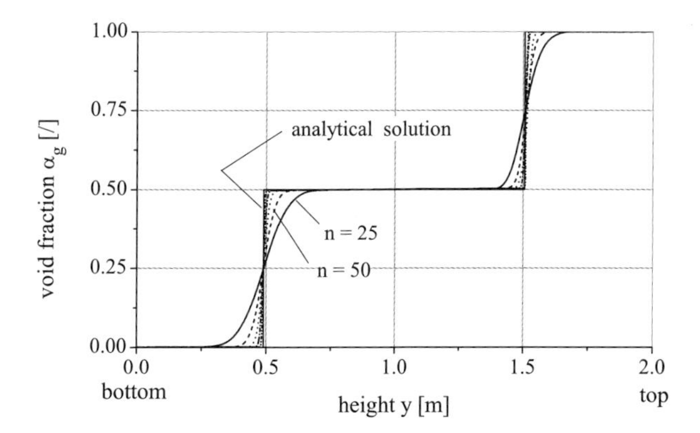

Fig. 9.3: Phase separation in a vertical pipe; convergence of solution in case of progressive grid refinement, $n = 25, 50, 100, 250, 500$ grid points, $t = 4$ s

9.2 U-tube oscillations

This idealized test case, first proposed by Ransom and published in [1], has become a standard benchmark test case to evaluate numerical diffusion inherent in numerical solution schemes for two-phase flow. It consists in calculating a gravity driven oscillation of a water column in a U-tube manometer. The geometry of the U-tube and the initial conditions are shown in Fig. 9.4.

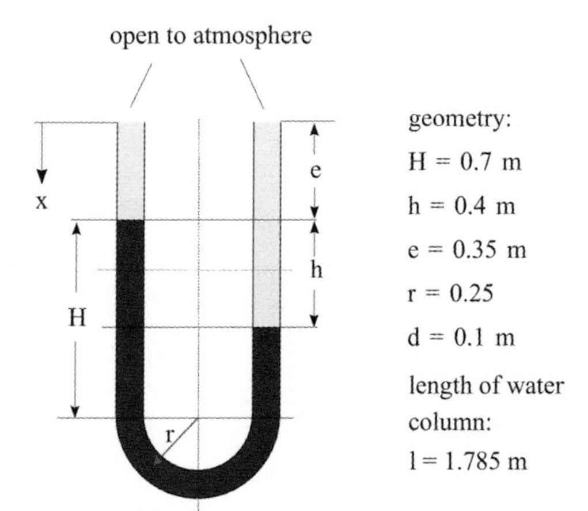

Fig. 9.4: Oscillating water column in a U-tube manometer; geometry, and initial conditions

The initial conditions start with the maximum displacement of the water column in the left leg and with an assumed zero velocity. In the particular case that the wall friction is zero, (inviscid flow), the liquid mass oscillated indefinitely as a "rigid body" and for the whole process a simple algebraic solution exists. Any damping of the predicted oscillation can be attributed to numerical diffusion or viscosity of the finite difference or the finite volume scheme applied.

9.2.1 Analytical solution

Neglecting viscosity and wall friction effects, the oscillation frequency ω and the maximum velocity of the water column $u^{\max}$ can be derived as follows:

$$\omega = \frac{1}{T} = \sqrt{\frac{g}{2\pi^2 l}} \quad \text{and} \quad u_l^{\max} = \sqrt{\frac{2g}{l}} h. \tag{9.9}$$

Assuming initial conditions of zero liquid velocity everywhere and a maximum liquid level displacement h, the velocity at the bottom of the U-tube at time t after the release is

$$u_l = u_l^{\max} \sin\left(\frac{t}{\omega}\right). \tag{9.10}$$

9.2.2 Numerical results

The results shown in the following are obtained with the second-order Flux Vector Splitting technique with 200 computational cells assuming a strictly one-dimensional flow conditions.

The predicted velocity at the bottom of the U-tube (solid line) is compared with the analytical solution (dotted line) in Fig. 9.5. The practical absence of (numerical) damping indicates the low numerical viscosity of the numerical method used. The still existing small attenuation might be (at least) partially attributed to the existing differences in phasic velocities close to the moving interface and the related dissipative effects.

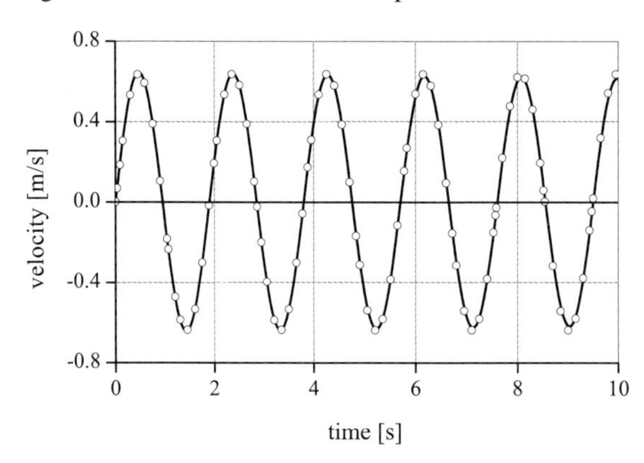

Fig. 9.5: Liquid velocity at the bottom of the U-tube as a function of time; comparison of CFD calculation (solid line) with analytical solution (dotted line with circles)

The predicted void fraction and pressure distribution along the U-tube for the first six oscillation cycles are given in Figs. 9.6 and 9.7. The figures show a high resolution of the moving liquid level where the discontinuities in void fraction and accompanied pressure gradient is represented by two to three grid points. The corresponding parameter profiles are maintained during various oscillation cycles, indicating a practical absence of numerical dispersion or viscosity effects. Since the liquid velocity is relatively low, the dynamic momentum contribution is negligible small and the pressure values shown in Fig. 9.7 are dominated by the gravity head determined by the actual liquid level elevation.

The successful prediction of the U-tube test case and the sedimentation case governed by counter-current two-phase flow conditions have been a milestone for the development of the hyperbolic two-fluid model as described in Chapter 5.

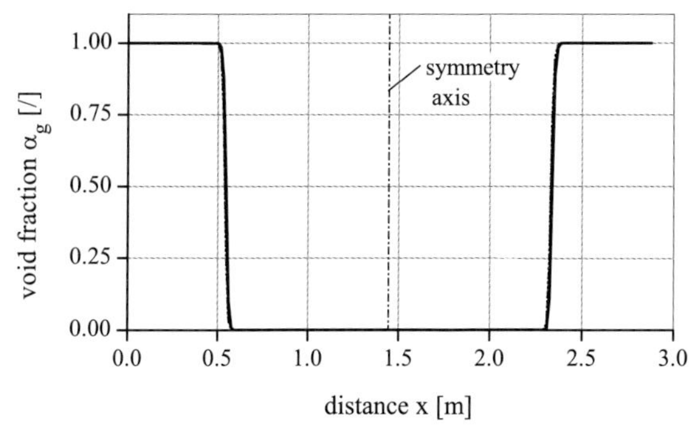

Fig. 9.6: Overlay of void fraction distributions for six consecutive oscillation cycles at times of equal level elevations on both legs

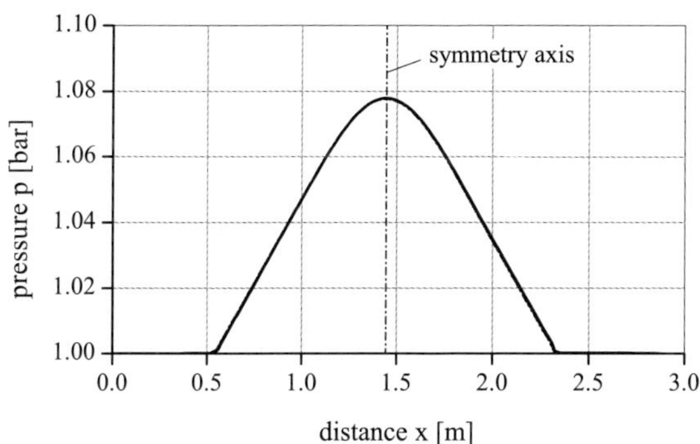

Fig. 9.7: Overlay of pressure distributions for six consecutive oscillation cycles at times of equal level elevations on both legs

9.3 Pressure wave propagation phenomena

Weak pressure disturbances ($\Delta p \rightarrow 0$) in a compressible homogeneous media propagate with the speed of sound which is determined by the compressibility of the fluid and inherent inertia as given by the density. In the case of absence of viscosity and nonequilibrium effects, the sound speed is independent of the sound frequency. For the specific case of single-phase gas the sound velocity can be calculated by the isentropic relation

$$a_0 = \sqrt{\left(\frac{\partial p}{\partial \varrho_g}\right)_s}. \tag{9.11}$$

Assuming an ideal gas law, expression (9.11) simplifies to

$$a_0 = \sqrt{\varkappa_g \frac{p}{\varrho_g}}, \qquad \text{or equivalently} \qquad a_0 = \sqrt{\varkappa_g R_g T_g}, \tag{9.12}$$

with the gas constant R_g and the isentropic exponent $\varkappa_g$. With an increased strength of the shock wave the propagation velocity u_{sw} exceeds the value of the sound velocity depending on the pressure rise $\Delta p = p_1 - p_0$

$$u_{\mathrm{sw}} = a_1 \sqrt{1 + \frac{\varkappa_g + 1}{2\varkappa_g}\left(\frac{\Delta p}{p_0}\right)}. \tag{9.13}$$

For two-phase flow conditions, analytical descriptions for wave propagation phenomena with finite wave strengths do not exist, apart from the cases where extreme simplified assumptions such as homogeneous equilibrium or complete "frozen" conditions (where the algebraic mass, momentum, and energy transfer terms are set to zero) are applied.

In the following, the propagation of one-dimensional (plane) pressure waves in two-phase media are studied numerically assuming the absence of wall friction and wall heat transfer. The initial and boundary conditions are shown schematically in Fig. 9.8. At the left side of the pipe a sudden pressure increase is imposed. The calculations are terminated before the waves have reached for the far right end of the 4 m long pipe.

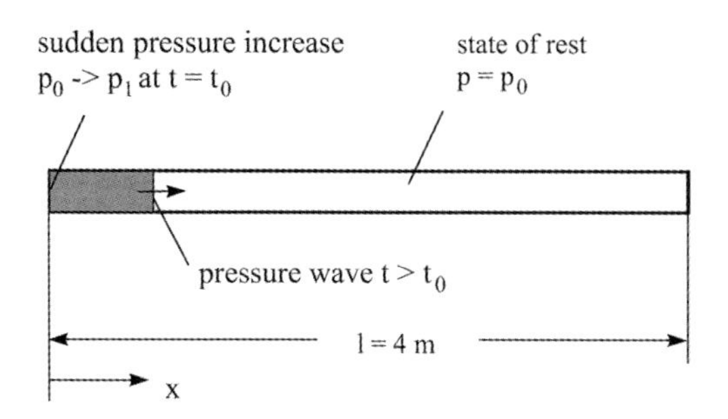

Fig. 9.8: Shock wave propagation in a straight pipe, pipe geometry, and boundary conditions

9.3.1 Single-phase gas flow

Before dealing with the more complex two-phase flow conditions, a few results are included here, mainly to demonstrate the capability of the Flux Vector Splitting (FVS) scheme, a numerical method that will be also used for the following two-phase cases. Predicted values for the shock wave propagation in a gas medium assuming ideal state equations are shown in Fig. 9.9 in comparison with the existing analytical solution. The calculation is performed with 1600 computational cells which corresponds to a grid spacing of 2.5 mm. The case selected represents a rather moderate shock strength with a shock velocity of $u_{\mathrm{SW}} = 473.05$ m/s, which corresponds to a Mach number of $M = 1.36$ based on the sound velocity in the undisturbed region in front of the wave.

The figure indicates a high resolution of the flow discontinuities and a perfect prediction of shock velocity. Within the shown scale, there is practically no difference between the CFD calculation and the analytical solution. The calculated results are also free of any numerically induced instability and, within the predicted time frame, there is no evidence of dispersive effects. All calculated parameters across the wave including, pressure, velocity, temperature, and entropy are in good agreement with the analytical values as can be seen from Fig. 9.10. The correct prediction of temperature and entropy could be achieved only when using the full energy conservation equation.

Of particular interest with regard to the numerical feature of the FVS scheme is how the selection of the grid spacing affects the numerical results compared with the analytical solution. This is demonstrated in Fig. 9.11 showing the effect of a progressive refinement of the grid from cell sizes of 8 mm to 1 mm. The figure indicates a clear convergence of the predictions toward the discontinuous analytical solution which might be fully reached for $\Delta x \rightarrow 0$.

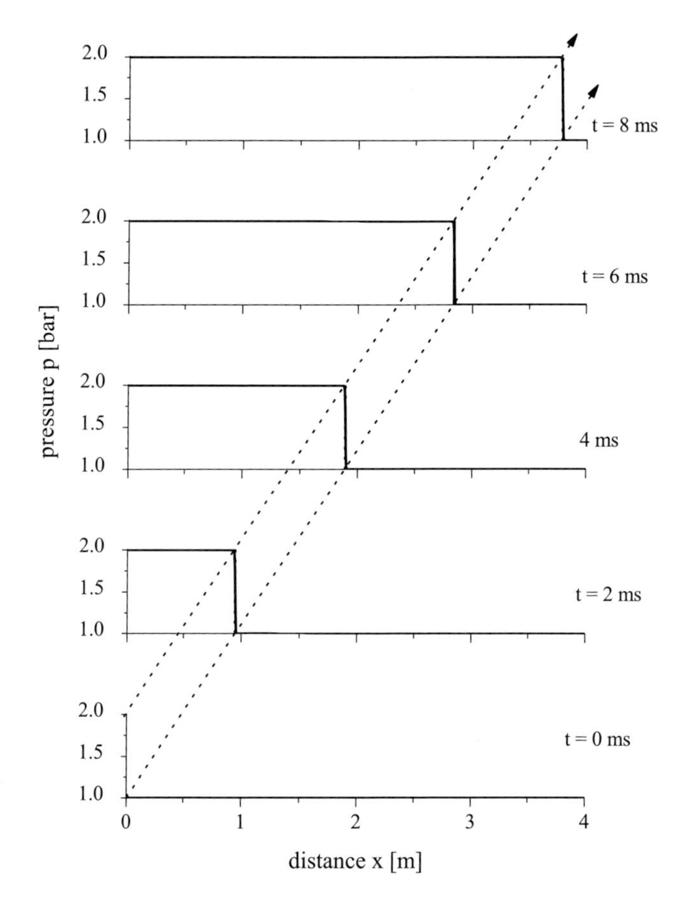

Fig. 9.9: Shock wave propagation in *single-phase gas* , second-order Flux Vector Splittting scheme with 800 cells $\Delta x = 2$ mm; comparison of CFD calculation (straight line) and analytical solution (dotted line)

9.3.2 Two-phase flow

For a two-phase test case, the pipe is assumed to be filled with a water/air mixture at equilibrium conditions with a temperature of 300 K and a void fraction of $\alpha_g = 0.05$. It is further assumed that the gas is homogeneously distributed in the form of equally sized bubbles having a bubble diameter of 4 mm. As for the previously described gas case, the shock wave is initiated by a sudden pressure rise from $p_0 = 1$ bar to $p_1 = 2$ bar. The calculations are performed with the same numerical methods as the previous case using the second-order FVS scheme. For the whole length of the pipe, 2000 computational cells are used which corresponds to a cells sizes of $\Delta x = 2$ mm. The calculated pressure wave propagation at five consecutive time values is shown in Fig. 9.12.

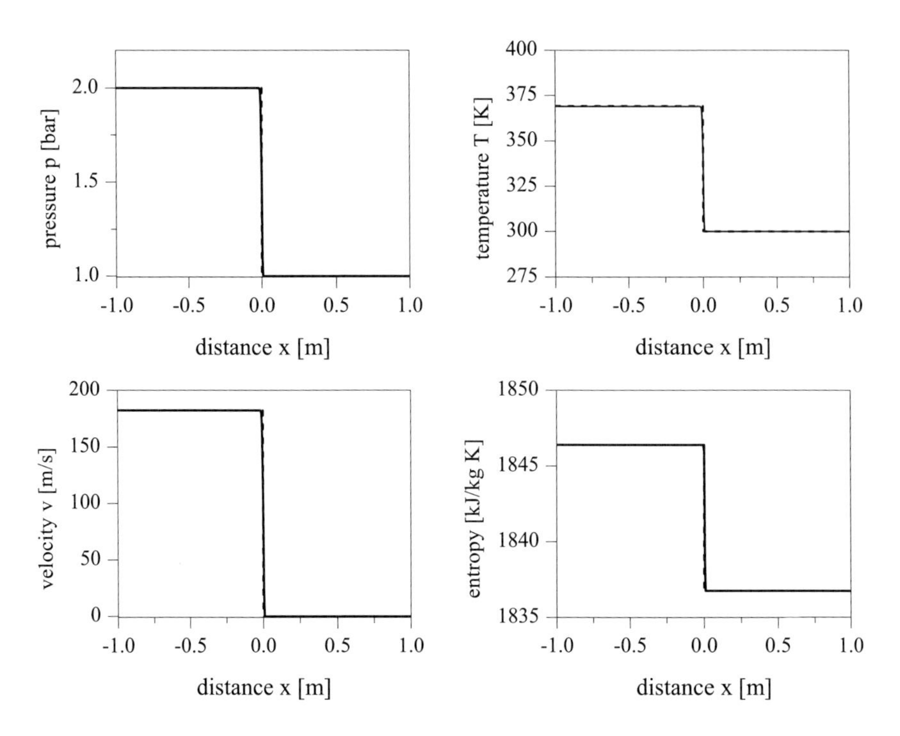

Fig. 9.10: Shock wave propagation in a *single-phase gas*, second-order FVS scheme with 800 cells ($\Delta x = 2$ mm); comparison of CFD calculation (straight line) and analytical solution (dotted line)

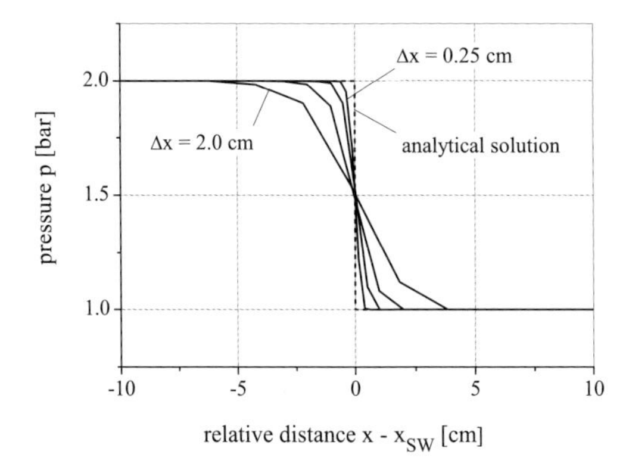

Fig. 9.11: Numerical simulation of a shock wave in *single-phase gas* media; convergence of solution in the case of grid refinement

The figure includes the results of three calculations which differ only in the specification of the algebraic source terms describing the momentum and heat coupling between the phases:

(a) this base calculation uses a standard modeling of the momentum and energy coupling resulting in finite values for the interfacial forces and interfacial heat transfer,

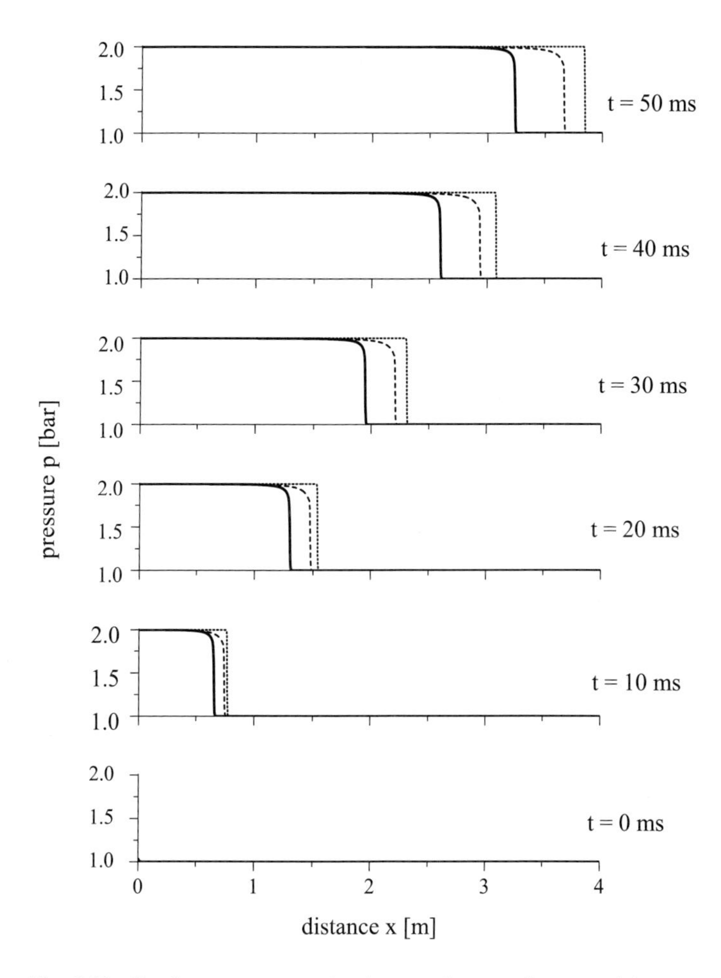

Fig. 9.12: Shock wave propagation in *two-phase* media, *water/air* at $p_1 = 1$ bar, $T_1 = 300$ K, $\alpha_g = 0.05$; second-order FluxVector Splitting scheme with 2000 cells

(b) here partial "frozen" conditions are applied where the heat transfer between the phases is neglected, the interfacial forces are modeled as in case (a),

(c) this prediction assumes complete "frozen" conditions where both the interfacial forces and the interfacial heat transfer are set to zero. However, the nonviscous interfacial forces described by exclusively differential terms are taken into account as in the calculations (a) and (b).

Although the same sudden pressure increase is applied to initialize the shock wave, there are remarkable differences between the results with regard to the predicted wave speed and pressure profile. For fully "frozen" conditions (a) the pressure wave propagates with a constant supersonic velocity ($u_{\mathrm{sw}} > a_0$) into the undisturbed region of the pipe, and at the same time, the initial discontinuous pressure change is (apart from very small numerical diffusion effects) maintained.

Specifying finite values for the interfacial heat transfer and interfacial friction as in the case (a) results in a further reduction of the wave speed and, due to the inherent physical dissipation, in a transfer of the initial pressure jump into a continuous pressure profile.

A more detailed picture of the wave structure is given in Fig. 9.13 showing the governing state and flow parameters across the wave at the time $t = 30$ ms. Two different regions can be distinguished:

(1) a wave front with rather steep (but finite) parameter gradients accompanied with strong excursions for the gas velocity and temperature and resulting strong deviations from thermal and mechanical equilibrium between the phases. Within this region the effect of the algebraic source terms is practically negligible ("frozen") and the continuous change of parameters is determined by the differential coupling terms between the phasic momentum and energy equations.

(2) a relaxation zone behind the shock front with a continuous change of the flow parameters toward a new equilibrium between the phases. The width of this region is governed by the finite values for the (algebraic) terms for the interfacial friction and heat transfer and the corresponding characteristic time values. As shown in Fig. 9.13, with the present modeling assumptions the thermal equilibrium is reached considerable faster as the mechanical equilibrium between the phases.

The variation of the pressure profile relative to the shock front is presented in Fig. 9.14 for different time values after initiation of the shock wave. The figure shows that already for $t \geq 10$ ms a nearly constant wave profile is reached where dissipative effects and dynamic forces are balanced resulting in a wave structure which remains practically constant during the subsequent propagation process.

To what extent the predictions might be affected by the computational grid is illustrated in Fig. 9.15 showing the results obtained for a progressively reduced grid spacing. As indicated in Fig. 9.15, there exists a clear convergence of solution toward the continuous pressure profile which is practically reached for a grid size of $\Delta x \leq 1$ mm.

The effect of the wave strength is illustrated in Figs. 9.16 and 9.17 showing the pressure and velocity profiles at $t = 30$ ms after initiation of the shock wave. Apart from the initiating pressure p_1 all other initial and boundary conditions are the same as those applied before. As for single-phase gas flow, the growing shock strength results in an increase of the wave velocity accompanied with a steepening of pressure and velocity at the shock front. The figures also indicate that the homogeneous equilibrium model, indicated by the dotted line, provides a good approximation for the wave speed and the new equilibrium state downstream of the shock wave.

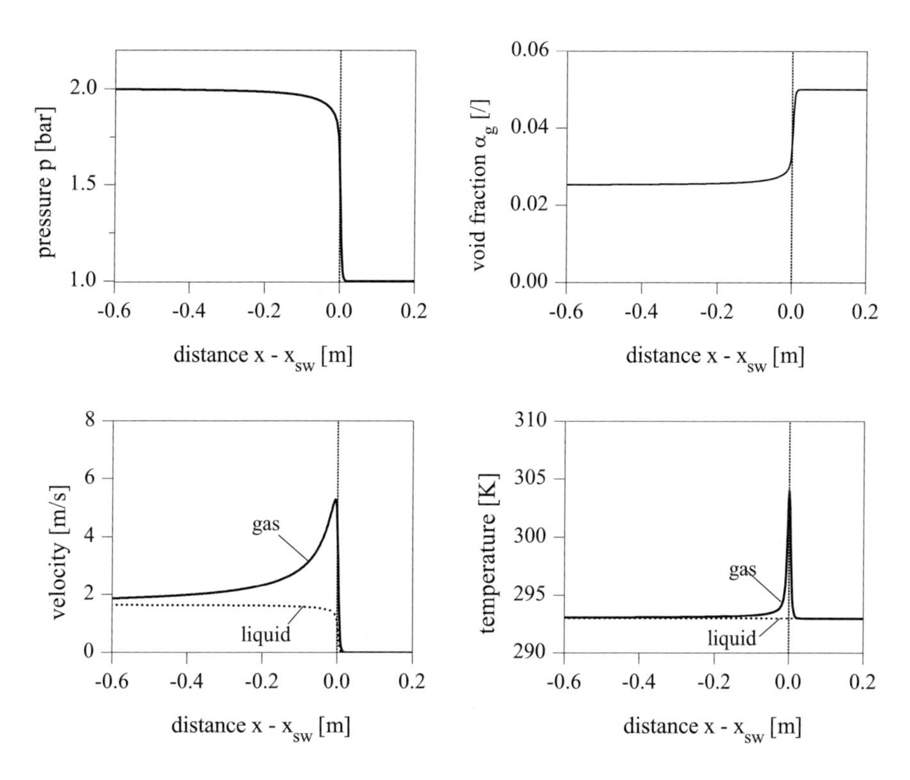

Fig. 9.13: Shock wave in *two-phase* media, *water/air*, $\alpha_g = 0.05$, parameter profiles at time $t \geq 10$ ms, second-order FVS scheme with 2000 cells, $\Delta x = 2$ mm

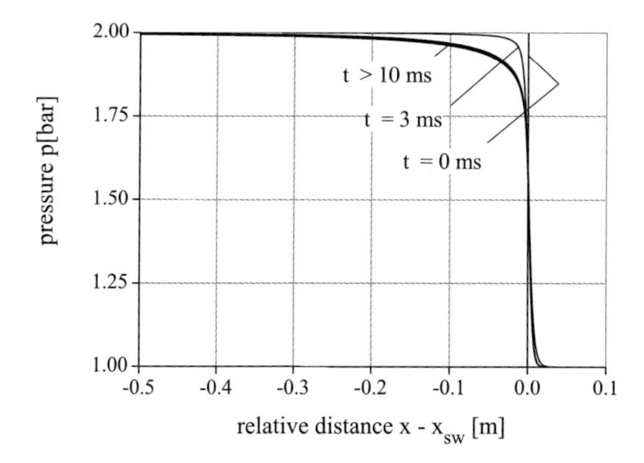

Fig. 9.14: Numerical simulation of a shock wave in *two-phase water/air* media, $\alpha_g = 0.05$, evolution of pressure profile during wave propagation, time: $0 \text{ ms} \leq t \leq 50$ ms

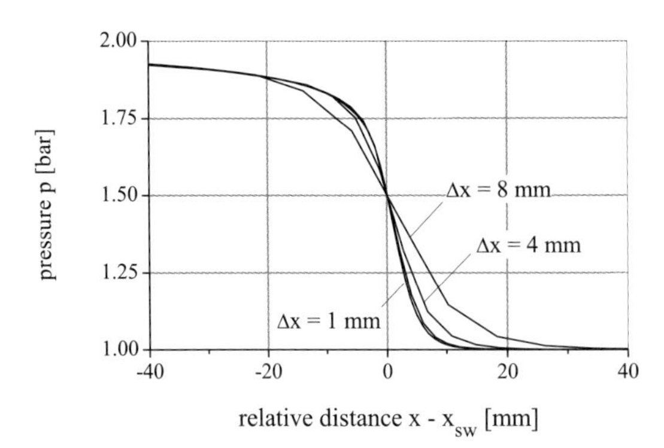

Fig. 9.15: Numerical simulation of a shock wave in *two-phase water/air* media, $\alpha_g = 0.05$; convergence of solution in the case of grid refinement, second-order FVS scheme

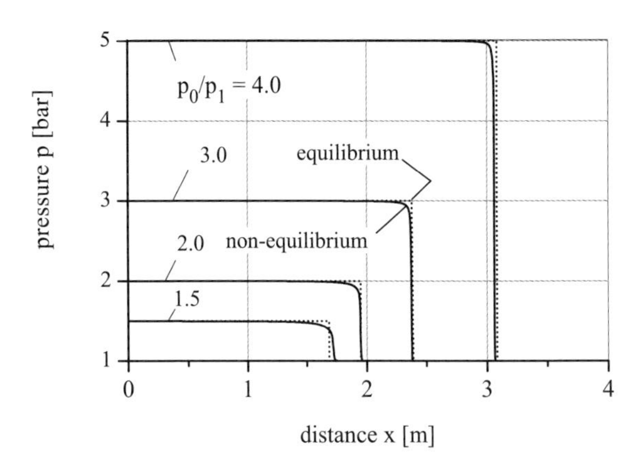

Fig. 9.16: Shock wave propagation in *two-phase water/air* media, $\alpha_g = 0.05$, pressure distributions at $t = 30$ ms, effect of shock strength; comparison of CFD calculation (straight line) with analytical solution for homogeneous equilibrium flow (dotted line)

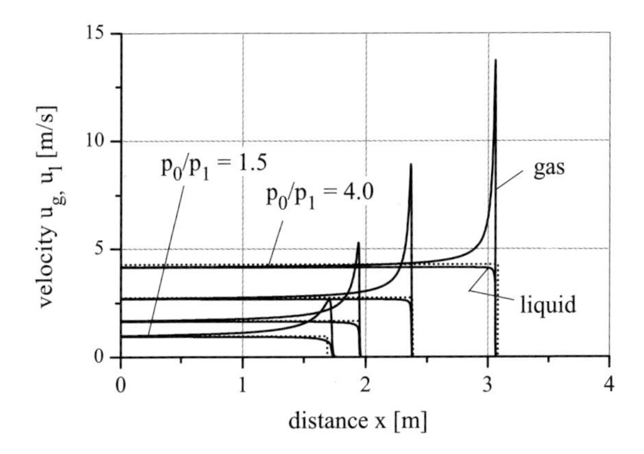

Fig. 9.17: Shock wave propagation in *two-phase water/air* media, $\alpha_g = 0.05$, phasic velocity distributions at $t = 30$ ms, effect of shock strength; comparison of CFD calculation (straight line) with algebraic solution for homogeneous equilibrium flow (dotted line)

9.4 Shock tube

Shock-tube devices have been extensively used to study shock wave propagation phenomena in compressible fluids like gases or gas–liquid two-phase mixtures. Usually a high (left) and a low (right) pressure region is separated by a diaphragm as schematically shown in Fig. 9.18. The transient is initiated by an instantaneous removal of the diaphragm resulting in a shock wave and rarefaction wave propagating toward the right and left ends of the pipe respectively. Assuming strictly one-dimensional flow conditions, the shock tube mathematically represents a "Riemann problem" where the initial flow velocities on both sides of the diaphragm have been set to zero.

9.4.1 Single-phase gas

For the single-phase gas case the geometry of the pipe and initial conditions are given in Fig. 9.18. For simplicity an ideal gas law is assumed with an isentropic exponent of $\varkappa = 1.4$.

The numerical results for the shock tube problem are obtained with the ATFM code using a second-order Flux Vector Splitting scheme with 500 cells which corresponds to a grid size of 0.8 mm. Figures 9.19 and 9.20 show a comparison of the CFD results with the existing analytical solution including the pressure distribution along the pipe length at various time values (Fig. 9.19) and the parameter distribution for a fixed time of $t = 8$ ms (Fig. 9.20). The figures indicate a nearly perfect agreement of the CFD calculation with the analytical solution including a perfect match of the timing for wave propagation phenomena and high resolution of the shock wave and contact discontinuity.

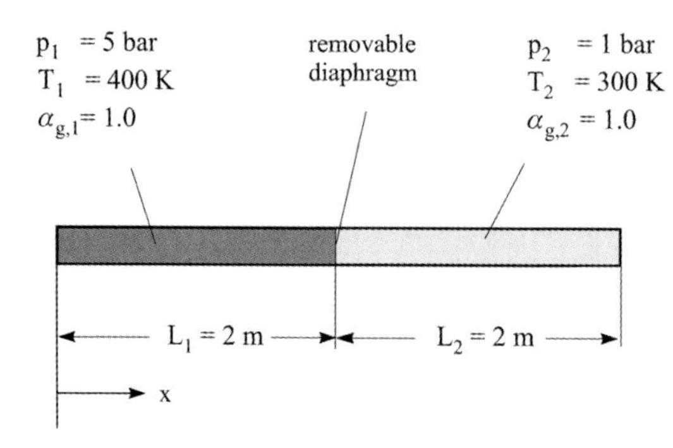

Fig. 9.18: Shock tube problem, for *single-phase gas*; geometry and initial conditions

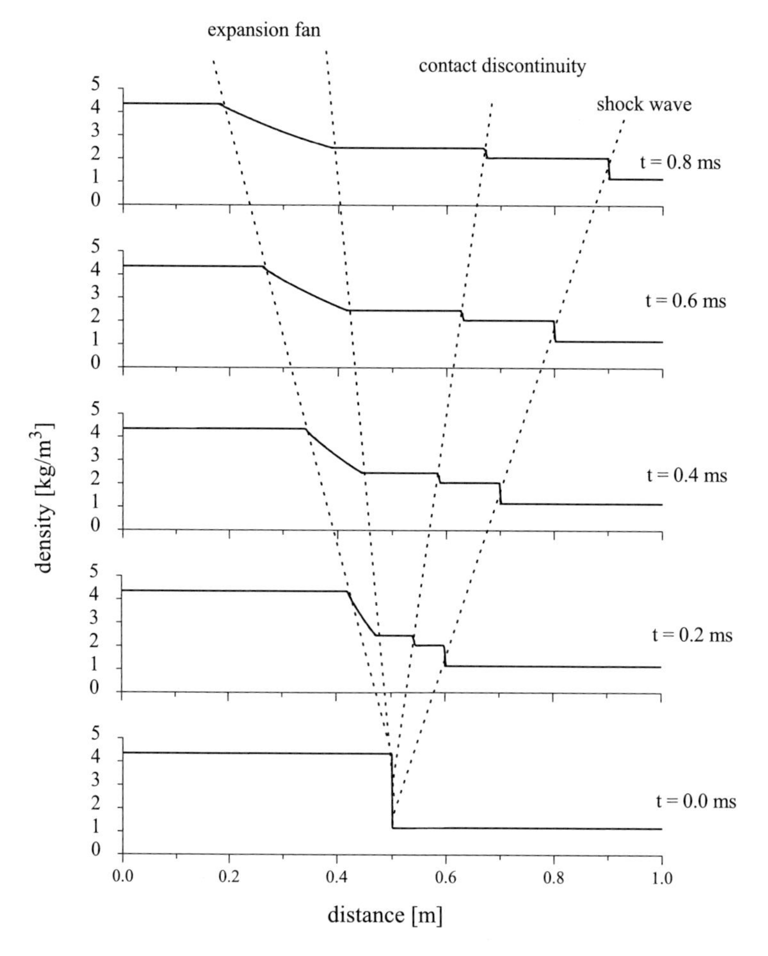

Fig. 9.19: Shock tube test problem for *single-phase gas*; density distribution at different time values, second-order Flux Vector Splitting scheme with 500 cells; comparison of CFD calculation (solid line) with analytical solution (dashed line)

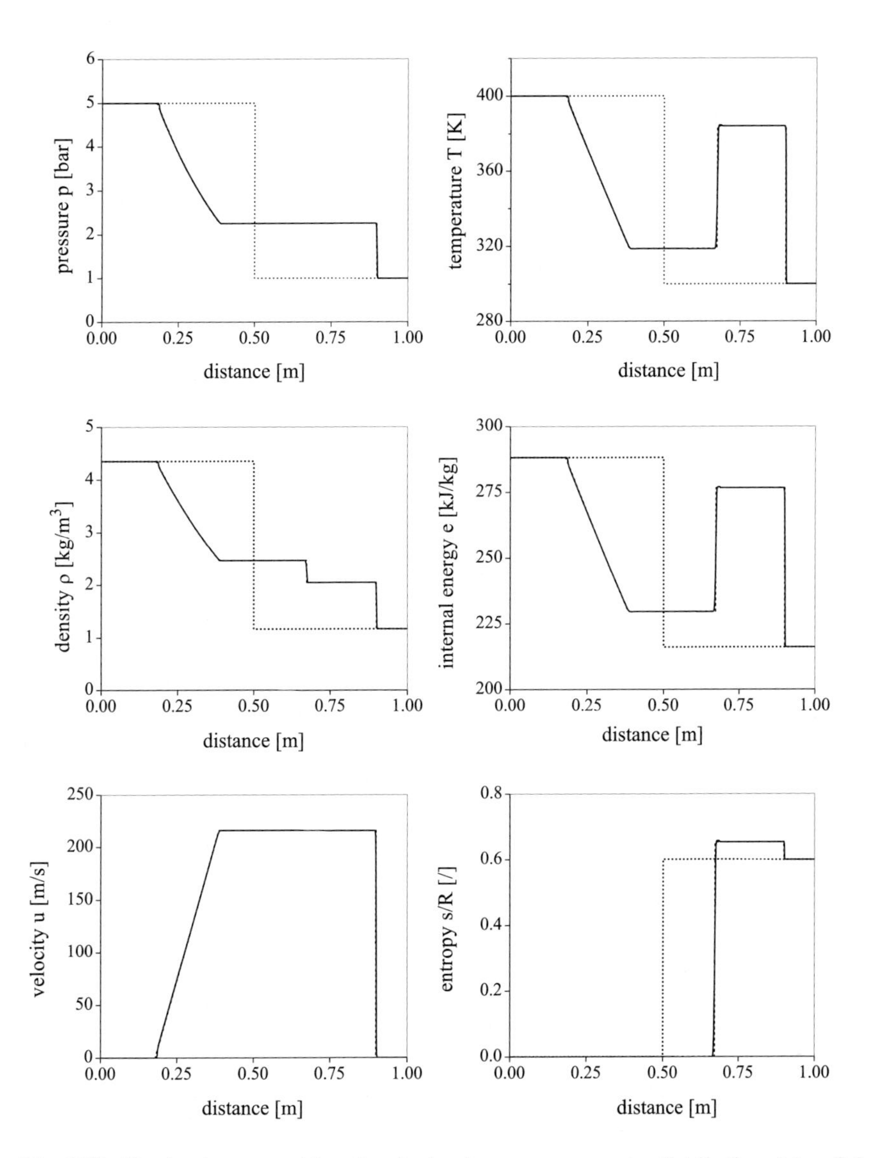

Fig. 9.20: Shock tube test problem for *single-phase gas*: parameter distribution at $t = 0.8$ ms; comparison of CFD prediction (solid line) with exact solution (dashed line), dotted line: initial conditions

9.4.2 Two-phase flow

Compared with the single-phase gas flow case, the shock tube problem becomes more complex for two-phase flow conditions. The reason for this is not only the increased number of governing flow parameters and related flow equations, but rather the presence of algebraic source terms controlling deviations between local phasic temperatures and flow velocities. As an example a two-component water/air mixture is chosen with a relatively high liquid content to exagerate the difference to the pure gas case. Due to the large complexity, no algebraic solutions exist for the general two-phase shock tube problem. Nevertheless, for the specific condition of homogeneous equilibrium flow an iterative algebraic solution can be derived in a similar way as was described for the pure gas case in Chapter 4. The tube geometry and initial conditions as used in the predictions are given in Fig. 9.21.

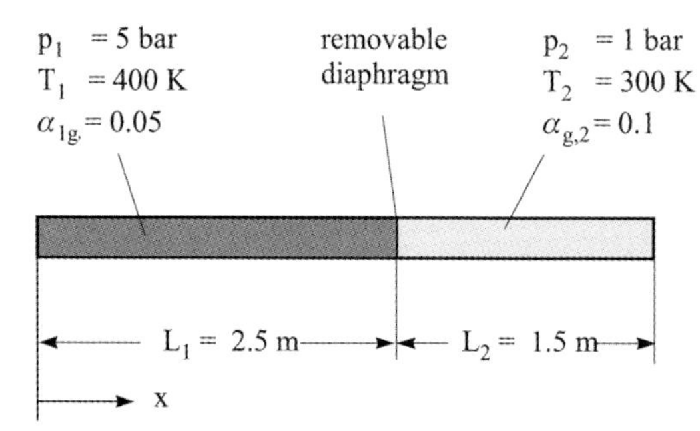

Fig. 9.21: Shock tube problem for *two-phase water/air* mixture; geometry and initial conditions

As in the previous case, the ATFM calculations were performed with the second-order Flux Vector Splitting scheme. A relatively large number of 1000 cells was used to guarantee a converged numerical solution. The predicted pressure distributions at various time values after the rupture of the diaphragm are shown in Fig. 9.22. On first glance the results look qualitatively very similar to what was predicted for the single-phase gas, including the presence of shock and expansion waves propagating with constant velocities into the low and high pressure regions. However, as a result of the increased inertia and reduced compressibility of the two-phase mixture the wave propagation velocities are considerably lower than for the pure gas case.

More details on the wave propagation processes can be seen in Fig. 9.23 showing the distribution of governing parameters at a constant time of $t = 18$ ms. As for the pure gas case three wave propagation phenomena can be distinguished:

1. A shock wave traveling into the low pressure region which is composed of a shock front characterized by steep (but finite) gradients for all involved flow parameters followed by more of a relaxation region with moderate changes. Due to the very small time scale for crossing the shock front, all interfacial transfer processes as described by algebraic source terms are practically "frozen", resulting in a strong deviation (overshoot) of phasic velocities and temperatures from the equilibrium conditions. The velocity of the shock front depends on the sound velocity within the undisturbed region as well as on the pressure ratio across the front which is not *a priori* known. The relaxation region behind the front is governed by the interfacial heat, mass and momentum coupling driving the phase

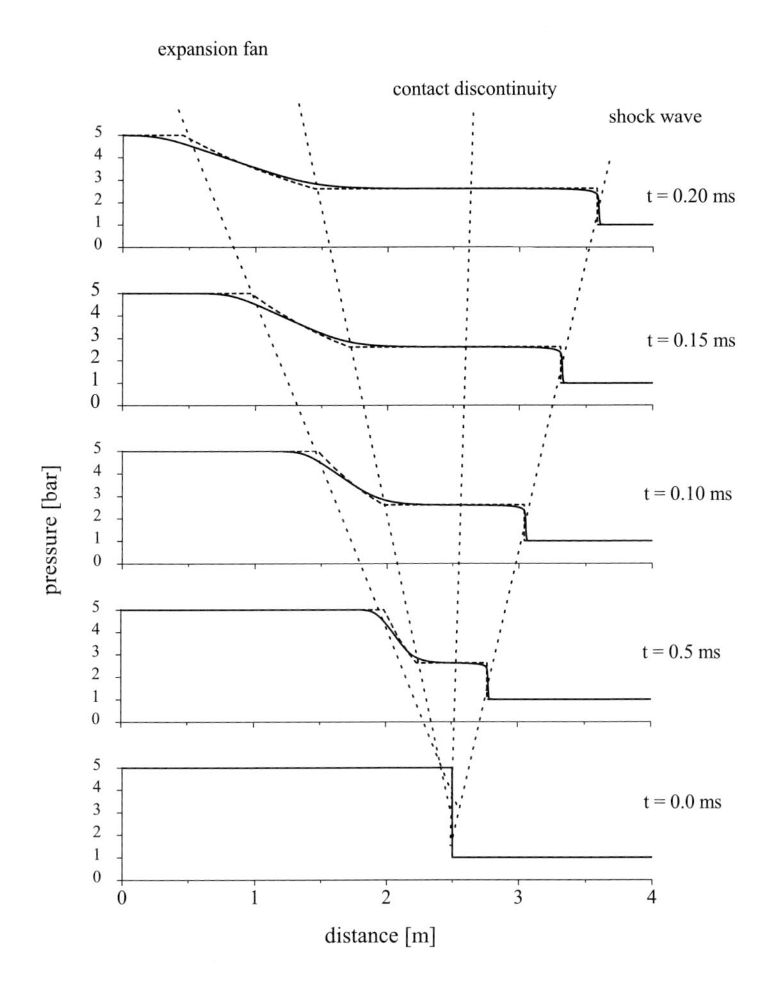

Fig. 9.22: Shock tube test problem for *two-phase flow of water/air* mixture; pressure distribution at different time values, second-order Flux Vector Splitting scheme with 1000 cells; comparison of CFD calculation (solid line) with analytical solution for homogeneous equilibrium condition (dashed line)

parameters toward a new equilibrium state. The length of the relaxation zone depends on the intensity of the interfacial coupling terms.

2. A pressure wave followed by a “contact discontinuity” which in an ideal case marks where the two fluids were initially separated by the diaphragm. Similar to the pure gas case, the predicted pressure and mixture (to a certain extent also phasic) velocities remain equal on both sides. All the other parameters including mixture density, void fraction and phasic temperature show abrupt changes across the “contact discontinuity”. However, contrary to the gas flow case, the parameter gradients remain at finite values due to diffusion effects resulting from different local phasic velocities.

3. A smooth expansion wave traveling into the high pressure region. The front of the waves propagates with the sound velocity in the undisturbed high pressure region. The continuous dispersion of the wave is a result of the decrease of sound velocity during the expansion of the fluid.

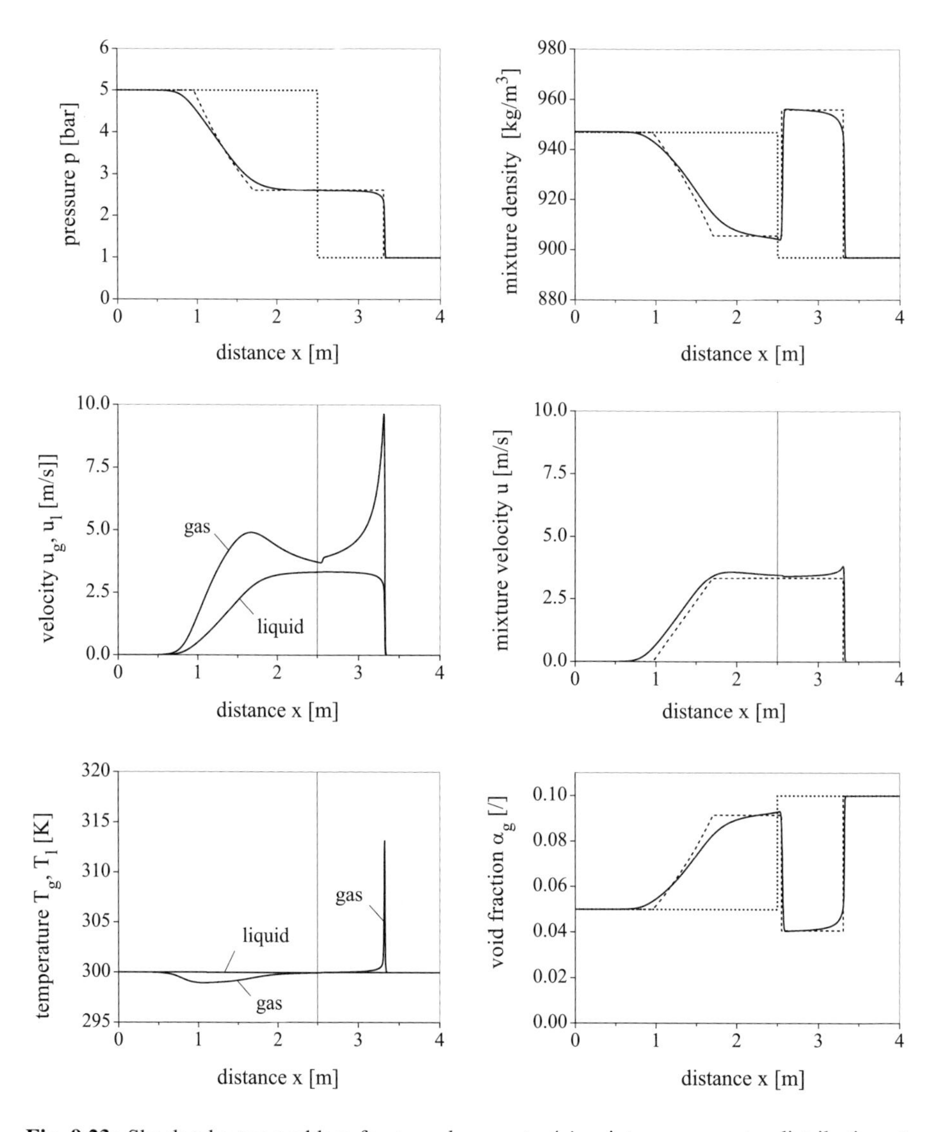

Fig. 9.23: Shock tube test problem for *two-phase water/air* mixture; parameter distribution at $t = 15$ ms, comparison of CFD prediction (solid line) with algebraic solution for homogeneous equilibrium conditions (dashed line); dotted line: initial condition

In addition to the CFD prediction, Figs. 9.22 and 9.23 also include the results of an iterative analytical solution for the simplified case of homogeneous equilibrium flow. It is evident that the assumption of equal local phase velocities and temperatures suppresses all thermal and mechanical disequilibrium effects. However, there still exists a remarkable agreement with respect to pressure, mixture density, and mixture (average) velocity as well as for propagation velocities of different wave propagation processes.

9.5 Multidimensional wave propagation and explosion phenomena

The shock tube problem as described above can be extended to multidimensional "explosion" test cases, where a cylindrical or spherical high pressure core region is surrounded by a constant pressure environment as schematically indicated in Fig. 9.24. As in the shock tube problem the initial velocities in the high and low pressure regions are set to zero. As long as the fastest waves have not reached the outer walls no further specific boundary conditions are needed.

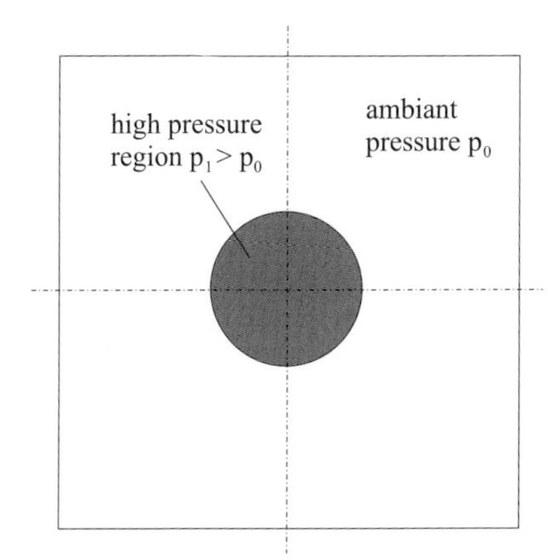

Fig. 9.24: Cylindrical explosion test case

For the following numerical analysis two different computational grids will be used as shown in Fig. 9.25: (1) an equally spaced Cartesian grid (left) where, due to the expected strictly axisymmetric behavior, only one quadrant is actually used in the prediction, and (2) a quasi-one-dimensional nonuniform grid (right) which takes a full advantage of the symmetrical feature of the test case.

9.5.1 Single-phase gas flow

Before dealing with the more complicated two-phase flow conditions the numerical approach will be first tested for a pure gas case. The initial pressure and temperature values in the high pressure region are $p_1 = 5$ bar and $T_1 = 400$ K, respectively. In the outer low pressure region pressure and temperature are specified as $p_0 = 1$ bar and $T_0 = 300$ K. The diameter of the high pressure core is $d = 1.6$ m.

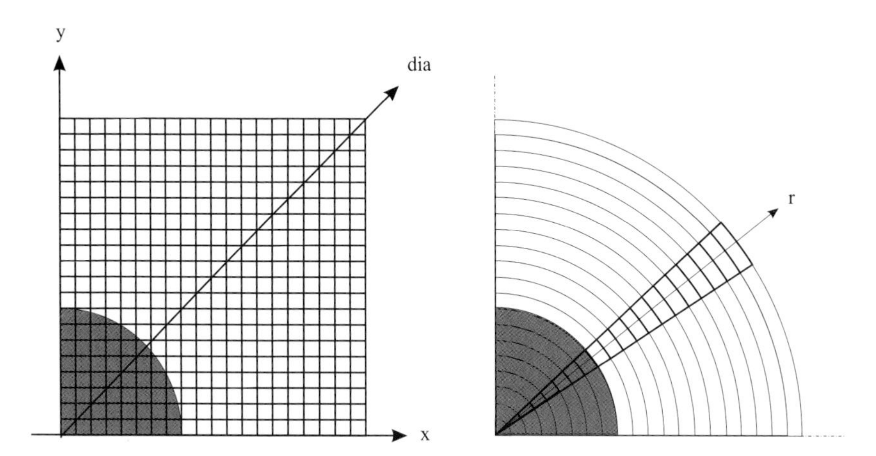

Fig. 9.25: Two-dimensional Cartesian grid (left) and quasi-one-dimensional grid (right) for explosion test case

The effect of the numerical approach on the explosion test case is shown in Figs. 9.26 and 9.27, comparing results for the pressure and density profiles obtained for different spatial resolutions. The figures include the results of three different calculations: (a) a two-dimensional calculation using a Cartesian grid of $100 \times 100 = 10000$ computational cells, (b) a quasi-one-dimensional calculation with 1000 cells (e.g., the same number of cells in the radial direction as used for the two-dimensional calculation) and (c) a quasi-one-dimensional "reference" calculation with 2000 cells. From the figures the following conclusions can be drawn:

1. in the two-dimensional calculation there is no noticable difference whether the parameters are taken along the x- (or y-) axis or along the diagonal axis. This means that, at least for the present test case, the grid orientation effect is largely negligible,

2. the results for the two-dimensional and one-dimensional calculations are nearly identical when using the same number of cells in the radial direction,

3. the predicted wave velocities and corresponding wave locations are largely independent on the spatial resolution of the computational grid,

4. the solution with only 100 cells in radial direction did not reach spatial convergence as indicated by the poor representation of the shock wave and contact discontinuity. For the quasi-one-dimensional calculation the convergence is almost reached when 2000 cells are used in radial direction.

The figures also demonstrate a dilemma of multidimensional simulation of pressure (explosion) wave propagation problems. For a high resolution of the parameter changes in the waves, a detailed fine grid spacing would be needed only in regions of large spatial parameter variation as in the vicinity of the moving contact surface and shock wave. Using a uniform Cartesian grid as in the present case is highly inefficient and may become impractical in the case of more complex geometries.

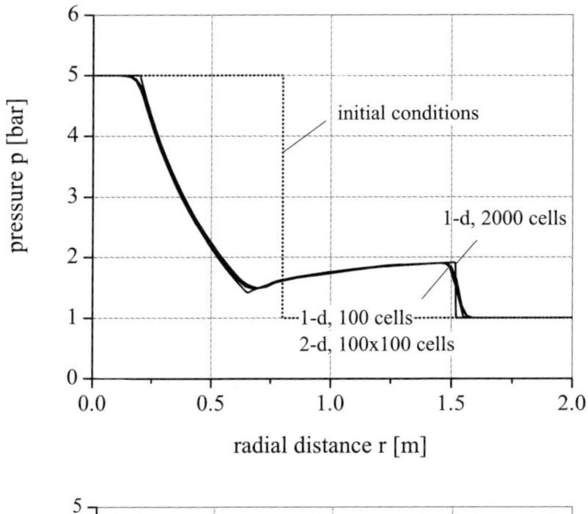

Fig. 9.26: Explosion test case, *single-phase gas* flow, pressure profile in the radial direction at $t = 1.5$ ms; comparison of one- and two-dimensional calculations

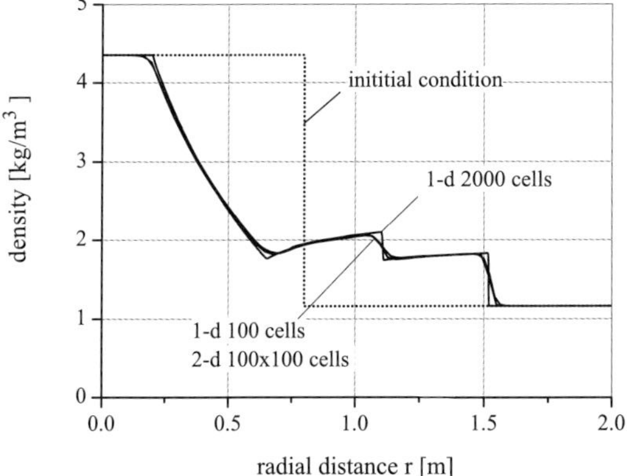

Fig. 9.27: Explosion test case, *single-phase gas* flow, density profile in the radial direction at $t = 1.5$ ms; comparison of one- and two-dimensional calculations

Results of a "reference" calculation using a quasi-one-dimensional nonuniform grid with 2000 cells in radial direction are given in Figs. 9.28 and 9.29 showing the radial pressure and density distributions for consecutive time values during the first 2 ms of the transient. Apart from the axis symmetry the behavior is qualitatively similar to the shock tube problem including the outward propagation of a (now circular) shock wave, a circular contact discontinuity traveling with some smaller velocity in the same direction, and a circular rarefaction wave traveling toward the origin. Contrary to the strictly one-dimensional shock tube case the shock strength becomes weaker while traveling in outward direction and the velocity profile between shock wave and contact discontinuity is no longer constant.

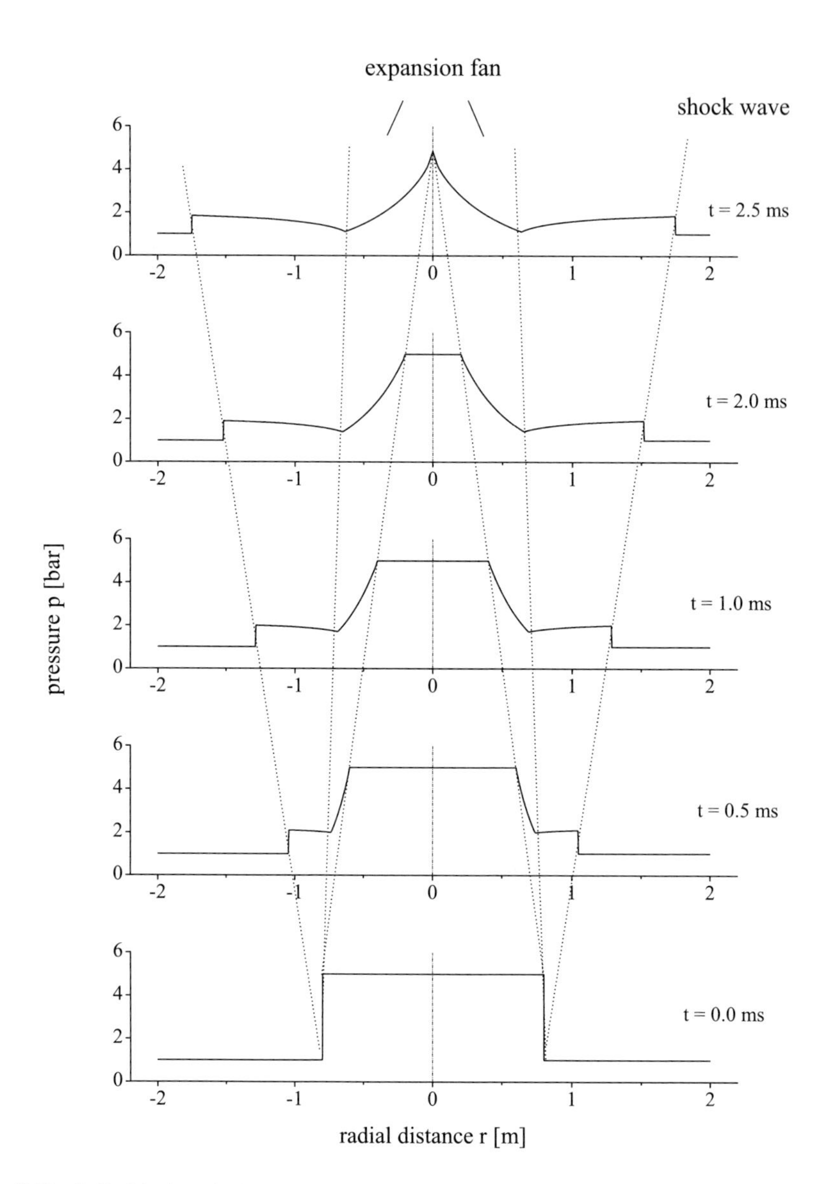

Fig. 9.28: Cylindrical explosion test case for *single-phase gas*, pressure distribution at different time values, second-order Flux Vector Splitting technique with 2000 cells

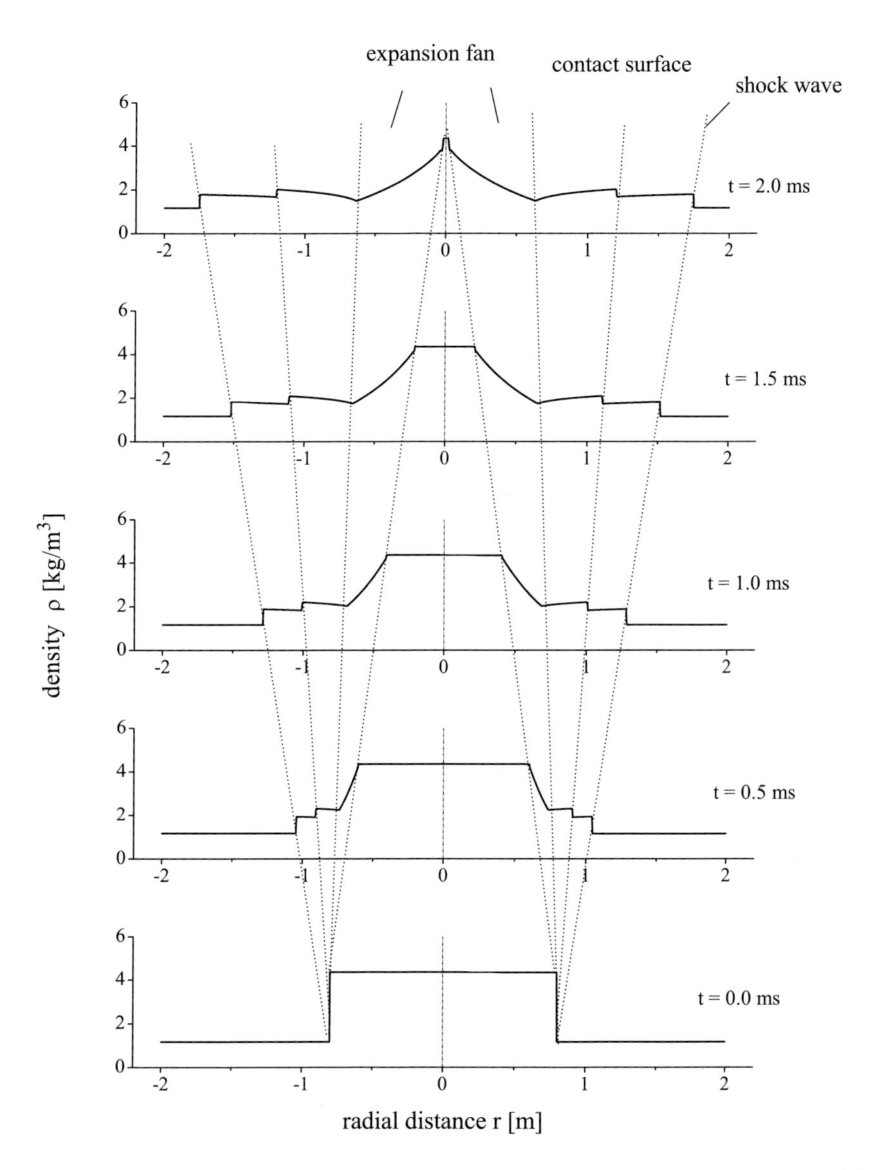

Fig. 9.29: Cylindrical explosion test case for *single-phase gas*, density distribution at different time values, second-order Flux Vector Splitting technique with 2000 cells

9.5.2 Two-phase flow

For the two-phase flow explosion test case a water/steam mixture is considered with the initial conditions as shown in Fig. 9.30. As in previous cases the pressure region with $p_1 = 5$ bar is surrounded by a constant ambient pressure $p_0 = 1$ bar. The gas (steam) volume fractions are $\alpha_{g,0} = 0.05$ in the high pressure region and $\alpha_{g,0} = 0.10$ on the low pressure side. In both regions thermal equilibrium is assumed with $T = T^{\text{sat}}(p)$.

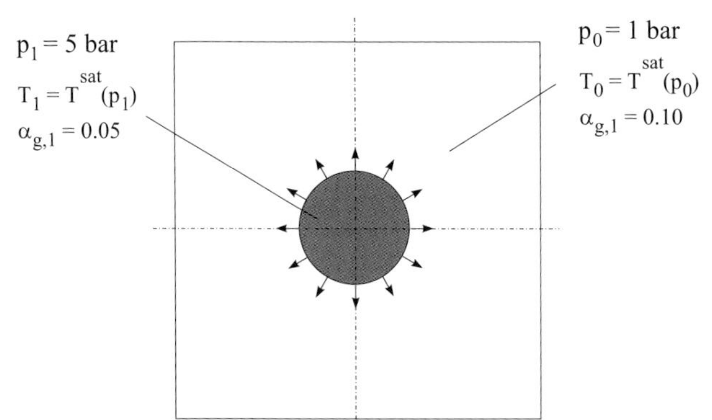

Fig. 9.30: Cylindrical explosion in water/steam media

How the chosen computational grid affects the calculation of the two-phase explosion test case is illustrated in Figs. 9.31 and 9.32. As for the single-phase gas case shown in Figs.9.26 and 9.27 the results using a Cartesian grid or a quasi-one-dimensional (nonuniform) grid are nearly identical as long as the same number of grid points are used in radial directions. A converged solution is practically reached for the quasi-one-dimensional calculation for 2000 cells in radial direction which corresponds to a grid size of 1 mm.

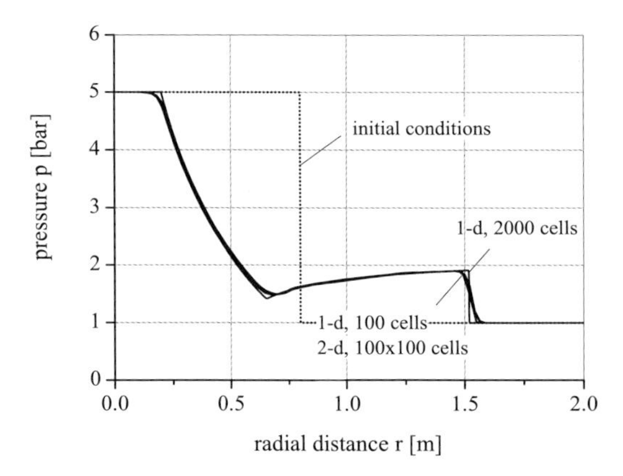

Fig. 9.31: Explosion test case, *twophase water/air* flow, pressure profile in the radial direction at $t = 40$ ms; comparison of one- and two-dimensional calculations

Results for the reference calculation using a quasi-one-dimensional grid with 2000 cells are presented in Figs. 9.34 and 9.35 showing the radial profiles for pressure and void fraction

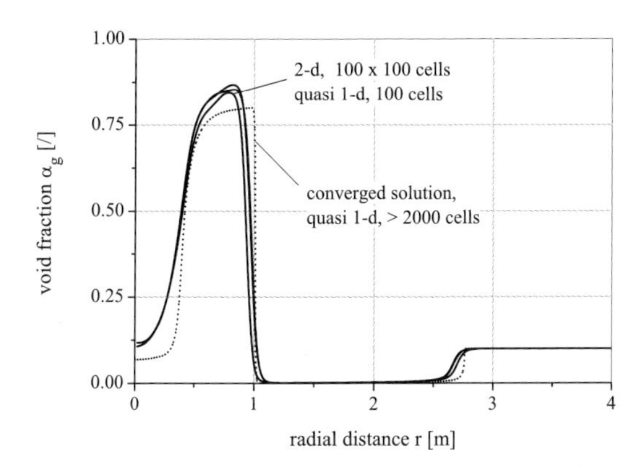

Fig. 9.32: Explosion test case, *two-phase water/air* flow, density profile in the radial direction at $t = 40$ ms; comparison of one- and two-dimensional calculations

at consecutive time values during the evolution of the transient. More detailed information on the parameter distributions at a fixed time of $t = 40$ ms are given in Fig. 9.33 also including phase velocities and evaporation/condensation rates. Although some common features still exist with the pure gas case as were described above, the inhomogeneity of the two-phase flow ($u_g \neq u_l$) and the large variation in mixture density due to the phase change processes add a considerable complexity to the wave propagation phenomena.

As in the gas case, a circular shock wave is traveling in an outward direction with a velocity of $u_{\mathrm{sw}} = 68.75$ m/s which corresponds to a Mach number (based on the "frozen" sound velocity in the upstream undisturbed region) of $M = 1.46$. The shock wave is comprised of a leading steep wave front creating a sudden disequilibrium between the phases, and a continuous downstream relaxation region where interfacial transfer processes for mass, momentum and energy drive the flow toward a new equilibrium state. With the initial conditions chosen, the shock wave results in a complete condensation of steam which is achieved about 0.5 m downstream of the shock front.

At the same time a circular rarefaction wave is propagating into the initial high pressure core region toward the origin. The propagation velocity of this wave is largely retarded by the onset of a strong evaporation as indicated by the large increase in void fraction at the outer core region. This leads to a prolonged holdup of the pressure in the core region up to the time when most of the liquid is evaporated (not shown here).

The third wave represents a type of "contact discontinuity" showing a nearly discontinuous change in void fraction (see Fig. 9.35) from pure liquid ($\alpha_g = 0$) to high void fraction ($\alpha_g > 0.8$) two-phase conditions. The wave practically marks the boundary between the two fluids which were initially present in either the high or low pressure region. As for the gas flow the pressure remains unchanged across the contact discontinuity. The major difference to the gas case is that due to the nonhomogeneous flow conditions ($u_g \neq u_l$) some mixing occurs across the wave which results in penetration of vapor into the adjacent subcooled region and associated condensation (see Fig. 9.33).

The results shown here represent typical examples to demonstrate some characteristic thermal-hydraulic features of explosion phenomena in two-phase media. Nevertheless, the present modeling and numerical approach can be easily extended to other initial and boundary conditions or to diabatic flow conditions with external heat sources.

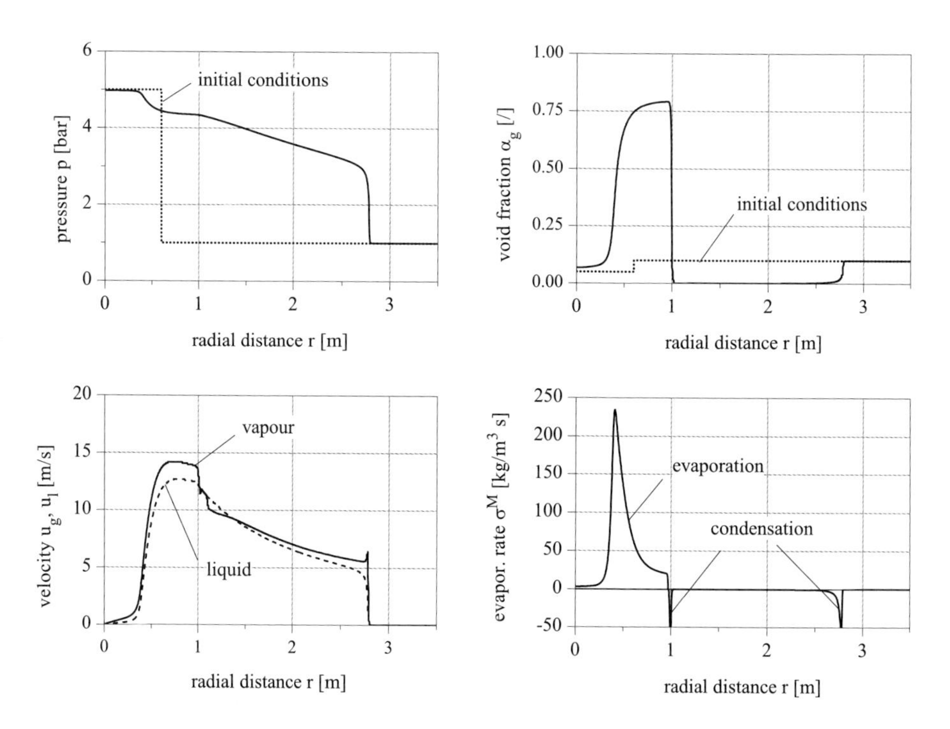

Fig. 9.33: Cylindical explosion test case for *two-phase water/steam* flow, paramter distributions in radial direction at time $t = 40$ ms

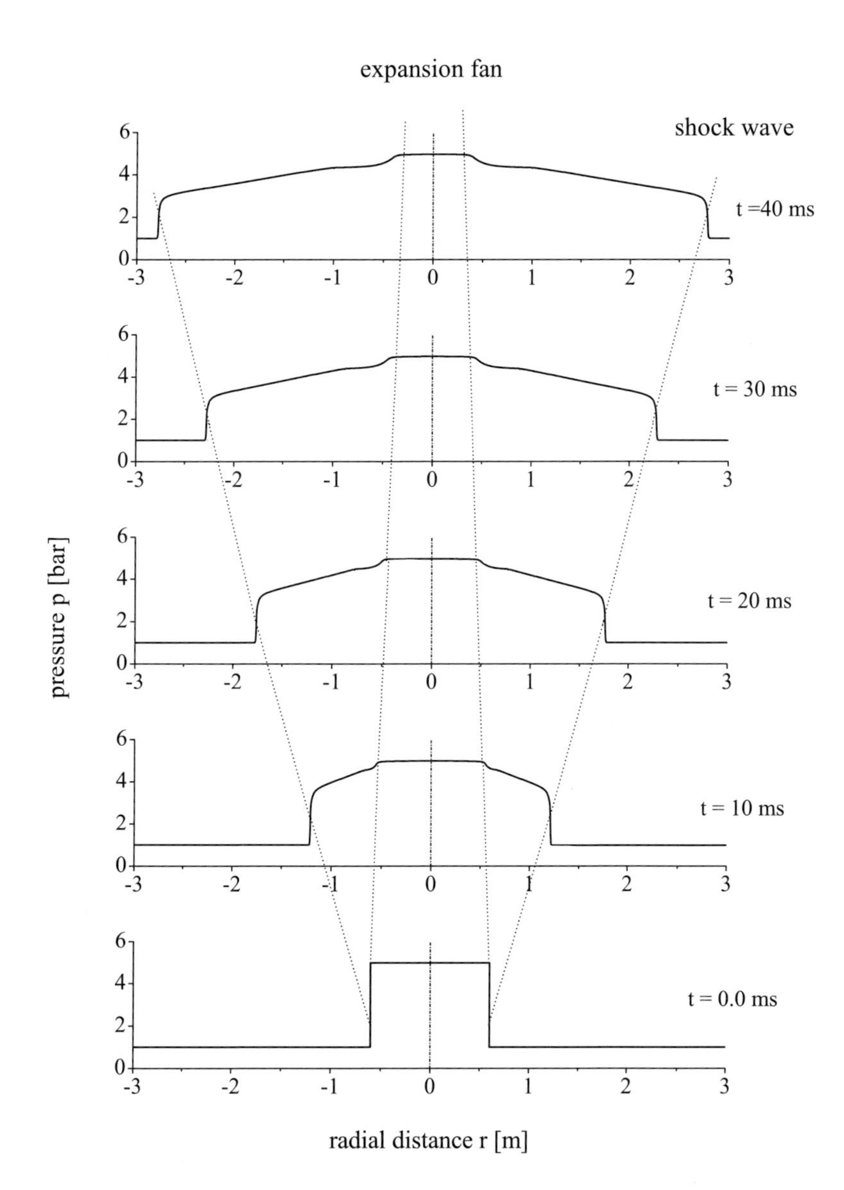

Fig. 9.34: Cylindical explosion test case for *two-phase water/steam* flow; pressure distribution at different time values, second-order Flux Vector Splitting technique with 2000 cells

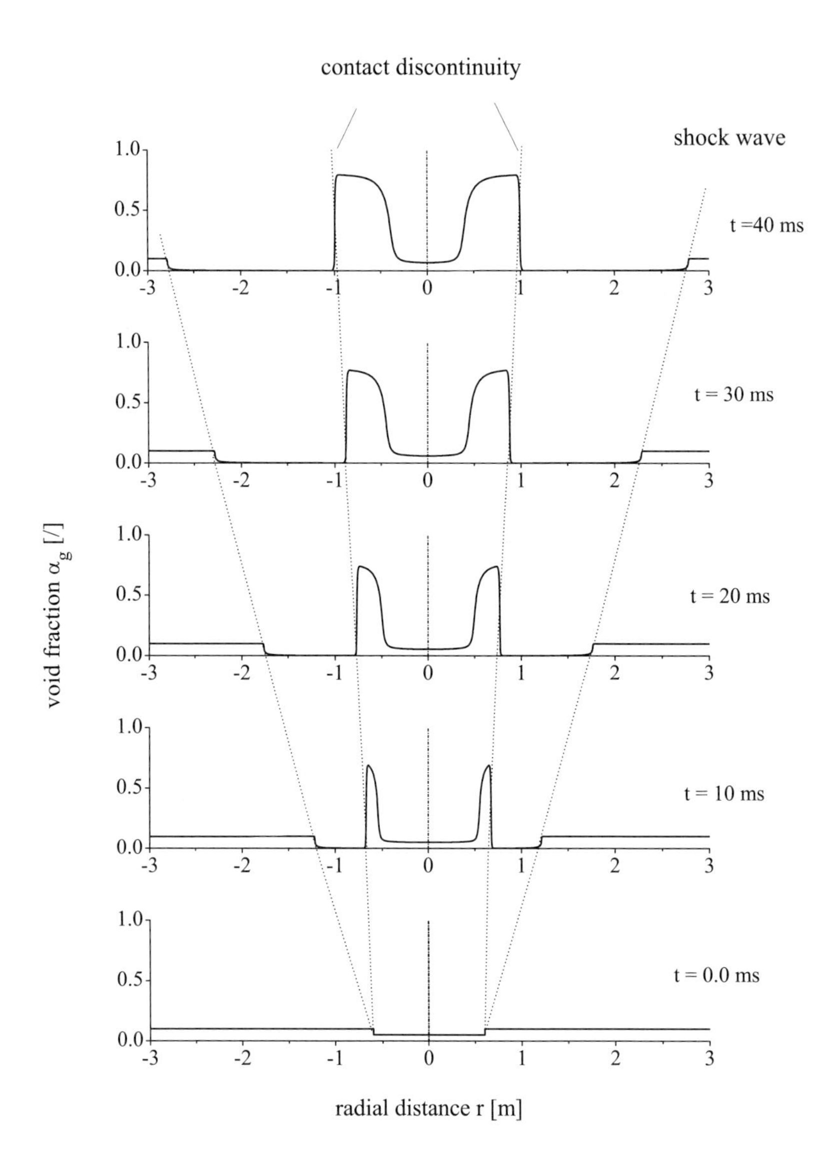

Fig. 9.35: Cylindical explosion test case for *two-phase water/steam* flow, void fraction distribution at different time values, second-order Flux Vector Splitting technique with 2000 cells

How the multidimensional effects influence the wave propagation and attenuation is demonstrated in Figs. 9.36 and 9.37 comparing pressure and void fraction distributions at a fixed time of $t = 40$ ms for plane, cylindrical, and spherical configurations. The initial conditions in all the three cases are the same as used before for the cylindrical explosion cases.

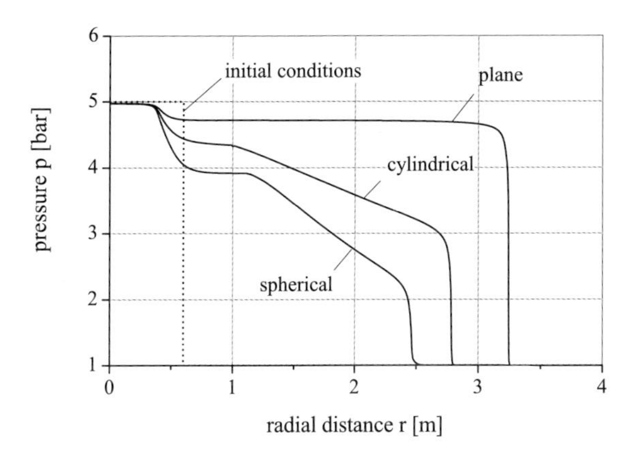

Fig. 9.36: Explosion test case for *water/steam* flow; pressure profiles in the radial direction at $t = 40$ ms for plane, cylindrical, and spherical configuration

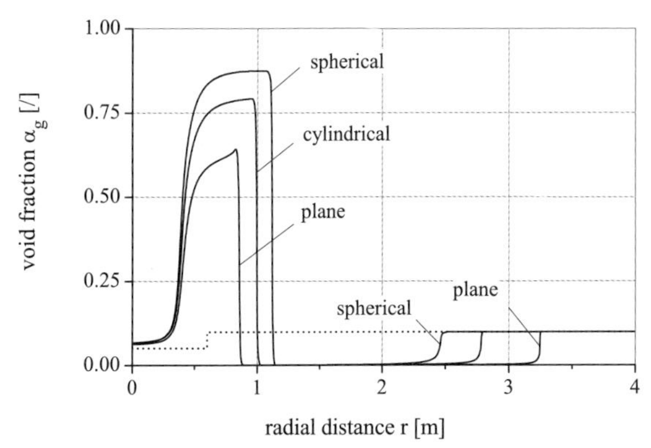

Fig. 9.37: Explosion test case, *water/steam* flow; void fraction profiles in the radial direction at $t = 40$ ms for plane, cylindrical, and spherical conditions

For the strictly *one-dimensional conditions* the "explosion case" becomes identical with the shock tube problem as was discussed in some detail in the previous section for water/air media. For the water/steam mixture, a strong shock wave is formed where in the trailing relaxation region the vapor is completely condensed. Due to the practical absence of viscosity effects (wall friction is neglected) the shock strength remains unchanged.

For the *cylindrical*, and even more for the *spherical configuration*, the continuous enlargement of the wave front while propagating in the outward direction results in a decrease of the wave intensity. On the other side, the spatial contraction to cylindrical or spherical geometries causes a faster depressurization of the core region and a more rapid evaporation as shown by the increased level of void fraction in this region.

The explosion test case as described above may also be reversed creating an implosion where an internal cylindrical or spherical core is surrounded by a high pressure region of finite thickness as schematically shown in Fig. 9.38.

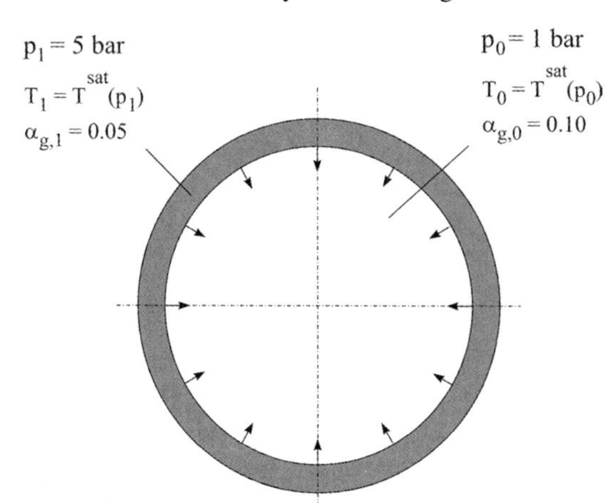

Fig. 9.38: Spherical *water/steam* implosion test case; basic configuration and initial conditions

Apart from the geometrical configuration, all other initial conditions for the high and low pressure regions are the same as in the previous explosion case. The quasi-one-dimensional prediction is done using the second-order Flux Vector Splitting scheme with 2000 grid points. Predicted parameter distributions for pressure and void fraction at five consecutive time values are shown in Figs. 9.39 and 9.40.

Similar to the explosion case, the onset of a strong evaporation in the initial high pressure (now outer) region creates a shock wave focusing toward the center of the sphere. The large thermal nonequilibrium in the wake of the shock leads to a complete condensation, which is practically reached within a distance of about 20 cm behind the shock front. Due to the convergence of the shock wave while propagating toward the origin the shock strength and propagation velocity continuously increase and, theoretically, an infinite pressure value would be achieved when the wave has shrunk to a single point. The prediction was terminated at $t = 30$ ms when the maximum pressure was $p = 46$ bar and the velocity of the shock wave was about $u_{\mathrm{sw}} = 200$ m/s.

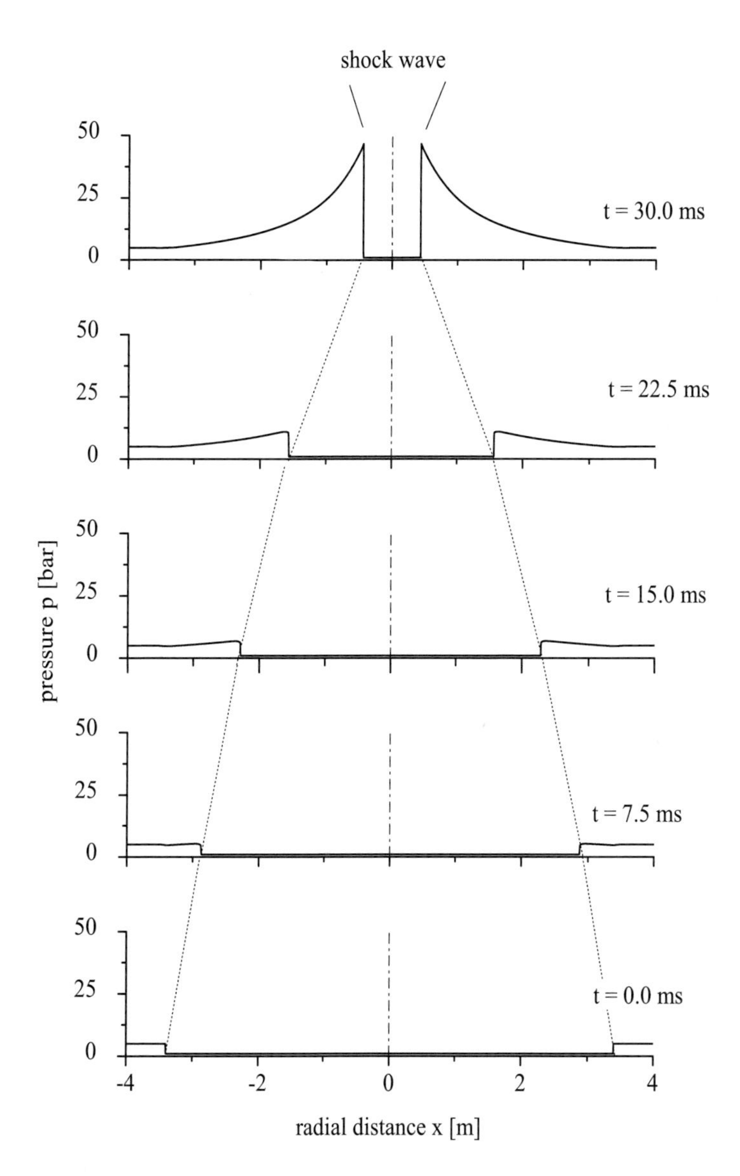

Fig. 9.39: Spherical implosion test case for *two-phase water/steam*flow, pressure distribution at different time values, second-order Flux Vector Splitting technique with 2000 cells

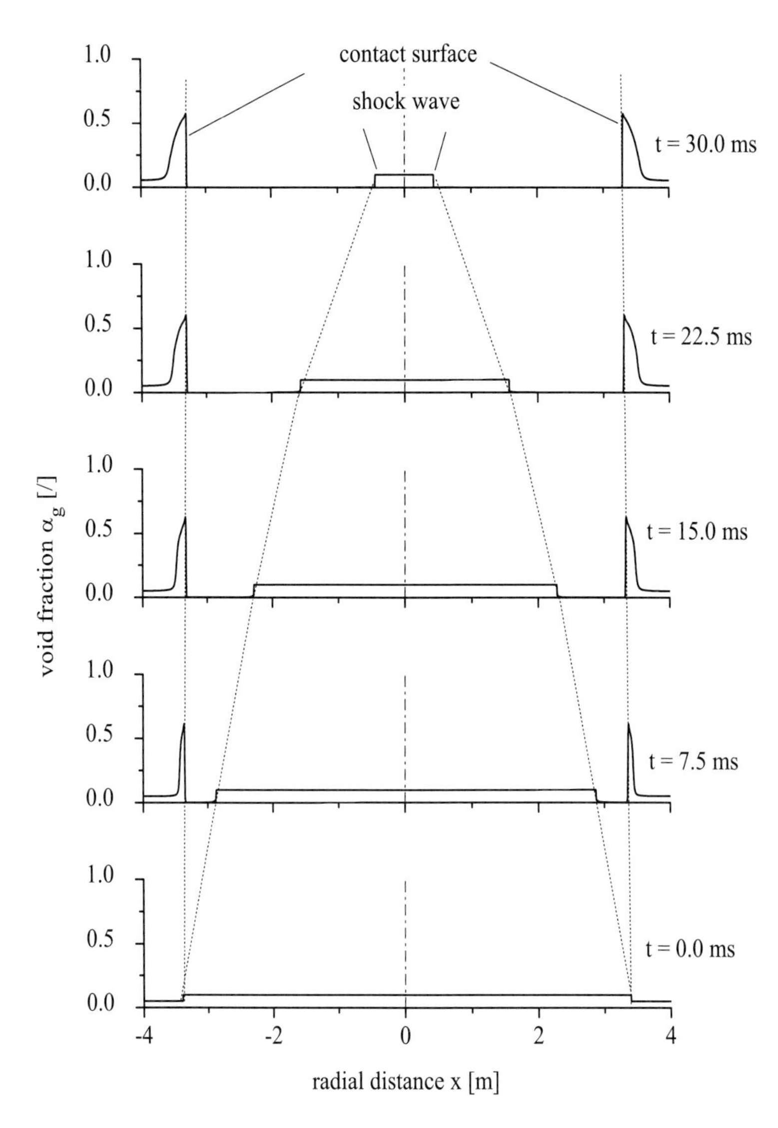

Fig. 9.40: Spherical implosion test case for *two-phase water/steam* flow, void fraction distribution at different time values, second-order Flux Vector Splitting technique with 2000 cells

9.6 Flow through convergent–divergent nozzles

The steady state flow of compressible fluid through convergent–divergent nozzles cover various important flow phenomena like the occurrence of critical flow conditions, transition from subsonic to supersonic flow or the occurrence of flow discontinuities. For the steady state quasi-one-dimensional nozzle flow of a single-phase gas, relatively simple algebraic solutions exist as described in many textbooks of gasdynamics. For two-phase flow conditions, iterative algebraic solutions can be derived only for the rather restrictive assumption of homogeneous flow (equal local phase velocities) and thermal equilibrium between the phases as presented in Chapter 4. For the more general case of nonhomogeneous and nonequilibrium conditions, usually only numerical solutions can be obtained where the large variety of Mach number and the possibility of flow discontinuities (shock wave) represent a major challenge.

In the following various types of nozzle flows are analyzed which include different nozzle geometries, one-component (water/steam) and two-component (water/air) fluids as well as the effect of different upstream reservoir conditions and back pressure values. Where available, measured data are included for comparison. In some cases also the results from homogeneous equilibrium are included to distinguish between the effects of fluid compressibility and the contributions resulting from mechanical and thermal disequilibrium between the phases.

9.6.1 The ASTAR nozzle

Within the framework of the EU sponsored project entitled "Advanced Simulation Tool for Application to Reactor Safety" (ASTAR) (Städtke et al. [2]) various benchmark test cases have been defined to assess different approaches for the numerical simulation of two-phase flow processes. This included the stationary flow in a "smooth" convergent–divergent nozzle with the geometry shown in Fig. 9.41.

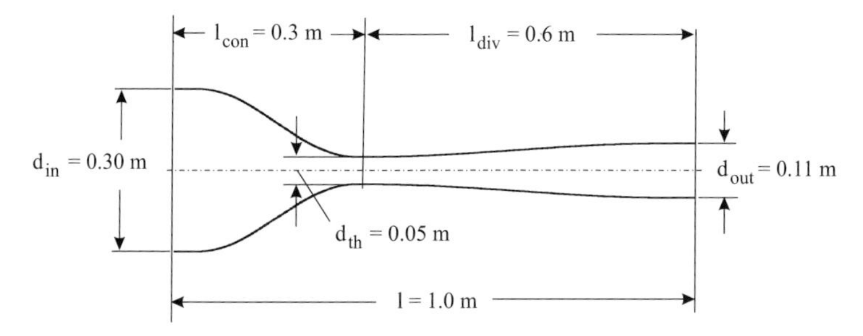

Fig. 9.41: ASTAR nozzle geometry

All calculations are related to the flow of a two-component (water/air) mixture with fixed upstream reservoir pressure and temperature values of $p_0 = 10$ bar and $T_0 = 570$ K. The upstream gas content (gas mass fraction) has been limited to $X_0 \geq 0.1$, or $\alpha_l < 0.10$ respectively, in order to guarantee dispersed droplet flow for all the test cases. To allow a comparison with analytical solutions (e.g., for single-phase gas or liquid flows) simplified state equations are assumed such as ideal gas law and pseudo-incompressible liquid with a constant value for the sound velocity in the liquid phase. For the prediction of heat and mass transfer between the phases the assumption of an ideal droplet flow regime is recommended based on uniform spherical droplets with a constant prescribed radius of $r_{dr} = 0.4$. The ATFM calculations shown in the following are performed using a nonuniform grid as schematically shown in Fig. 9.42.

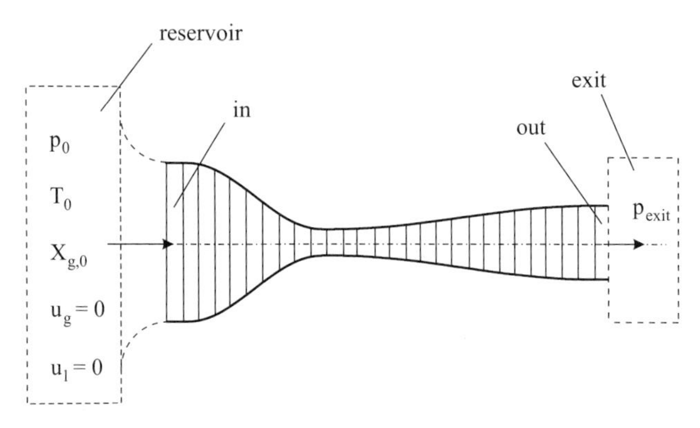

Fig. 9.42: ASTAR nozzle: computational grid and boundary conditions for quasi-one-dimensional flow

For the prediction of the boundary conditions a one-dimensional homogeneous equilibrium flow is assumed between the reservoir and the nozzle inlet using the actual mixture velocity u_{in} and mixture entropy s_{in} at the first cell at the nozzle inlet

$$h_{in} = h_0 - \frac{u_{in}^2}{2} \tag{9.14}$$

$$s_{in} = s_0. \tag{9.15}$$

With the assumption of thermal equilibrium the pressure at the nozzle inlet is updated at each time step from the state equation

$$p_{in} = (h_{in}, s_{in}). \tag{9.16}$$

For the outlet boundary condition a constant ambient (exit) pressure is applied; for all other parameters the spatial gradients at the nozzle exit are assumed to be zero. The initial conditions in the nozzle are identical with the upstream reservoir conditions which implies that the transient calculation starts with a strong discontinuity at the nozzle exit. These initial boundary conditions are also applied qualitatively for the other nozzle test cases as described in this section.

Single-phase liquid flow

For a numerical scheme based on characteristic information, the flow of pure liquid represents a significant challenge, due to the large differences between the flow and the sound velocities. In order to test the Flux Vector Splitting technique for such conditions the flow of pure water through the ASTAR nozzle was predicted for a reservoir pressure of $p_0 = 10$ bar and an exit pressure value of 9.99 bar $\geq p_{\text{exit}} \geq 9.80$ bar. For these conditions the maximum Mach number at the nozzle throat remains below 0.02. Predicted flow pressure values and flow velocities using 500 computational cells shown in Fig. 9.43. As indicated in the figure the CFD

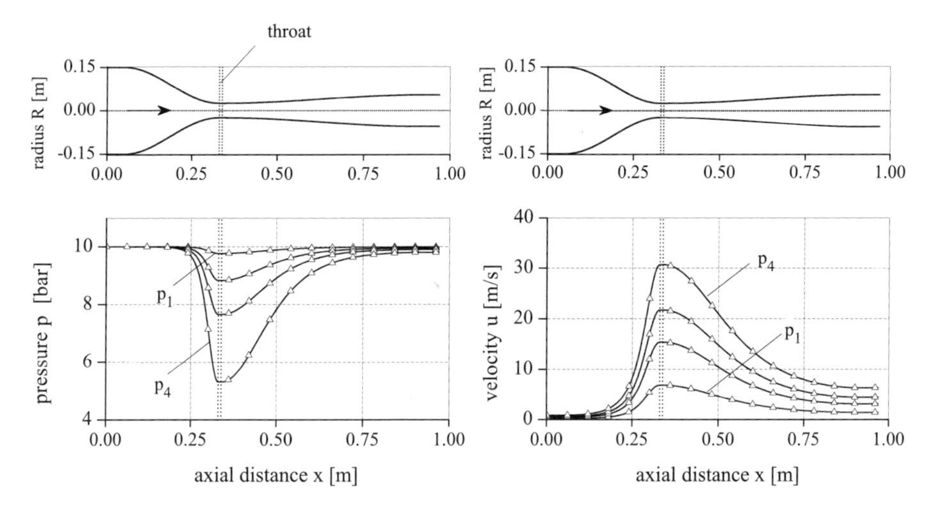

Fig. 9.43: ASTAR nozzle: stationary flow of *single-phase liquid* through a convergent–divergent nozzle, exit pressure: $p_{\text{exit}} = p_i$, with $p_1 = 9.99$ bar, $p_2 = 9.95$ bar, $p_3 = 9.9$ bar, $p_4 = 9.8$ bar; comparison of CFD calculation (straight line) with the results of the Bernoulli equation (triangle)

calculations are in nearly perfect agreement with the analytical solution given by the Bernoulli equation. The capability to handle such low Mach number flow is a necessary prerequisite for the numerical simulation of two-phase nozzle flow with subcooled liquid conditions, as will be described later.

Single-phase gas flow

For the second limiting case, the flow of pure gas, the specific numerical difficulties are related to the transition through the sonic point (saddle-point singularity for $M = 1$) at the nozzle throat and the occurrence of flow discontinuities (shock waves) in the divergent part of the nozzle depending on the back pressure at the nozzle exit. The ATFM predictions are performed as for the previous liquid case with a second-order Flux Vector Splitting scheme with 500 computational cells to provide a high degree of convergence.

Figure 9.44 shows the predicted parameter distributions along the nozzle axis for various pressure values at the nozzle exit. In all the cases a nearly perfect agreement is obtained between the CFD calculation and the analytical solution, including the correct position of shock waves, and a high resolution of the corresponding flow discontinuities.

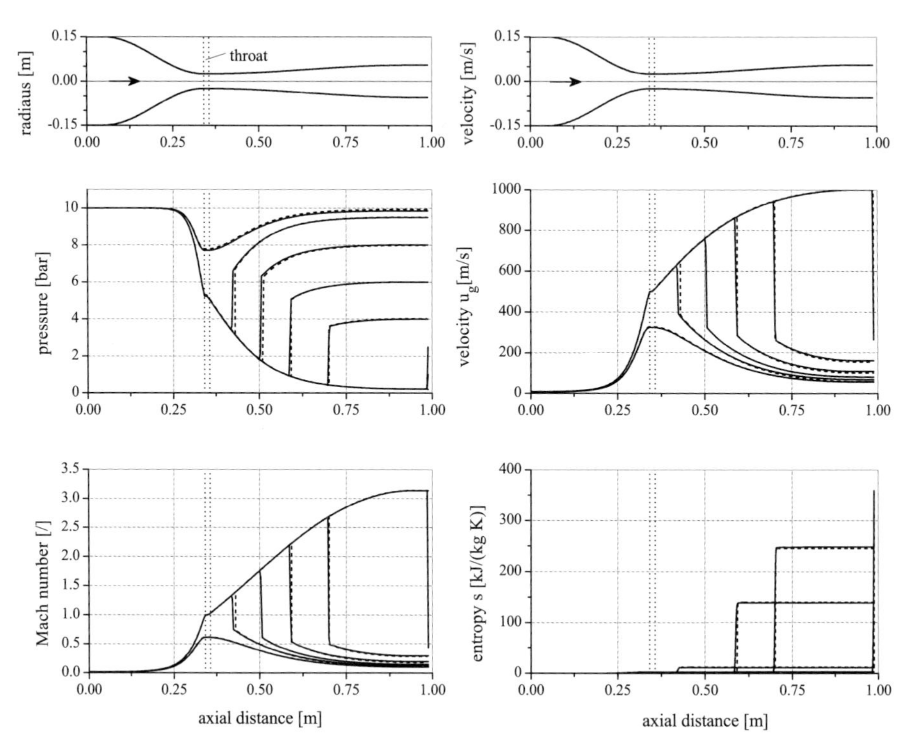

Fig. 9.44: ASTAR nozzle: flow of *single-phase gas* through a convergent–divergent nozzle, exit pressure $p_1 \geq p_{\text{exit}} \geq p_7$, $p_1 = 9.85$ bar, $p_7 = 0.2$ bar; comparison of results from CFD calculation (straight line) with analytical solution (dashed line)

Two-phase flow

The two-phase flow calculations are performed for two-component water/air mixtures with a gas mass fraction in the region $1.0 \geq X_0 \geq 0.10$. The predicted results for the steady state pressure and Mach number distributions for different nozzle back pressure values are shown in Fig. 9.45.

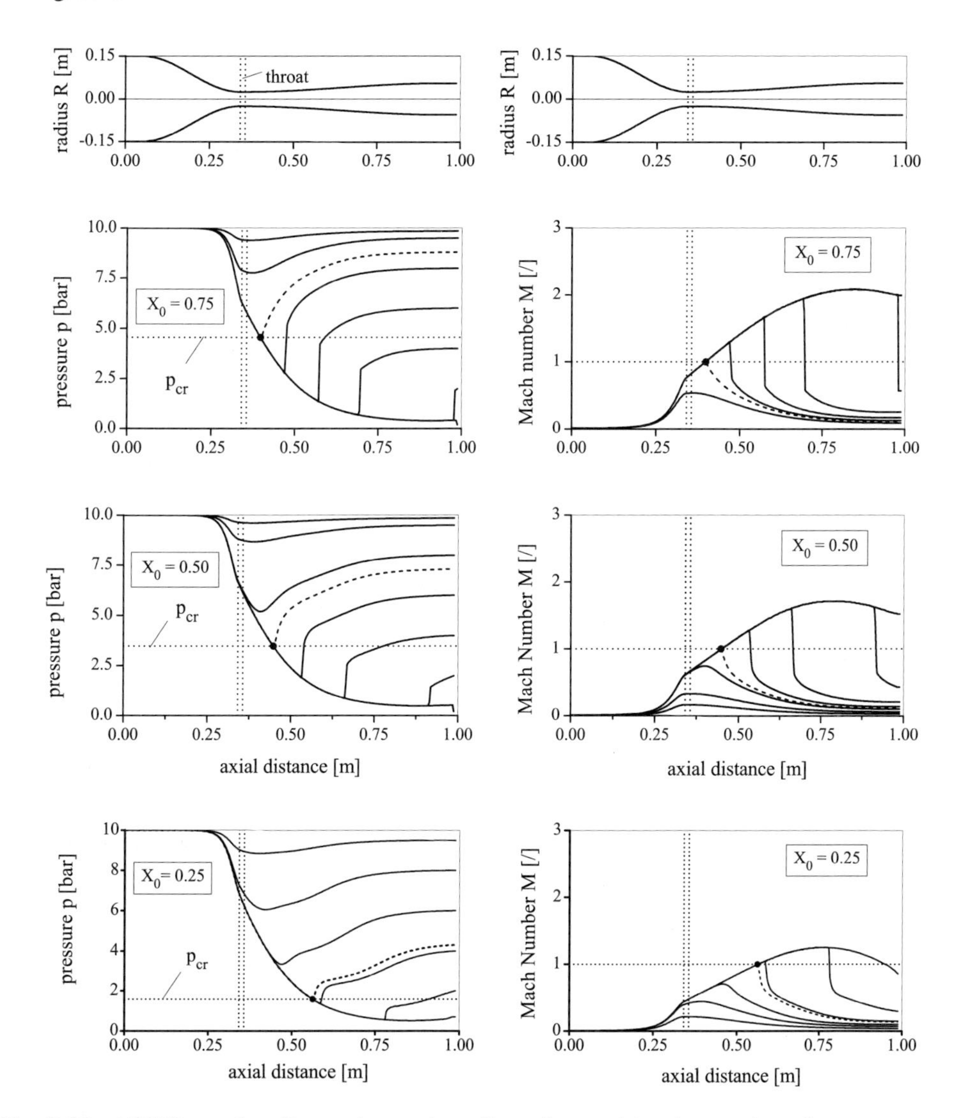

Fig. 9.45: ASTAR nozzle: dispersed *two-phase* flow of *water/air* mixtures through a convergent–divergent nozzle for different exit pressures 0.2 bar $\leq p_{\text{exit}} \leq$ 0.9 bar, effect of gas content X_0 on the location of the sonic point

As for the single-phase gas flow, the transition through the sonic point ($M = 1$) is characterized by a saddle-point singularity with a bifurcation of solution into a subsonic and supersonic branch separating the subsonic ($M < 1$) and supersonic ($M > 1$) regions in the divergent section, respectively. However, different to the pure gasdynamic case, the location of the singularity has moved downstream of the throat into the divergent section of the nozzle accompanied by a reduction of the critical pressure. The exact position of the sonic point depends strongly on the actual interfacial coupling between the phases as resulting from the algebraic source terms is analyzed in detail in Section 5.1.7.

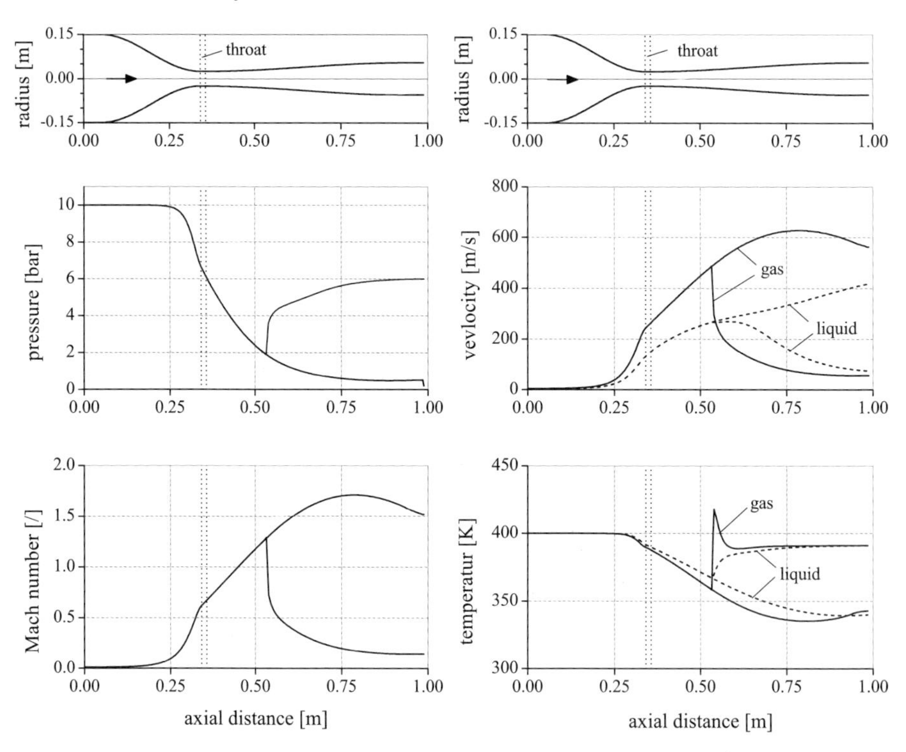

Fig. 9.46: ASTAR nozzle: dispersed *two-phase of water/air*, gas content $X_0 = 0.50$, exit pressure $p_{\mathrm{exit}} = 6.0$ bar and 0.2 bar

Depending on the back pressure values, shock waves occur in the supersonic region downstream of the sonic point. As already explained in Section 8.3 the shock wave structure is characterized by steep but finite parameter gradients. This "smoothing" effect results from the nonconservative terms in the phasic momentum equations which becomes more pronounced for increased liquid volume fraction. If the sonic point has moved to the nozzle exit, which in the present case is reached for $X_0 = 0.1$, the flow in the divergent section is free of any shock wave and the flow remains continuous up to the prescribed exit pressure.

For the pure gasdynamics case, the occurrence of critical flow conditions ($M = 1$) implies a maximum flow rate through the nozzle (choking condition) independent of a further reduction of the pressure at the nozzle exit. This is principally the same for two-phase nozzle flows as shown in Fig. 9.47. For the progressive reduction of the nozzle exit pressure, two major trends are visible: (1) a continuous reduction of the exit pressure becomes necessary to achieve critical flow conditions ($M = 1$) in the nozzle and (2) an increased retardation of the flow to asymptotically reach the maximum flow rate. This means, in particular, that a type of pre-choking might occur where the maximum flow rate is approached much earlier before the critical conditions are reached within the nozzle.

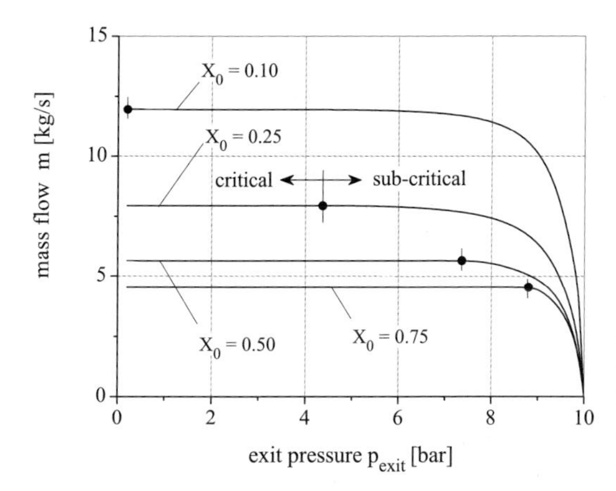

Fig. 9.47: Flow of *water/air mixtures* through a convergent–divergent nozzle, mass flow rates as function of back pressure at nozzle exit, gas content $X_0 = 0.75, 0.50, 0.25, 0.10$

The specific peculiarities of the two-phase nozzle flow described above are largely determined by the mechanical and thermal disequilibrium as illustrated in Fig. 9.46 for a gas mass fraction of $X_0 = 0.50$ and exit pressure values of $p_{\mathrm{exit}} = 0.2$ bar and 6.0 bar.

The strong pressure gradient and related acceleration of the flow results in an increase of the "slip velocity" which is governed by the forces acting on the dispersed droplet field. A maximum value for the velocity difference of approximately 100 m/s occurred at the middle of the divergent section. Due to a more moderate pressure decrease at the end of the nozzle, the flow shows a trend toward a mechanical equilibrium which, however, is not achieved at the nozzle exit. As a result of the intense heat transfer from the dispersed liquid to the gas phase, the thermal disequilibrium is less pronounced and toward the nozzle exit a new thermal equilibrium is practically achieved.

For an exit pressure of $p_{\mathrm{exit}} = 6$ bar, a moderate shock wave is formed in the divergent section of the nozzle with an abrupt decrease of gas velocity and increase of gas temperature. Resulting from the continuous heat transfer to the gas phase, the gas temperature at the shock wave is much higher than the value predicted by an isentropic gas flow, and therefore, it is not surprising that the shock wave exhibits a peak gas temperature above the temperature at the reservoir. Due to the thermal and mechanical inertia, the liquid velocity and liquid temperature are practically "frozen" across the shock wave which leads to a reverse of thermal and mechanical disequilibrium immediately downstream of the shock. The region behind the shock is characterized by a relaxation region superimposed with the effect resulting from the change of the nozzle cross section in the flow direction. Within the relaxation zone, the flow is approaching a new thermal and mechanical equilibrium which is achieved at the nozzle exit.

As already mentioned the deviations from thermal and mechanical equilibrium between the phases are largely determined by the modeling details for the algebraic source terms describing interfacial transfer processes for momentum and energy. The validity of the coupling terms will be indirectly tested for the following nozzle test case by comparing with experimental data.

Two-dimensional nozzle flow

For the relatively large L/D ratio of the ASTAR nozzle geometry essentially one-dimensional flow conditions are expected for the single-phase flow of gases. However, this may no longer be the case for dispersed droplet flows due to the large density ratio between the liquid and gas phases as will be shown in the following. In order to investigate multidimensional effects, two-dimensional calculations have been performed for a planar nozzle with a body-fitted computational grid as schematically shown in Fig. 9.48.

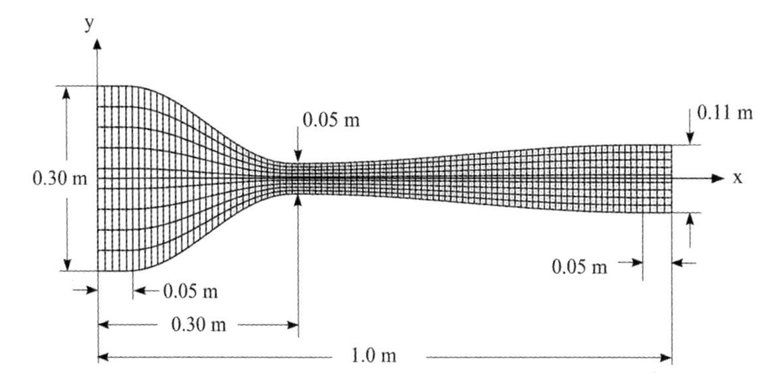

Fig. 9.48: ASTAR nozzle: computational grid for two-dimensional calculations (schematically)

Actually a computational grid of $20 \times 200 = 4000$ cells are used for the second-order Flux vector Splitting scheme. At the nozzle inlet a gas content of $X_0 = 0.5$ is assumed which corresponds with a void fraction of $\alpha_{g,0} = 0.98$. The calculations were performed for two different exit pressure values of $p_{\text{exit}} = 1.0$ bar and $p_{\text{exit}} = 6$ bar.

Calculated values for the liquid volume fraction α_l and corresponding vector fields of liquid mass flow densities $\alpha_l \varrho_l \vec{u}_l$ are shown in Fig. 9.49. The figure indicates a large enrichment of liquid near the wall region in the convergent section of the nozzle which, due to the large curvature near the nozzle throat, detaches from the wall and penetrates toward the nozzle axis. This creates a layer structure for the liquid fraction in the divergent section which remains evident up to the nozzle exit as indicated in Figs. 9.50 and 9.51.

Figure 9.50 provides spectral plots for the pressure distributions as were obtained for two different pressure values at the nozzle exit of $p_{\text{exit}} = 0.2$ bar and $p_{\text{exit}} = 6$ bar resulting either in a continuous depressurization to the ambient pressure ($p_{\text{exit}} = 1$ bar) or in the formation of a shock wave in the divergent section followed by a continuous pressure increase up to the exit pressure value of $p_{\text{exit}} = 6$ bar. The cross-sectional distributions of liquid volume fraction α_l as shown in the lower part of Fig. 9.50 clearly reflect the redistribution of liquid concentration

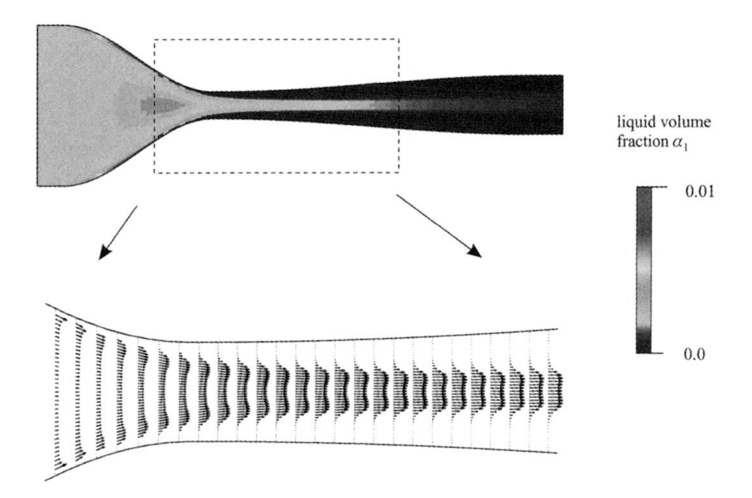

Fig. 9.49: ASTAR nozzle: liquid volume fraction distribution and vector field for liquid mass flow density

from a liquid enriched wall region in the convergent part to a gas enriched wall region in the divergent section of the nozzle.

The nonhomogeneous phase distribution over the nozzle cross-sectional area also affects the Mach number distribution as indicated in Fig. 9.51. This results from the large sensitivity of the sound velocity with regard to changes in the volume concentration for high void fraction (as in the present case) or low void fraction.

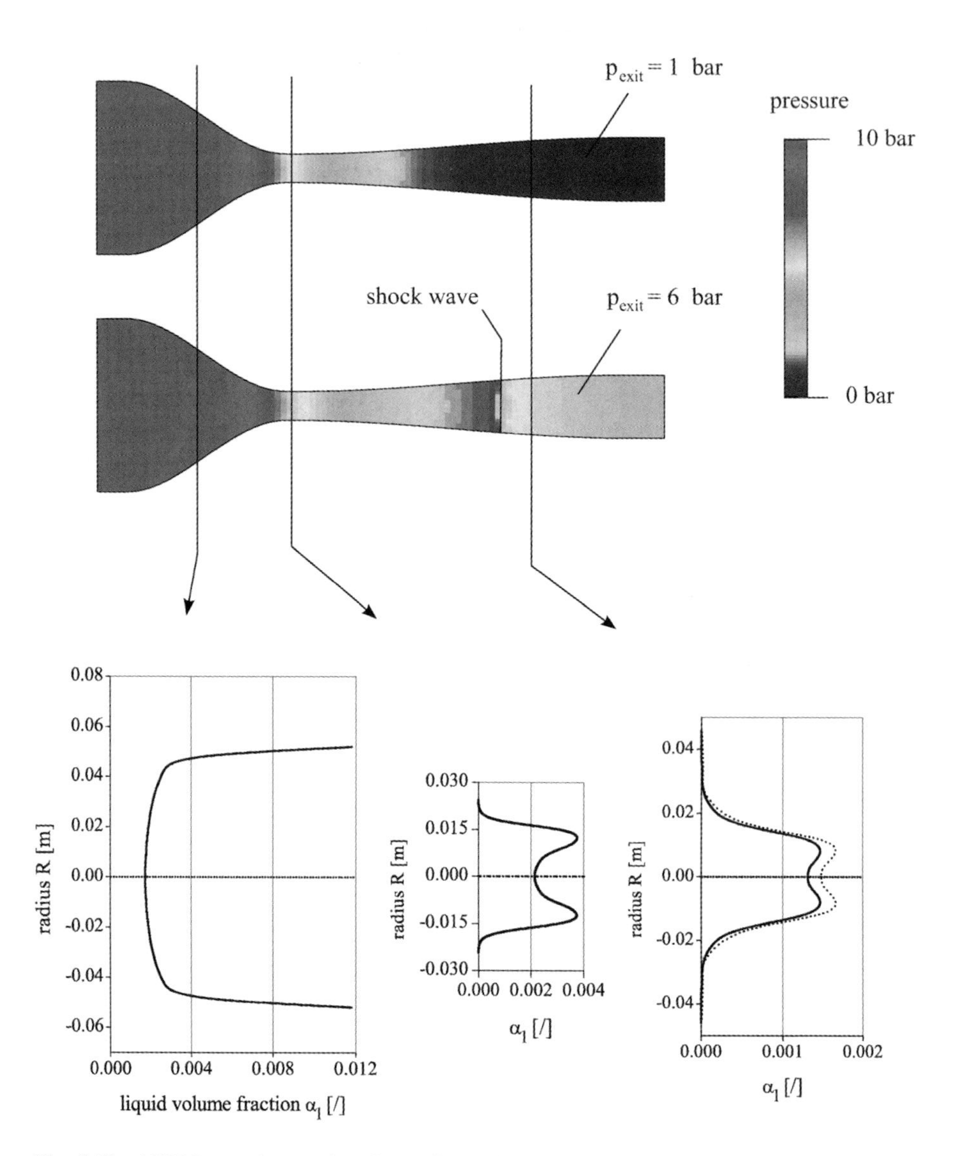

Fig. 9.50: ASTAR nozzle: results of two-dimensional calculation for $p_0 = 10$ bar and $X_0 = 0.5$; pressure distribution (top) and liquid volume fraction profiles at the convergent section, nozzle throat, and divergent section (bottom); second-order FVS scheme with $20 \times 200 = 4000$ cells

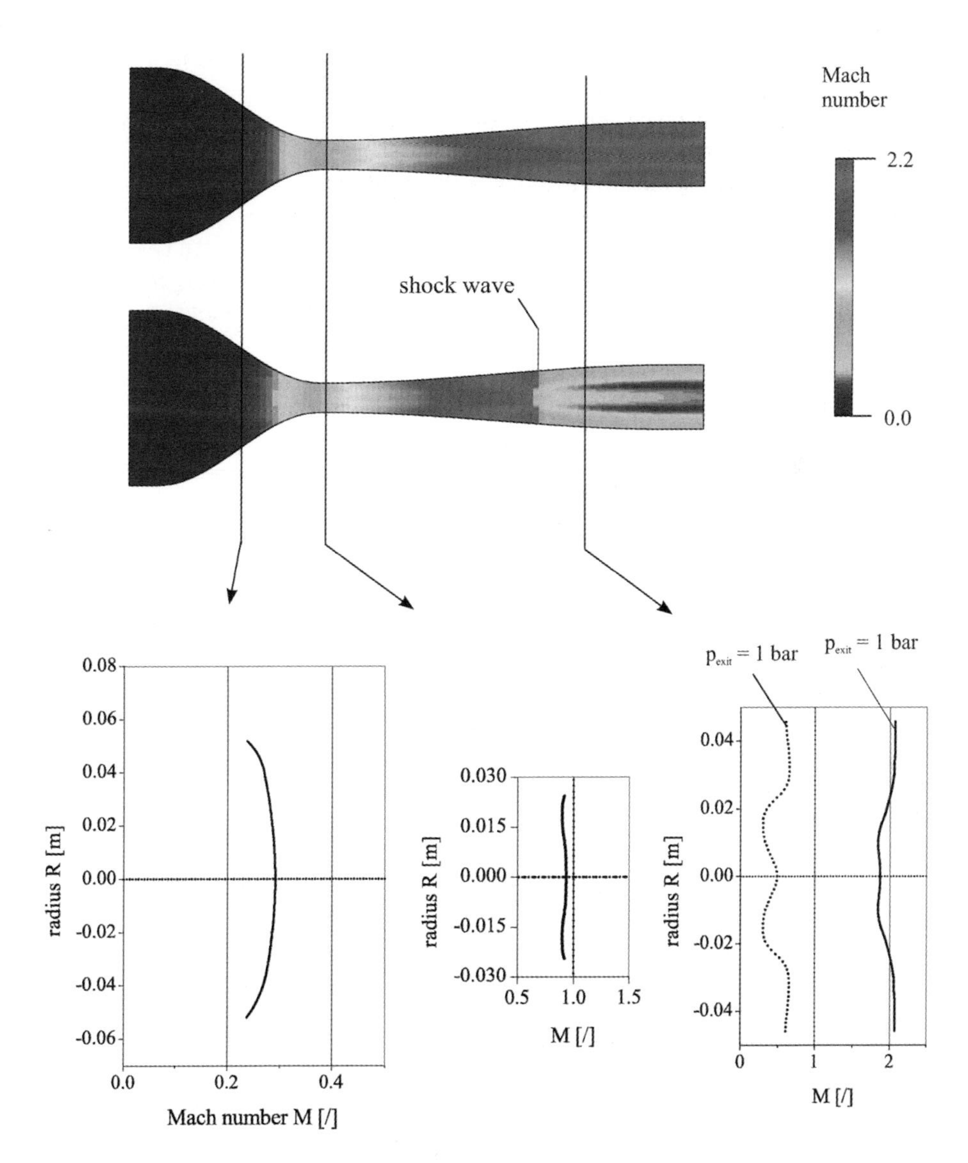

Fig. 9.51: ASTAR nozzle: results of two-dimensional calculation for $p_0 = 10$ bar and $X_0 = 0.5$; Mach number distribution (top) and Mach number profiles at the convergent section, nozzle throat, and divergent section (bottom); second-order FVS scheme with $20 \times 200 = 4000$ cells

9.6.2 Deich nozzle tests

The experimental program for the investigation of single component water/steam mixtures through a convergent–divergent (naval) nozzle as reported by Deich *et al.* [3] covers a wide spectrum of gas contents and exit pressure values. The experiments were performed with wet steam for a region of liquid contents Y_0 (wetness) ranging from $Y_0 = 0.0$ (slightly superheated vapor) up to a maximum liquid content of $Y_0 = 0.83$. The geometry of the nozzle as shown in Fig. 9.52 consists of a circular inlet section with a radius of 28 mm, followed by a cone with a constant angle of aperture $\Phi_i = 3.3°$ and a length of 122 mm. The nozzle inlet pressure was fixed at 1.2 bar, the outlet pressure was progressively reduced down to 0.05 bar. The measured data including static pressure values along the nozzle axis and mass flow rates of liquid water and vapor should be preferably considered as a qualitative measure for the flow behavior.

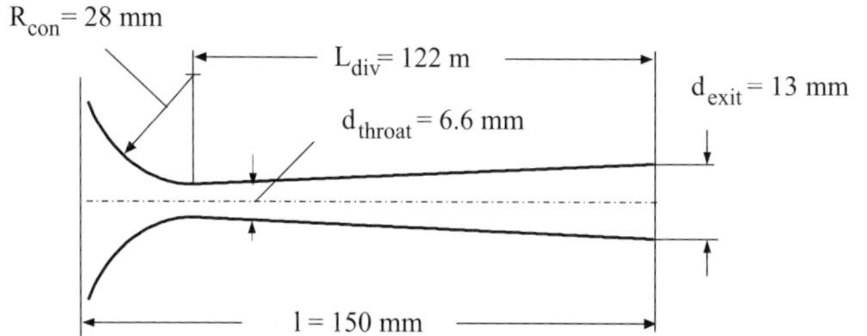

Fig. 9.52: Deich nozzle geometry

As an example a comparison of measured and predicted pressure values along the nozzle and corresponding Mach numbers are shown in Fig. 9.53 for an inlet liquid content (wetness) of $Y_0 = 0.83$ and progressively reduced pressure values at the nozzle exit.

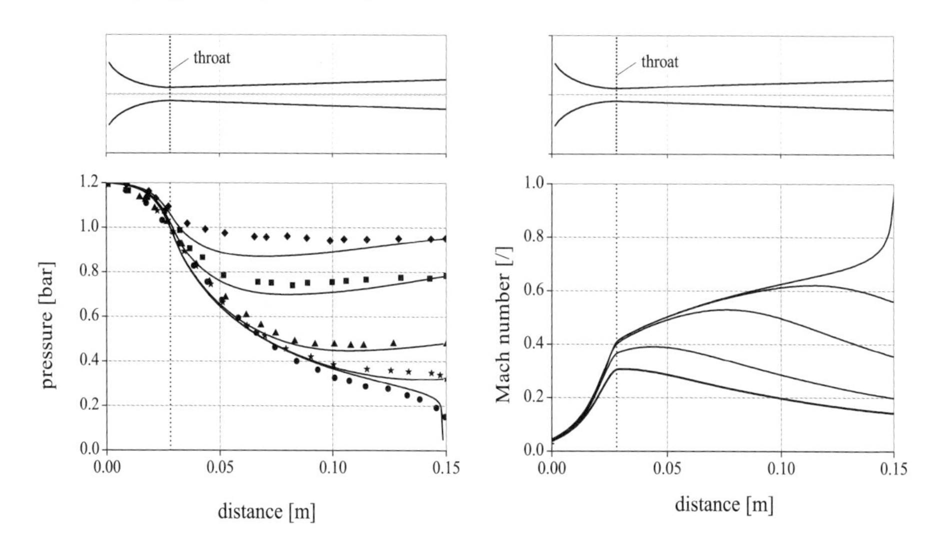

Fig. 9.53: Deich nozzle: parameter distribution along the nozzle axis for different exit pressure values 0.1 bar $\leq p_{\mathbf{exit}} \leq$ 0.95 bar; liquid mass fraction (wetness) at the reservoir $Y_0 = 0.83$

For all exit pressure values the pressure and the Mach number distributions change continuously and are free of any discontinuity. The steep pressure and Mach number gradient at the nozzle exit for $p_{\text{exit}} = 0.1$ bar as shown by the predictions and the measured pressure values suggest that critical conditions have been reached for an exit pressure of $p_{\text{exit}} = 0.1$ bar. Nevertheless, as also shown in Fig. 9.53, the conditions at the nozzle throat no longer change for $p_{\text{exit}} < 6.0$ bar, which indicates that a constant (maximum) flow rate through the nozzle already occurred before critical conditions were achieved in the nozzle exit area. This is confirmed by Fig. 9.54, where the calculated mass flow rate is given as a function of the exit pressure.

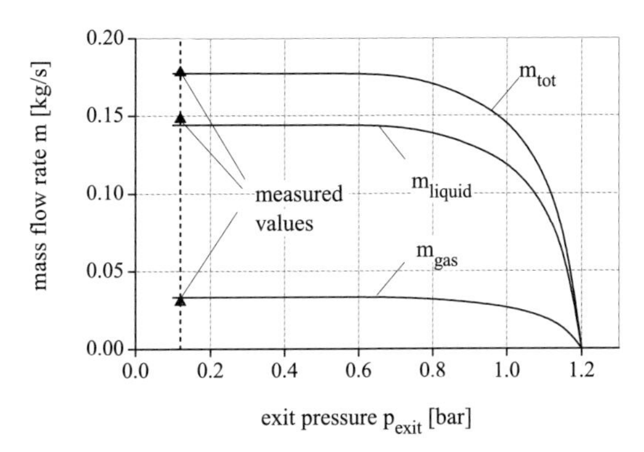

Fig. 9.54: Deich nozzle: comparison of measured and predicted mass flow rates as a funtion of the back pressure at the nozzle exit

In order to investigate the sensitivity of the nozzle flow with respect to the nozzle geometry various calculations have been performed for a wetness of $Y_0 = 0.43$ where the angle of aperture for the divergent section is varied in the region of $0° \leq \Phi_i \leq 6°$. The exit pressure for all calculations was $p_{\text{exit}} = 0.1$ bar to guarantee that critical conditions are achieved in all cases.

The predicted pressure and Mach numbers as shown in Fig. 9.55 indicate that for all cases a critical state has been achieved in the divergent section of the nozzle. Any reduction of the angle Φ_i results in downstream movement of the critical cross-section from a position close to the throat for $\Phi_i = 6°$ toward the nozzle exit which is reached for values $\Phi_i \leq 1°$. For the experimental nozzle configuration of $\Phi_i = 3.3°$ the predicted pressure distribution is in good agreement with the corresponding measured data which is also the case for the critical mass flow as given in Fig. 9.56.

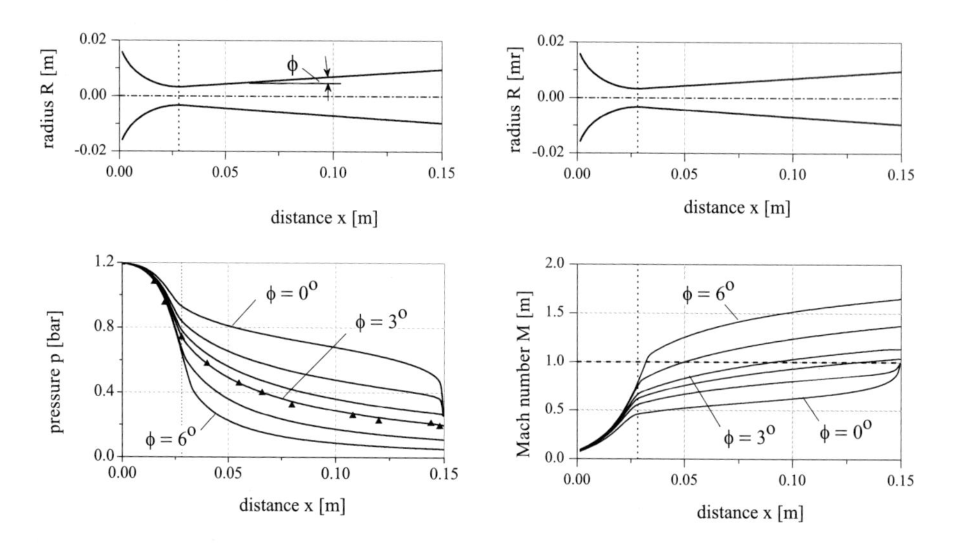

Fig. 9.55: Deich nozzle: critical flow of water/steam mixture; liquid mass fraction at the reservoir $Y_0 = 0.57$, effect of angle of aperture in divergent section, triangles represent measured data for $\Phi_i = 3.3°$

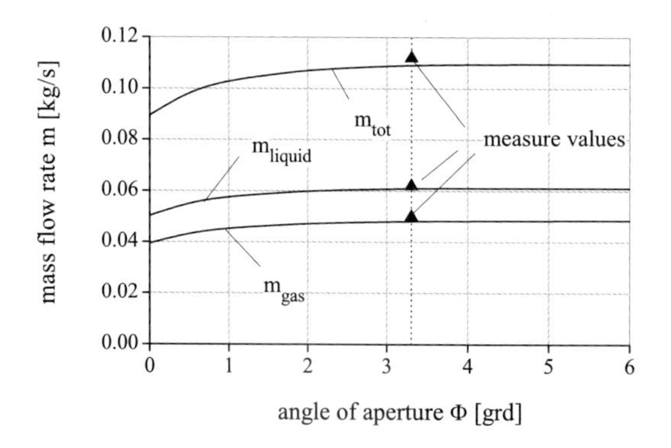

Fig. 9.56: Deich nozzle: mass flow rate as a function of the angle of aperture Φ in the divergent section, liquid mass fraction at reservoir $Y_0 = 0.83$

9.6.3 Moby–Dick nozzle tests

The Moby–Dick nozzle test program [4] was performed at the Centre d'Etude Nucleaire (CEA) de Grenoble as part of the qualification of the French Nuclear Thermal Hydraulic code CATHARE [5]. The tests were designed to study two-phase critical flow conditions which are of particular importance for the analysis of hypothetical Loss of Coolant Accidents (LOCA) in Pressurized Light Water reactors. Such accidents might be initiated by a structural failure of the high pressure primary system of a PWR resulting in a fast depressurization of the coolant system and in a degradation of the heat removal from the reactor core.

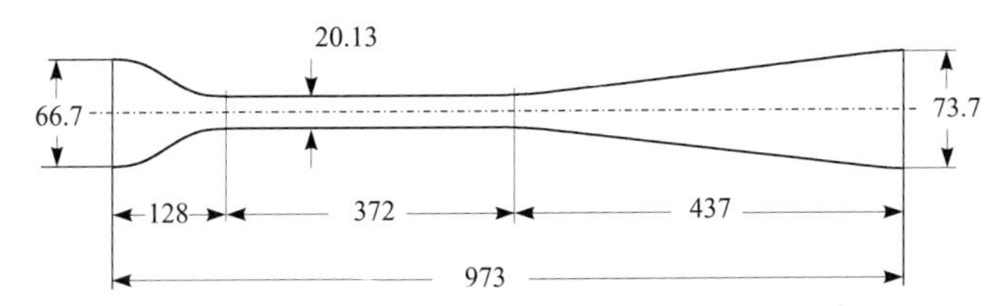

Fig. 9.57: Moby–Dick nozzle geometry, all dimensions in mm

The Moby–Dick nozzle as shown in Fig. 9.57 has a total length of about 1 m and consists of a smooth convergent section, a relatively long cylindrical throat and a conical divergent section with an angle of aperture of 7°. The nozzle inlet conditions range from subcooled liquid with different degrees of subcooling to saturated conditions with different vapor mass fractions. From the large experimental program, two tests are selected with subcooled conditions at the nozzle entrance. The upstream reservoir pressure is in both cases $p_0 = 20$ bar with a degree of subcooling of $\Delta T_{\text{sub}} = 2$ K and 25 K, respectively. In the experiments the pressure downstream of the nozzle was continuously reduced up to a point (or even below) where a maximum flow rate was obtained through the nozzle. In the ATFM calculation the measured pressure at the nozzle exit is used as a boundary condition. Predicted parameter distributions for steady state conditions are given in Figs. 9.58 and 9.59 including experimental data for pressure and void fraction (not for all tests available) as well as analytical results based on the homogeneous equilibrium assumptions.

For the low degree of subcooling (Fig. 9.58) the flow in the convergent section remains pure liquid. The evaporation starts immediately at the entrance to the cylindrical throat section followed by a moderate acceleration of the fluid and a related drop of pressure. At the inlet to the divergent section the further expansion of the fluid becomes more pronounced leading to an increased acceleration of the fluid with maximum flow velocities for gas and water of $u_g = 240$ m/s and $u_l = 150$ m/s. The flow remains always subsonic and, therefore, is free of any discontinuous (or near discontinuous) parameter change. The good agreement with measured pressure and void fraction data suggests that the prediction gives a fair picture of the flow behavior. The analytical solution assuming homogeneous equilibrium flow largely differs from the experimental data and shows the presence of an unrealistic shock wave in the divergent section of the nozzle.

For the high degree of subcooling (Fig. 9.59) the behavior of the nozzle flow becomes slightly different. The flow remains pure liquid up to the near end of the cylindrical section where a strong evaporation (flashing) starts, possibly triggered by the frictional pressure drop.

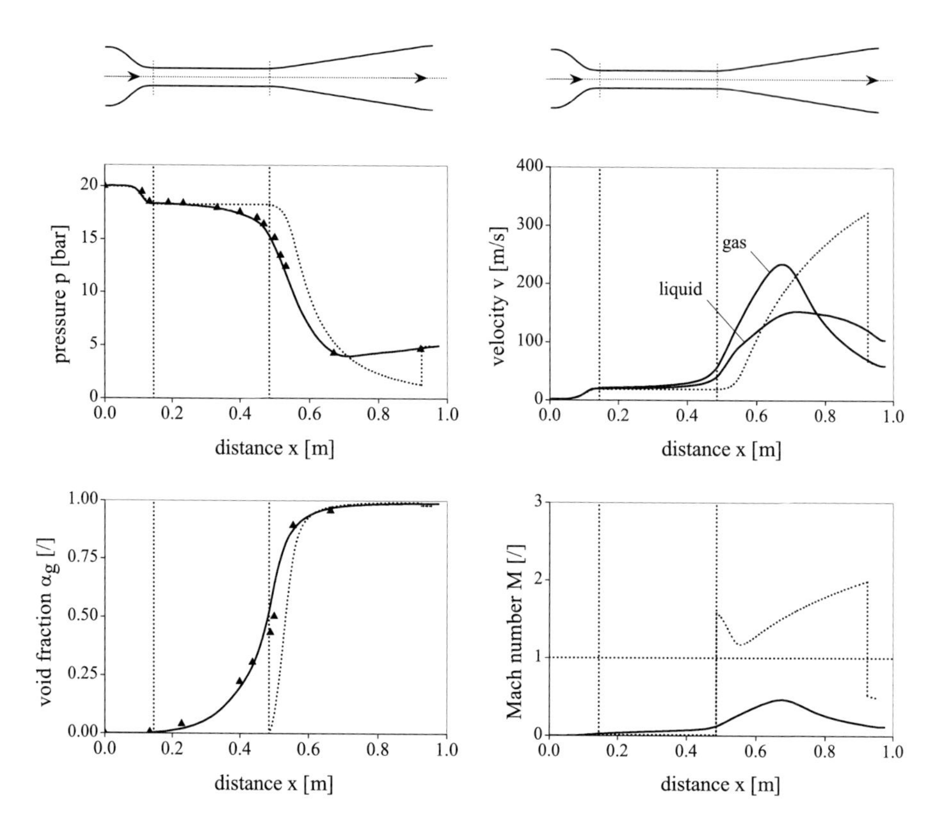

Fig. 9.58: Moby–Dick nozzle: $p_0 = 20$ bar, $\Delta T_{sub} = 2$ K; comparison of CFD calculation (solid line) and algebraic solution for homogeneous equilibrium conditions (dotted line) with experimental data (triangles)

Nevertheless, the void fraction is somewhat lower as for the previous case whith near saturated inlet conditions and consequently the flow shows a more moderate acceleration with the maximum phase velocities $u_g = 90$ m/s and $u_l = 70$ m/s. Also in this case the flow in nozzle remains subsonic and, therefore, is free of shock waves.

The fact that in both of the Moby–Dick test cases shown above the flow remained everywhere subsonic and therefore, the question arises whether the condition for "choking" has been obtained where the flow through the nozzle becomes independent of the exit pressure. This is demonstrated in Fig. 9.60 for the high degree of subcooling ($\Delta T = 25$ K) showing the distributions of governing flow parameters for progressively reduced pressures at the nozzle exit p_{exit}. The figure indicates that for $p_{exit} = 16$ bar any further reduction has no effect on the flow parameter in the convergent part and in cylindrical throat section of the nozzle, indicating that the mass flow through the nozzle has practically reached a maximum value. Any further reduction of back pressure pushed the region of influence toward the nozzle exit and at $p_{exit} = 1$ bar, the flow in the whole nozzle is unaffected by any further change of the back pressure.

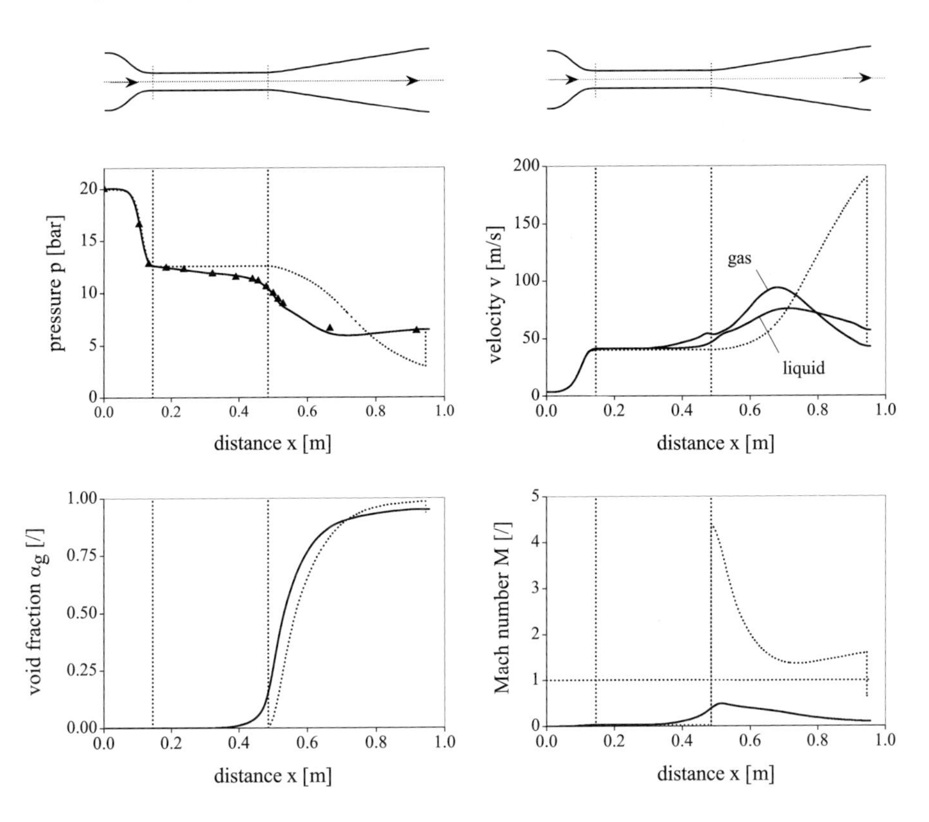

Fig. 9.59: Moby–Dick nozzle: $p_0 = 20$ bar, $\Delta T_{\text{sub}} = 25$ K; comparison of CFD calculation (solid line) with algebraic calculation for homogeneous equilibrium conditions (dotted line) and experimental data (triangles)

The effect of "pre-choking" is also seen on the total mass flow rate through the nozzle as a function of the exit pressure as shown in Fig. 9.61. Already at a rather moderate pressure reduction at the nozzle exit a practically constant mass flow is achieved for values much higher than those used in the experiment. The figure also indicates the correct prediction of the measured mass flow for both cases of subcooling at the nozzle entrance.

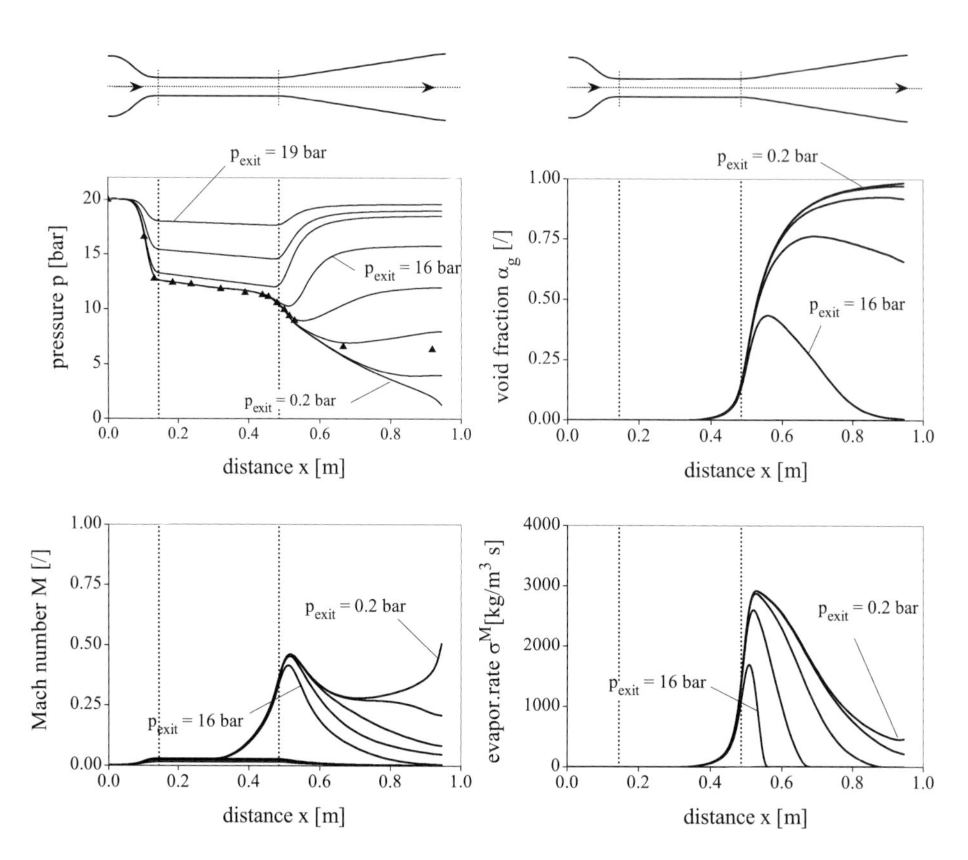

Fig. 9.60: Moby–Dick nozzle: $p_0 = 20$ bar, $T_0 = 461$ K, effect of back pressure on the nozzle flow, 19 bar $\geq p_{\rm exit} \geq$ 0.2 bar

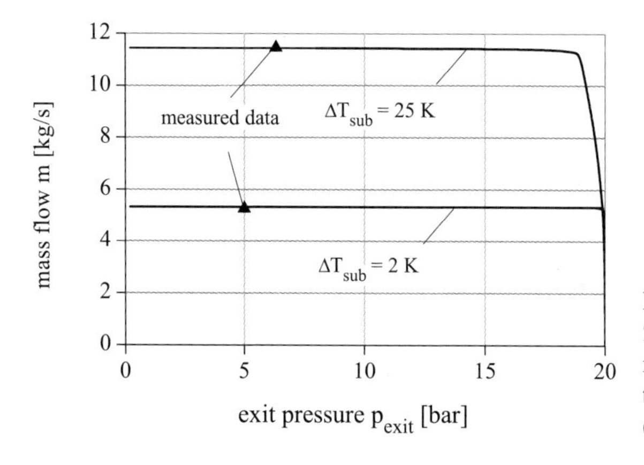

Fig. 9.61: Moby–Dick nozzle: p_0 = 20 bar; effect of back pressure on mass flow rate; comparison of prediction (straight line) with measured data (triangles)

9.7 Blowdown phenomena

The fast depressurization of pressure vessels or piping systems containing subcooled or saturated liquids is of large interest for the safety analysis of industrial installations. Such "blowdown" processes might originate by a structural failure or by an operational opening of safety valves to prevent damage to the plant. In particular for the safety analysis of Light Water Reactor (LWR) safety analysis, blowdown phenomena have been extensively studied. This included the event of the rupture of a main coolant pipe in the primary system of a pressurized LWR often postulated as a most severe credible accident for the design of emergency cooling systems and accident management procedures. Since for obvious reasons, full scale experiments are not feasible, complex thermal-hydraulic computer codes have been developed to describe such phenomena and their consequences for a safe shutdown of the plant. In order to assess these codes a number of standard test cases were defined by the Committee for the Safety of Nuclear Installation (CSNI) [6] which cover a wide spectrum of physical phenomena involved at different geometrical scales.

9.7.1 Edwards' pipe blowdown

A standard test case for thermal-hydraulic codes has been the blowdown of an initially hot pressurized liquid from a pipe of approximately 4 m length, known as the CSNI standard problem No. 1, also known as Edwards' pipe blowdown [7] (Fig. 9.62). The water in the pipe has an initial pressure of 7.0 MPa and a temperature of 502 K which corresponds to an initial subcooling of 56.8 K. The geometrical configuration is given in Fig. 9.63. The transient is initiated by the rupture of a bursting disk allowing the rapid discharge to the environment at atmospheric pressure.

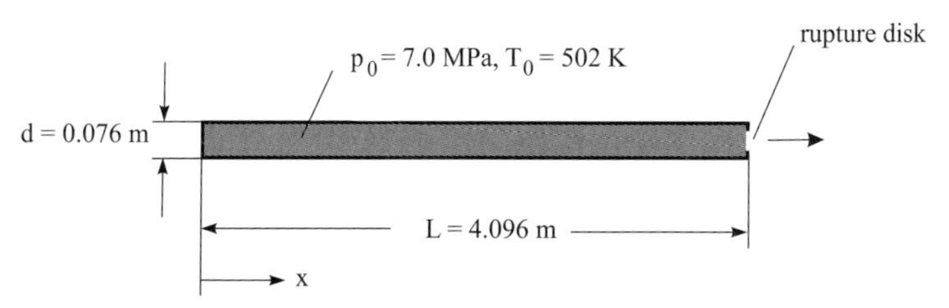

Fig. 9.62: Edwards' pipe blowdown: pipe geometry and initial conditions

Most of existing calculations for this test use a constant (atmospheric) pressure as a boundary condition immediately downstream of the pipe. This seems to be doubtful, especially in the cases where the flow in the pipe remains subsonic and the high pressure difference between the pipe exit area results in continuation of the evaporation process downstream of the exit. In the calculation presented here, the specification of boundary conditions at the very sensitive area at the pipe exit is avoided by enlargement of the numerical simulation to include the expansion of the two-phase mixture and jet formation downstream of the pipe. This is done by the modeling as an axisymmetric, quasi-two-dimensional flow process near the pipe exit with a constant (atmospheric) far-field pressure boundary. The computational scheme used in the following calculations is shown in Fig. 9.63.

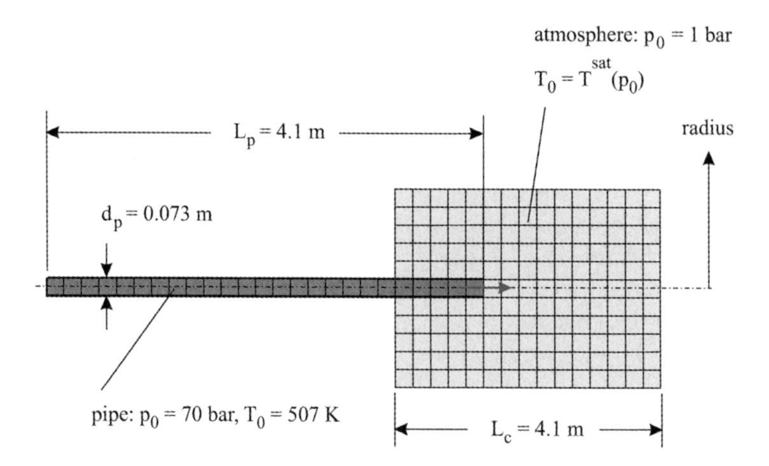

Fig. 9.63: Edwards' pipe blowdown: computational grid (schematically) and initial conditions

With regard to the governing phenomena of the blowdown process two different time periods can be distinguished:

(1) A *short time period* mainly characterized by wave propagation and reflection phenomena shown in Fig. 9.64. Immediately after the removal of the rupture disk, a sudden pressure drop occurs at the pipe exit resulting in the onset of a violent evaporation of liquid which limits the pressure decrease to a value slightly below the saturation pressure according to the initial liquid temperature. This pressure value is nearly maintained during the first 10 ms period of the transient characterized by the propagation of a rarefaction wave propagating with the speed of sound of the liquid phase into the pipe. The reflection of this wave at the left closed end of the pipe at about 3 ms results in pressure undershoot limited by the accompanied evaporation process. This forms a moderate pressure wave which travels back toward the pipe exit which is reached by about 6 ms. After the pressure wave has returned to the exit, the transient continues with the bulk evaporation over the full pipe length leading to a more moderate depressurization.

(2) A *long time period* showing a more steady transient governed by a continuous bulk evaporation as shown in Fig. 9.65. As shown in Figs. 9.64 and 9.65, the flow in the pipe remains subsonic over the whole transient, however, supersonic ($M > 1$) conditions occur temporarily in a region slightly downstream of the pipe. The governing process controlling the discharge from the pipe is the short region with extremely large evaporation rates close to the exit as created by the steep pressure gradient in this region.

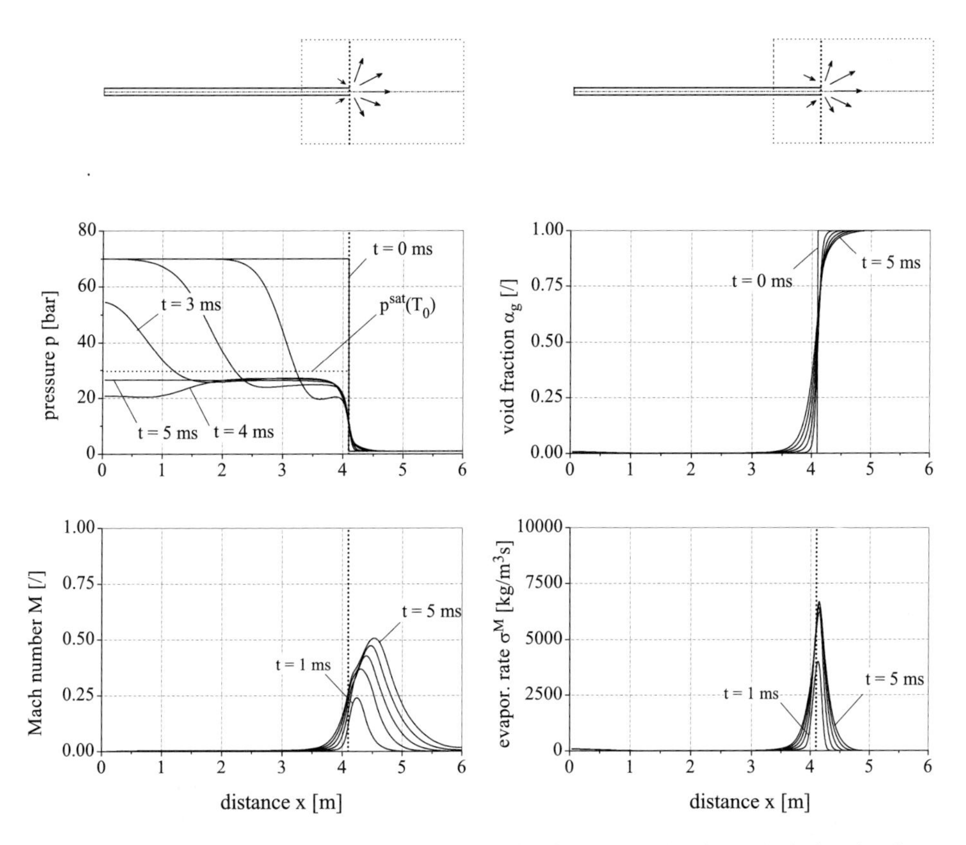

Fig. 9.64: *Edwards' pipe* blowdown: parameter distributions along the pipe axis during the *short time* period at time values 0 ms $\leq t \leq$ 5 ms

The long time blowdown behavior is governed by the discharge from the pipe, the continuous evaporation of liquid and to a lesser extent by the frictional forces at the pipe walls. With a further decline in pressure the flow velocities start to decrease up to the end of the blowdown at $t = 0.5$ s when atmospheric pressure is reached in the pipe.

A comparison of the calculated values for the pressure at the pipe head and for the void fraction at the pipe middle section is given in Fig. 9.66. The rather good agreement with the corresponding measured data suggests that the governing phenomena are correctly described.

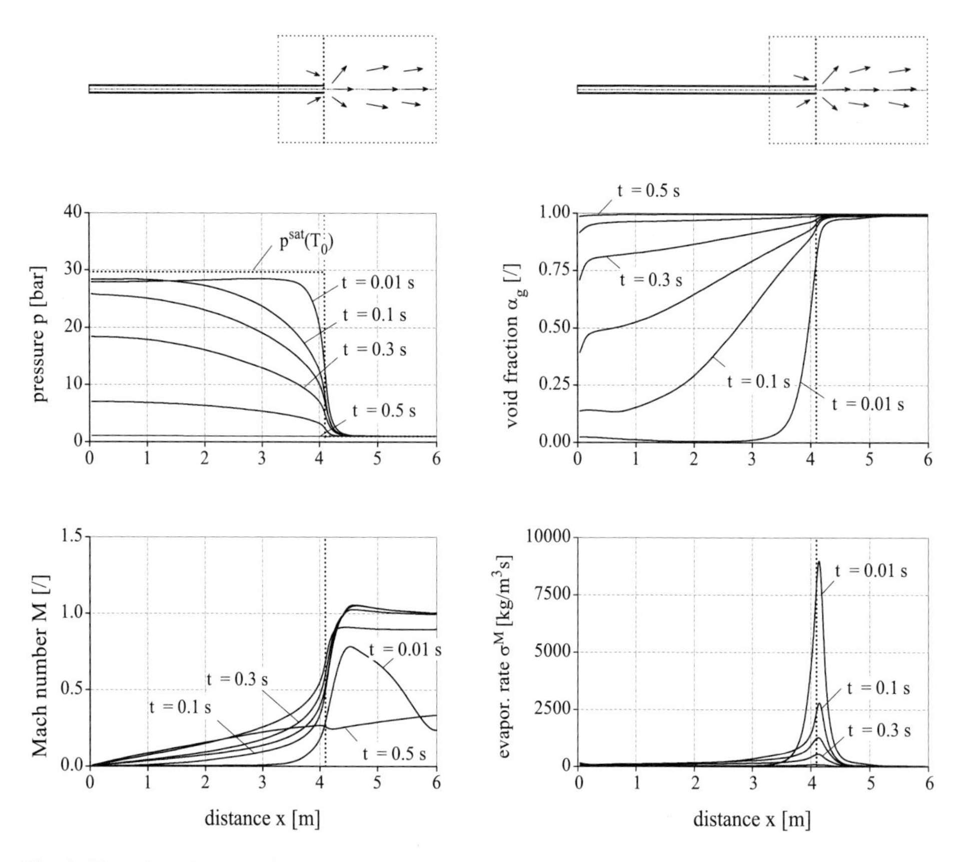

Fig. 9.65: *Edwards' pipe* blowdown: parameter distributions during the *long time* period at time values $0.01\ \text{s} \leq t \leq 0.5\ \text{s}$

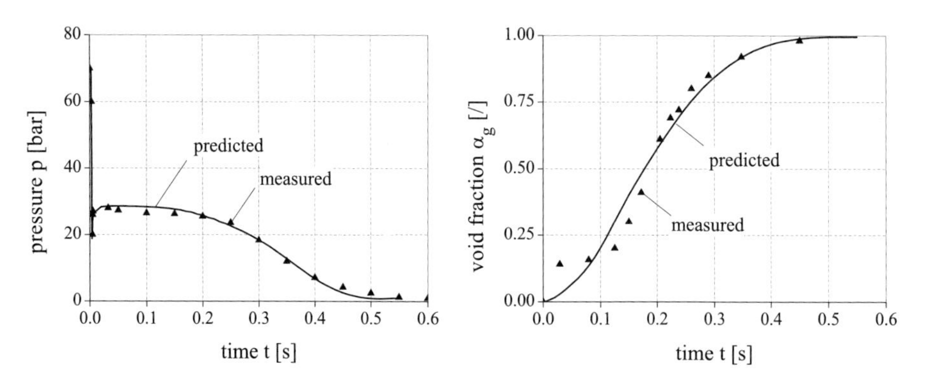

Fig. 9.66: Edwards' pipe blowdown: comparison of predicted and measured values; pressure at pipe head (left) and void fraction at pipe middle section (right)

9.7.2 Canon experiment

The Super-Canon test program was performed at the Centre d'Etude Nucleaire (CEA) de Grenoble with the aim to enlarge the experimental database for the assessment of thermal-hydraulic computer codes developed for the safety analysis of Light Water Reactors. The experiments were performed in a similar way as already described for Edwards' pipe, using the horizontal pipe of 4.39 m length and an internal diameter of 0.1 m. Compared with Edwards' pipe a more detailed instrumentation was used providing information on pressure and temperatures at different positions of the pipe and to a lesser extent on the void fraction. The geometry of the pipe and the measurement positions are given in Fig. 9.67.

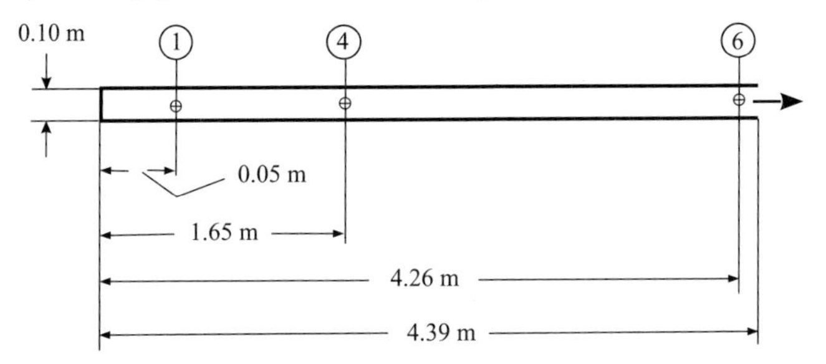

Fig. 9.67: Canon pipe geometry and location of measurement points

From the Super-Canon test program [8], an experiment has been selected with an initial pressure of $p_0 = 15.0$ MPa and a temperature of $T_0 = 507$ K (equivalent to a subcooling of 42 K). Due to the extremely high initial pressure, critical flow conditions are expected to exist at the pipe exit during most of the blowdown period. Therefore, it seems to be justified to assume atmospheric pressure at the exit of the pipe. The computational grid and initial conditions as used in the calculations are schematically shown in Fig. 9.68.

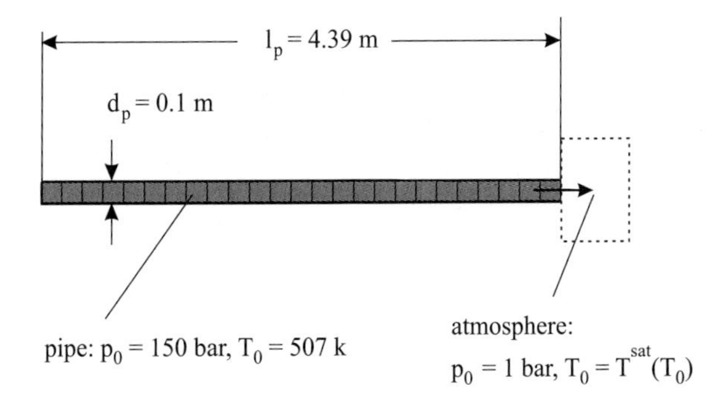

Fig. 9.68: Canon blowdown experiment: computational grid and initial conditions

The predicted behavior during the blowdown as shown in Figs. 9.69 and 9.70 is qualitatively very similar to that obtained for Edwards' pipe. This includes the wave propagation and reflection phenomena during the first 10 ms of the transient when the strong evaporation (flashing) upstream of the pipe exit prevents the pressure from droping below the saturation pressure $T^{sat}(p_0)$. Different to Edwards' pipe, critical flow conditions ($M = 1$) occurred immediately after the removal of the rupture disk and were also maintained as long as the pressure in the pipe considerably exceeded the atmospheric pressure.

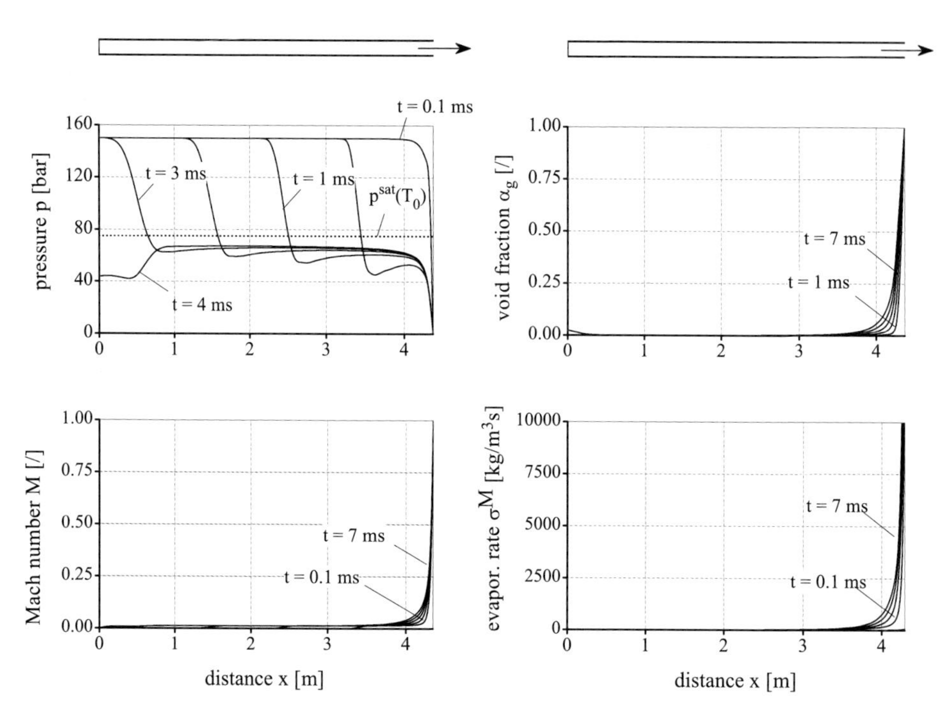

Fig. 9.69: *Canon blowdown test* case; parameter distribution for the *short term* period, time values; 0.1 ms $\leq t \leq$ 7 ms

A comparison with the measured data for pressure and void fraction is given in Fig. 9.71. Although the figure indicates that the general trends of the experiment are reasonable well predicted, a more qualitative evaluation is only partially possible due to the large scatter in the measured data (in particular for the void fraction).

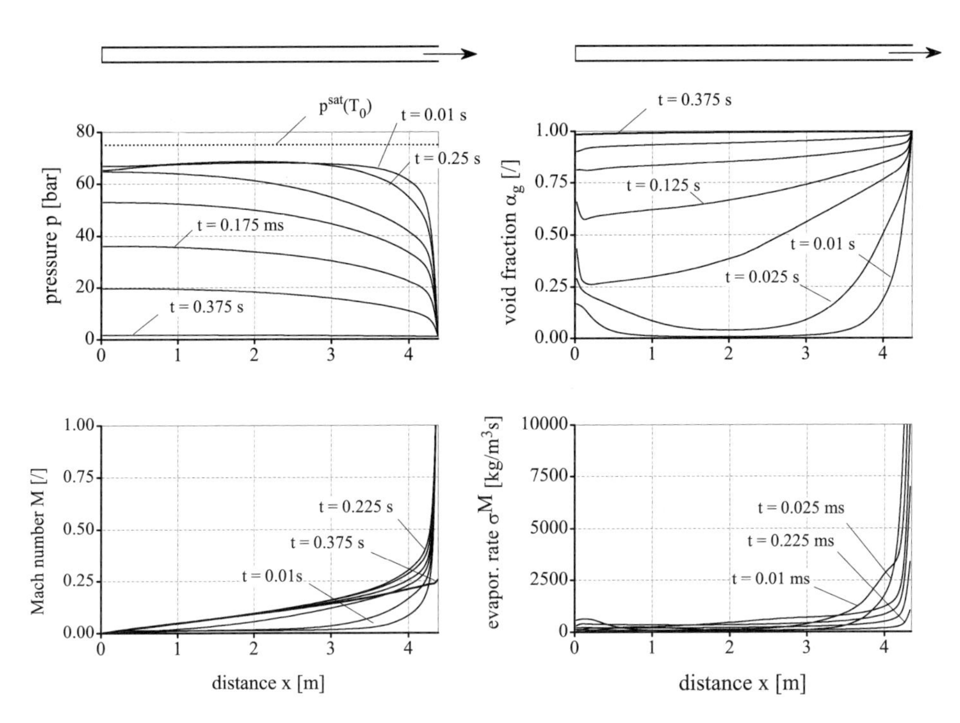

Fig. 9.70: *Canon blowdown test*; parameter distribution for the *long term* period, time values: $0.01 \text{ s} \leq t \leq 0.375 \text{ s}$

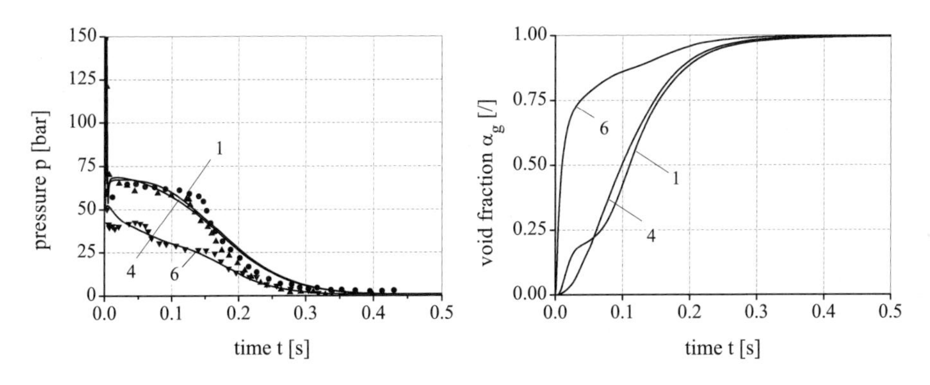

Fig. 9.71: *Canon blowdown test*: comparison of prediction with measured data; *left*: pressure at the pipe head (1), pipe middle section (4), and near pipe exit (6), *right*: corresponding void fraction values

9.7.3 Two-vessel test case

This purely hypothetical test case is included to demonstrate the capabilities of the presented modeling and numerical strategies for the numerical simulation of more complex multidimensional two-phase processes as are also of interest for many industrial applications. The assumed facility consists of two cylindrical vessels connected by a horizontal stand pipe schematically shown in Fig. 9.72. The high pressure container on the left side is partially filled with saturated liquid with a pressure of 10 bar whereas the right vessel contains pure vapor at atmospheric pressure. Both vessels are separated by a diaphragm at the exit of the connecting pipe which is assumed to be removed instantaneously at time zero.

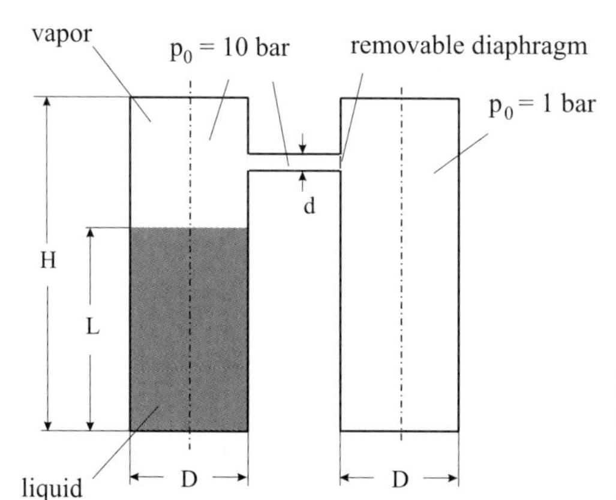

Fig. 9.72: Two-vessel test case: geometrical configuration and initial conditions, dimensions: $H = 1.8$ m, $D = 0.6$ mm, $L = 1.0$ m, $d = 0.01$ m

Since the major parameter changes are expected in the x-y plane, the problem is treated as a pseudo two-dimensional case where some three-dimensional effects are taken into account by a variable "depth" in the z-direction. The corresponding computational grid as used in the calculation is shown schematically in Fig. 9.73.

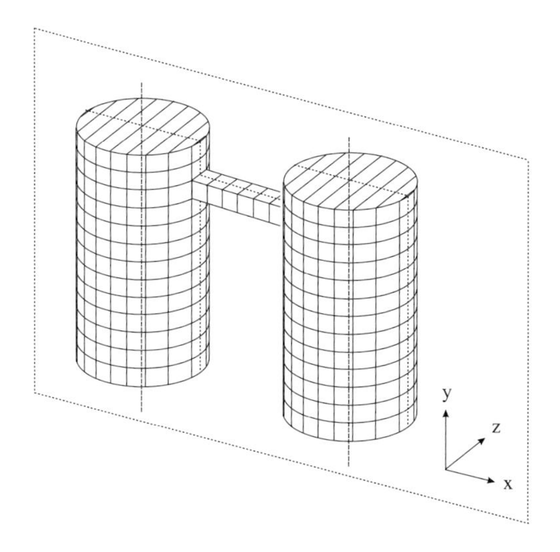

Fig. 9.73: Two-vessel test case: computational grid for a quasi-two-dimensional calculation (schematically)

In the actual calculations a hexagonal grid is used within the x-y plane as indicated in Fig. 9.74. This has the advantage that every cell has common interfaces with all neighboring cells, which provides a reduction of the grid dependence of solution compared with a Cartesian grid.

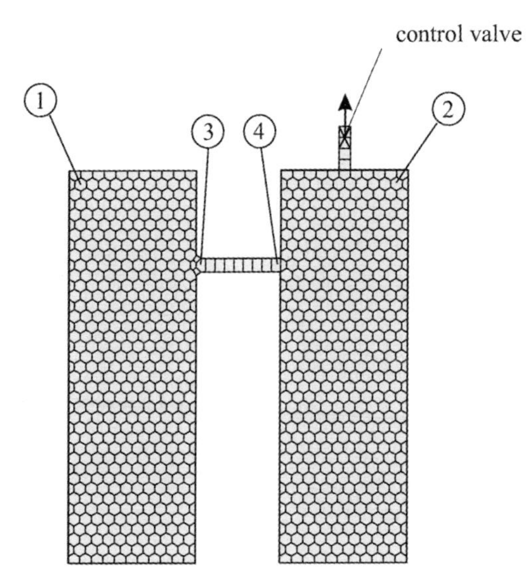

Fig. 9.74: Two-vessel test case: hexagonal grid at symmetry plane and reference data positions

In the following the results of two calculations are presented including a completely *closed system* where the valve on top of the right (low pressure) vessels remains closed during the whole transient and (b) and a *vented system*, where the valve is opened simultaneously with the removal of the diaphragm in the connecting pipe at time zero.

Closed system

A qualitative picture of the transient might be obtained from Figs. 9.79 and 9.80 showing the void fraction distribution and the vector field for the gas mass flow at various consecutive time values. A more detailed information is presented in Figs. 9.75 and 9.76 for the pressure values and Mach numbers in the vessels and in the pipe during the short time ($0.0 \text{ s} \leq t \leq 0.1 \text{ s}$) and long time ($0.1 \text{ s} \leq t \leq 10 \text{ s}$) periods:

(a) a rapid boil-off and swelling of the water pool in the left vessel due to fast evaporation,

(b) transition from single-phase vapor to two-phase flow and choking in the interconnecting pipe,

(c) jet formation in the right (low pressure) vessel, jet impingement at the vessel wall and a strong re-circulating flow pattern,

(d) gravity-induced phase separation,

(e) liquid collapse and a formation of residual liquid pools in the two vessels.

The transient is terminated at about 12 s when a new equilibrium state is reached in the whole system and the vapor and liquid phases in both vessels are completely separated. As can be seen from Fig. 9.76 a large amount of liquid is finely transported into the right vessel.

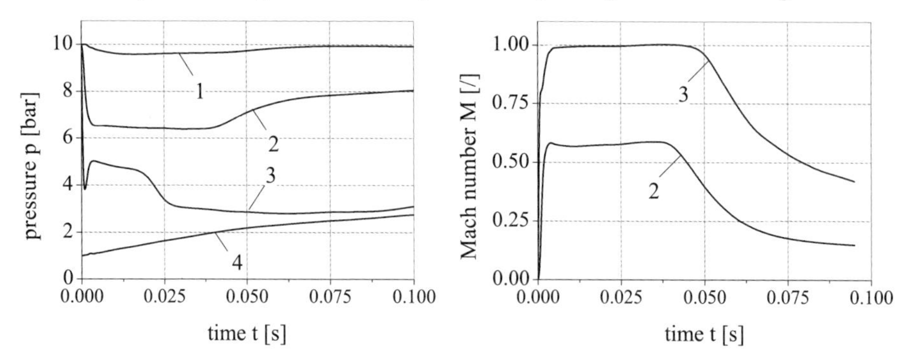

Fig. 9.75: Two-vessel test problem, *closed conditions*: pressure and Mach number during the *short time* period at top of the left vessel (1), inlet (2), and outlet (3) of the connecting pipe, and at the top of the right vessel (4)

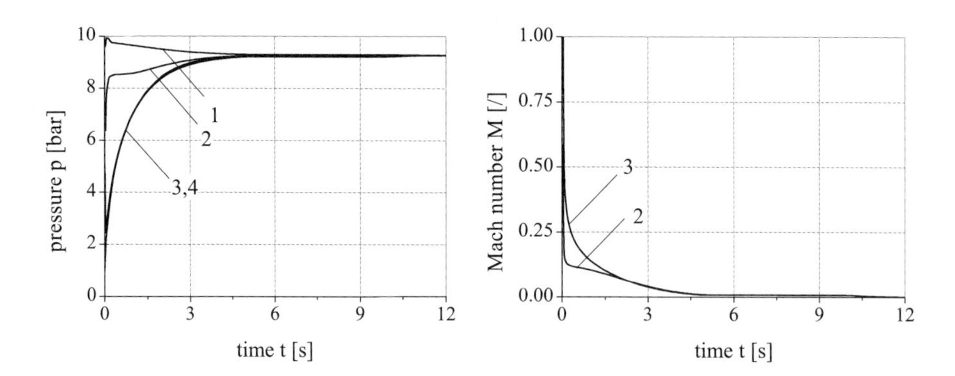

Fig. 9.76: Two-vessel test problem, *closed conditions*: pressure and Mach number during the *long time* period at the top of the left vessel (1), inlet (2) and outlet (3) of the connecting pipe, and at the top of the right vessel (4)

Open system

As long as there exists critical flow conditions in the connecting pipe during the short time period, the behavior of the vented system is nearly identical with those predicted for the closed system and, therfore, is not explicitly shown here. Shortly after the flow in the connecting pipe turns to subsonic condition the pressure in the right vessel reached a maximum value of about $p_4 = 3$ bar when the volumetric flow in the connecting pipe and the discharge through the valve are at the same order of magnitude (see Fig. 9.77). This pressure then remains nearly constant for a certain period of time before, due to the dominating effect of discharge to the atmosphere, a continuous depressurization of both vessels occurs up to the time when the atmospheric pressure is reached at about $t = 12$ s.

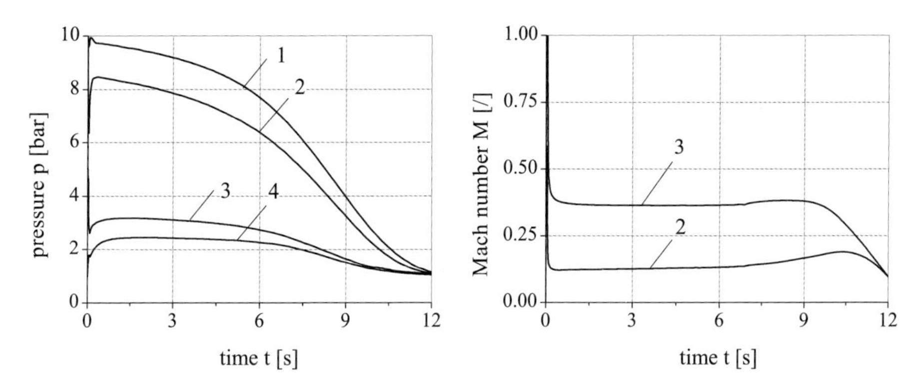

Fig. 9.77: Two-vessel test problem, *open conditions*: pressure and Mach number during the *long time* period at the top of the left vessel (1), inlet (2) and outlet (3) of the connecting pipe, and at the top of the right vessel (4)

Predicted values for pressure and mass inventories in both pressure vessels are compared in Fig. 9.78 for closed and open conditions. The figure indicates that in the case of *closed conditions*, the transient results mainly in a redistribution of the mass inventory and the final equilibrium pressure appears only slightly below the initial pressure of the left vessel. For the *open conditions*, a large amount of water is finally ejected from the system and only small liquid pools remained in both vessels at the end of the transient.

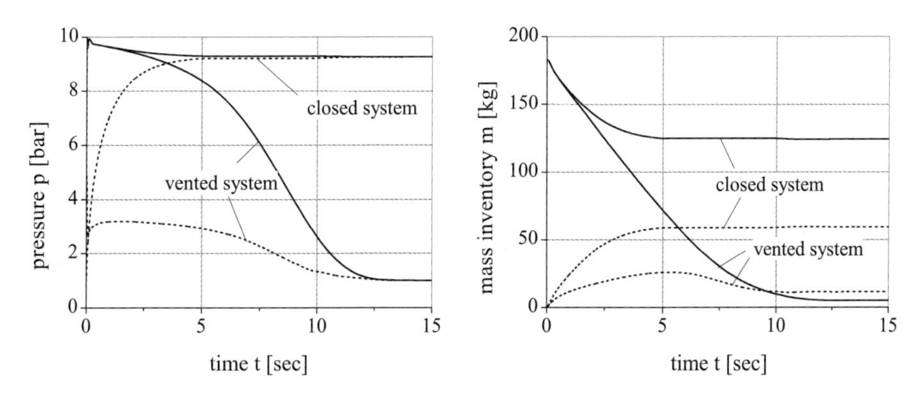

Fig. 9.78: Two-vessel test problem, comparison of results for *closed and open conditions:* pressure and mass inventory for the left (solid line) and right (dashed line) vessel as a function of time

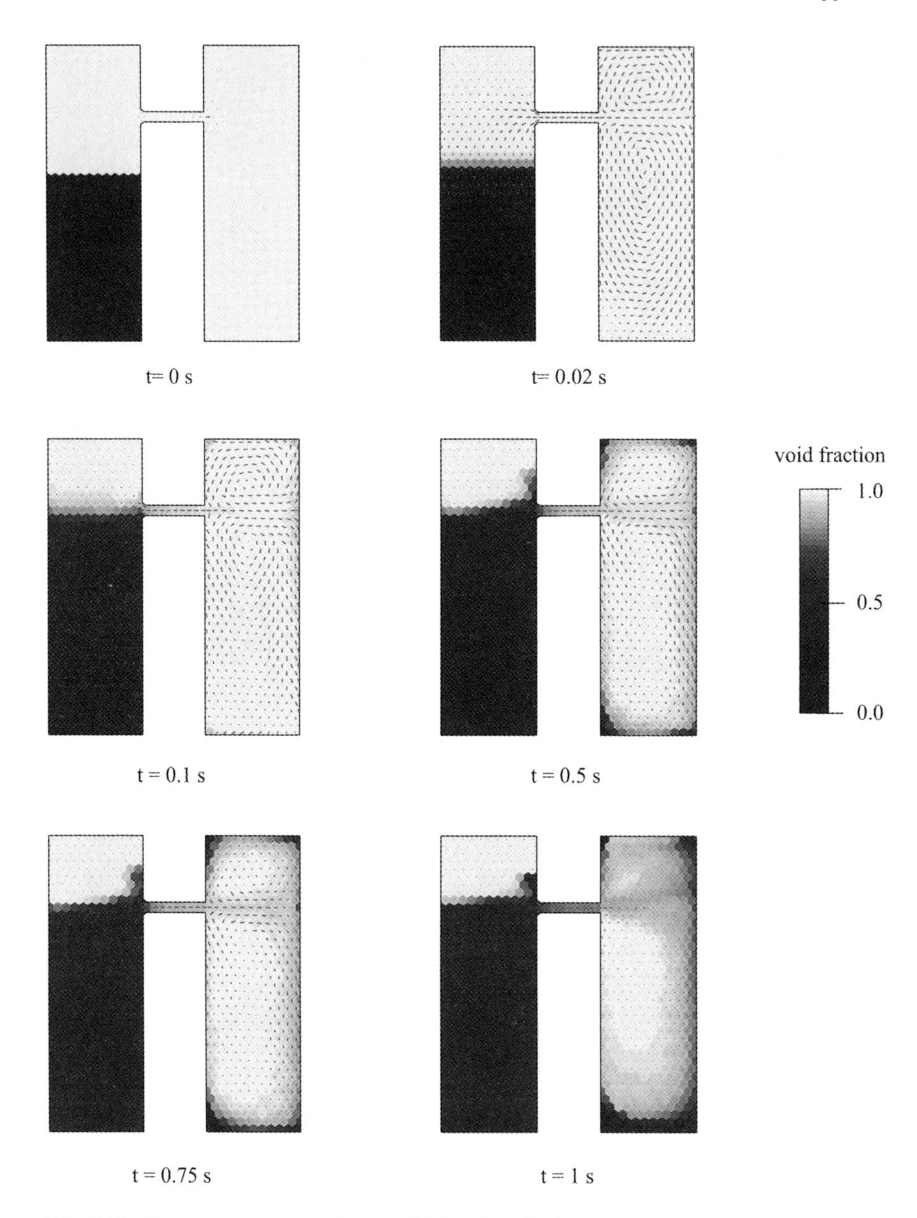

Fig. 9.79: Two-vessel test problem: void fraction distribution and vector field for gas velocity for short time period

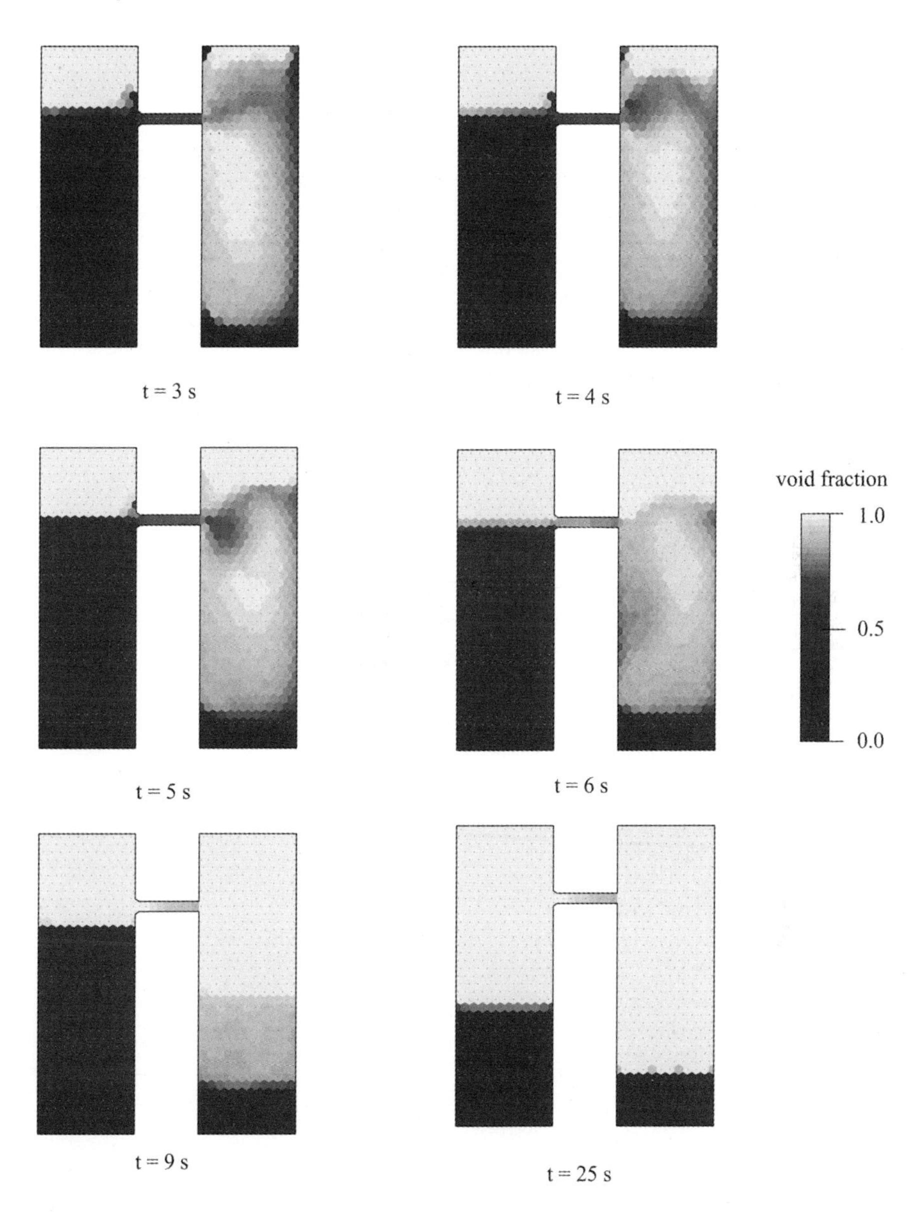

Fig. 9.80: Two-vessel test problem: void fraction distribution and vector field for gas velocity during long time period

References

[1] G.F. Hewitt, J.M. Delhaye, and N. Zuber, *Multiphase Science and Technology*, Vol. 6, 591–609, 1996.

[2] H.Städtke et al., *Advanced Three-Dimensional Two-Phase Flow Simulation Tool for Application to Reactor Safety (ASTAR),* Nuclear and Engineering Design 235, 379–400, 2005.

[3] M.E. Deich, V.S. Danilin, V.N. Shanin, G.V. Tsiklauri, *Critical Conditions in Laval Nozzles Operating in a Two-Phase Medium*, Teploenergetika 16 (6), 76–79, 1969

[4] A. Asaka, *CATHARE – Qualification of Critical Flow Experiments*, Centre d'Etudes Nucleaires de Grenoble, report STR/LML/EM/92-108, 1992

[5] D. Bestion, *The Physical Closure Laws in the CATHARE Code*, Nuclear and Engineering Design 124, 229–245, 1990.

[6] *CSNI International Standard Problems (ISP)*, Report NEA/CSNI/R(2000)5.

[7] A.R. Edwards, and T.P. O'Brien, *Studies of Phenomena Connected with the Depressurization of Water Reactors*, J. Br. Nucl. Energy 9, 125–135, 1970

[8] B. Riegel, *Experimental Data Report on Super Canon Test Programme - Brief Descriptions (1975–1999)*, CEA : TT/SETRE/79-2-B/BR, 1979.

10 Summary and Concluding Remarks

The objective of this book has been to provide a comprehensive review of two-phase flow processes strongly affected by the compressibility of two-phase media. This includes, in particular, the formation and propagation of weak or strong pressure waves, the occurrence of critical flow conditions associated with the flow through nozzles or the fast depressurization of initially subcooled liquids in pipes or vessels. The basis for the analysis of these processes is a newly developed hyperbolic two-fluid model which allows a complete algebraic evaluation of the eigenspace of the governing equations including eigenvalues and related sets of right and left eigenvectors. For integration of the resulting governing equations, a generalized second-order Flux Vector Splitting technique is used providing a high resolution of local flow processes such as steep parameter gradients or flow discontinuities. The accuracy of the chosen approach has been demonstrated in numerous test cases also including a comparison of predicted results with analytical solutions or measured data.

As long as homogeneous and equilibrium conditions are assumed, the two-phase flow behaves qualitatively as the single-phase gas counterpart. However, compared with the single-phase gas flow, the high density and the usually large compressibility of the homogeneous two-phase mixture result in rather low values for the sound velocity and, hence, supersonic flow conditions might be obtained already at low or moderate flow velocities. A further peculiarity exists for one-component media where, due to the thermal equilibrium assumption, the sound velocity changes discontinuously when crossing the saturation line between regions of subcooled liquid or superheated steam and two-phase conditions.

For a more realistic representation of two-phase flows, deviations from the mechanical and thermal equilibrium between the phases have to be taken into account such as unequal local phase velocities ($u_g \neq u_l$) and unequal temperatures ($T_g \neq T_l$), or deviations of the phasic temperatures from the corresponding saturated value ($T_g \neq T^{\text{sat}}, T_g \neq T^{\text{sat}}$). These nonequilibrium conditions are a consequence of the usually large differences in state and transport properties between the phases and the finite rates of the interfacial exchange processes for mass, momentum, and energy. The consequences of these nonequilibrium effects are manifold:

- The sound velocity in two-phase media is no longer a single value determined by simple thermodynamic state properties as in single-phase gas media. Instead, it becomes a function of the sound frequency with an upper (a^{fr}) or lower (a^{eq}) limiting value for very high or very low frequencies where either "frozen" or equilibrium conditions are reached. The "frozen" sound velocity resulting from the dispersion analysis is identical with corresponding values obtained from the characteristic analysis of the governing equations for the propagation velocity of pressure waves: $\lambda_{1,2} = u \pm a^{\text{fr}}$.

Gasdynamic Aspects of Two-Phase Flow. Herbert Städtke

ISBN: 3-527-40578-X

- Depending on the boundary conditions shock waves might occur for supersonic flows $u > a^{\mathrm{fr}}$, however, different to the pure gas flow or to homogeneous equilibrium two-phase flow, all parameters change continuously across the wave. Usually two different regions for the shock structure can be distinguished: (1) a shock front characterized by a steep (but finite) pressure gradient resulting in a sudden deviation from equilibrium between the phases, followed by (2) a relaxation region where flow is driven asymptotically toward a new equilibrium state. The differences between the shock front and the relaxation zone becomes more pronounced with increasing Mach number and related shock strength.

- In stationary flow situations, critical flow conditions are reached when the fastest wave speed becomes stationary for $u = a^{\mathrm{fr}}$. For the flow through a convergent–divergent nozzle, the occurrence of such critical conditions is characterized by a saddle-point singularity which, different to single-phase gas flow, appears in the divergent part of the nozzle downstream of the nozzle throat. The exact location of the critical state is *a priori* not known and can be determined only together with the integration of the complete system of flow equations.

- The presence (or absence) of critical flow conditions is not *a priori* an indication whether or not a maximum value for the mass flow (choking) in a nozzle or pipe is reached. As a consequence of strong interfacial heat and mass transfer processes, a maximum discharge mass flow might occur much earlier at subsonic conditions ($u < a^{\mathrm{fr}}$).

The effects of nonequilibrium conditions on pressure wave propagation or nozzle flows as described above are not unique to two-phase media. A qualitatively similar behavior also exists for high temperature gas flows exposed to nonequilibrium conditions resulting from vibrational relaxation or ionization processes, or for flow processes undergoing chemical reactions.

For the numerical simulation of transient two-phase flows the hyperbolicity of the flow equations represents a necessary requirement for the correct prediction of governing wave propagation phenomena. Nevertheless, the advantage of having a hyperbolic system of flow system of equations might be totally lost if improper numerical methods are applied characterized by large nonphysical (numerical) diffusion or viscosity effects. Therefore, the use of numerical methods, which make explicit use of hyperbolic features of the flow equations becomes indispensable.

Another aspect to be considered is that the "hyperbolic" numerical schemes require a complete evaluation of the eigenspectrum for all computational cells at any time step. This represents a large computational effort, however, this is reduced considerable whenever explicitly algebraic expressions can be derived as in the present two-fluid model. Having a fully algebraic solution for the eigenspectrum has some other advantages which might be worth mentioning. The model provides a consistent set of real eigenvalues and related independent eigenvectors which is free of any numerically-related approximation or uncertainty. Specific attention has to be given to conditions under which the basic flow equations degenerate as indicated by the occurrence of singularities in the split coefficient or Jacobian matrices , e.g., for instantaneous equal phase velocities $(u_g = u_l)$ or for the transition between single- and two-phase conditions where either the gas or the liquid phase appears or disappears. How such conditions can be handled is explained in Chapter 5.

In most of the numerical results shown, the viscous and heat conduction effects are taken into account only in the vicinity of the interface or at the pipe wall in the form of algebraic source terms describing interfacial coupling due to mass, momentum and energy transfer processes. This seems to be justified for most of the wave propagation processes described, in particular as far as strictly one-dimensional conditions are concerned. Nevertheless, there are many other conditions where the inclusion of bulk viscosity and heat conduction effects might become essential. As experienced with the ATFM code, the inclusion of bulk viscosity terms in the flow equations is a rather straightforward process and a finite volume equivalent to the centered finite difference technique is adequate to handle such terms.

A Basic Flow Equations for Two-Fluid Model of Two-Phase Flow

A.1 Flow topology

A.1.1 Phase distribution function

The whole spatial flow domain is considered to be divided into two subdomains where either only liquid or vapor is present as schematically shown in Fig. A.1. Both regions are separated by an infinitesimal thin layer, in the following called an interface. This situation can be described by a distribution function γ_i with $i = g, l$ defined as

$$\left.\begin{array}{ll} \gamma_i = 1 & \text{where phase } i \text{ is present} \\ \gamma_i = 0 & \text{where phase } i \text{ is not present} \end{array}\right\} \tag{A.1}$$

with the condition $\gamma_g + \gamma_l = 1$.

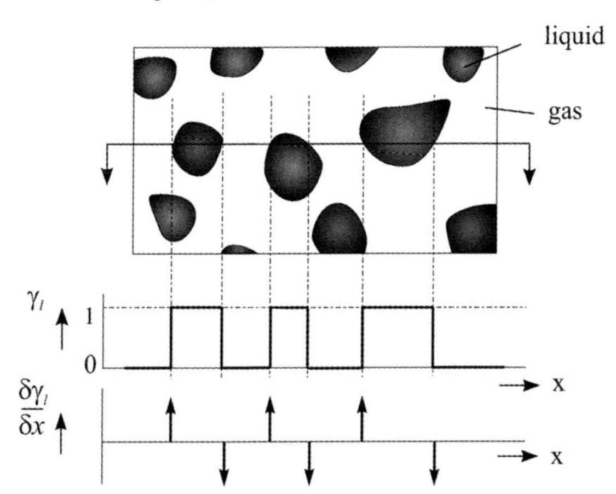

Fig. A.1: Two-phase flow topology

Using the definition for the distribution function (A.1) the volumetric concentration of the phase i in an an arbitrary finite volume V_ξ can be written as

$$\alpha_i = \frac{1}{V_\xi} \int\limits_{V_\xi} \gamma_i \, dV = \frac{A_i^{\text{int}}}{V_\xi}, \tag{A.2}$$

where α_g is known as the void fraction.

Gasdynamic Aspects of Two-Phase Flow. Herbert Städtke

ISBN: 3-527-40578-X

In a similar manner two types of volume averaged quantities as schematically shown in Fig. A.2 can be introduced for phase parameters ψ_i related to either the total value of the control volume V_ξ or the volume occupied by the phase V_i, respectively

$$\langle\psi_i\rangle = \frac{1}{V_\xi}\int\limits_{V_\xi} \psi_i\,\gamma_i\,dV \tag{A.3}$$

and

$$\langle\psi_i\rangle^i = \frac{1}{V_i}\int\limits_{V} \psi_i\,\gamma_i\,dV. \tag{A.4}$$

Both values are connected through the relations

$$\langle\psi_i\rangle = \alpha_i\langle\psi_i\rangle^i. \tag{A.5}$$

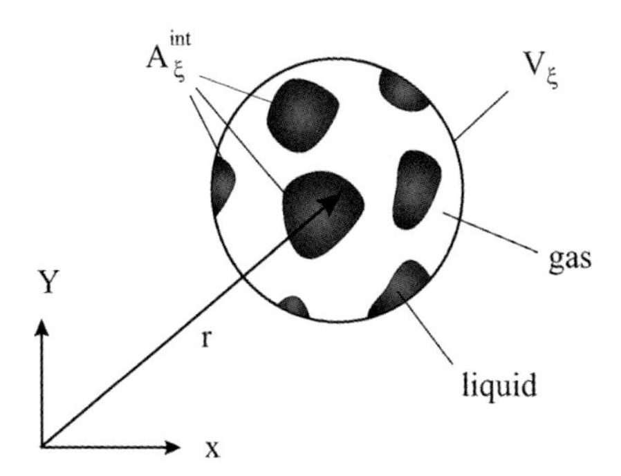

Fig. A.2: Volume averaged quantity

A.1.2 Interfacial properties

The gradient of the distribution function $\nabla\gamma_i$ as schematically shown in Fig. A.1 represents a vector field which is defined by $\nabla\gamma_i = 0$, everywhere, except at the interphase, where $\nabla\gamma_i$ is a vector directed into the phase i normal to the interface with the absolute value $|\nabla\gamma_i|^{\text{int}} \Rightarrow \infty$. This allows us to define a unit vector for the interface related to the phase i pointing outward with respect to phase i

$$\vec{n}_i^{\text{int}} = -\left(\frac{\nabla\gamma}{|\nabla\gamma_i|}\right)^{\text{int}}, \tag{A.6}$$

as schematically shown in Fig. A.3.

Assuming that $|\nabla\gamma_i|$ has the property of a Dirac delta function, the interfacial area within a certain control volume V_ξ is determined as

$$A_\xi^{\text{int}} = \int\limits_{V_\xi} |\nabla\gamma_i|\,dV = \int\limits_{A_\xi} dA. \tag{A.7}$$

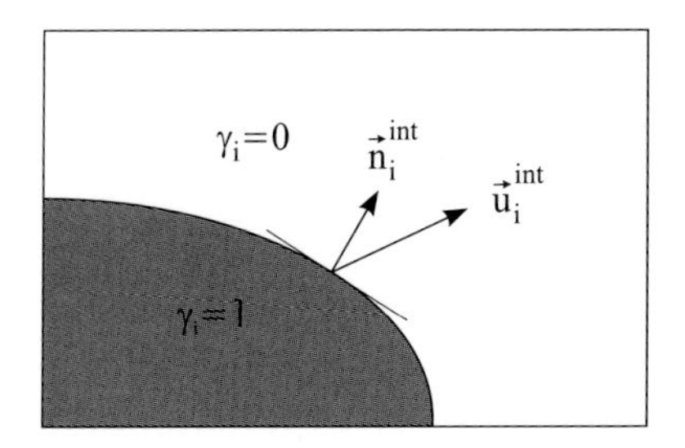

Fig. A.3: Interfacial velocity and normal unit vector

The concentration of the interfacial area per volume is therefore defined as

$$a^{\text{int}} = \frac{A_\xi^{\text{int}}}{V_\xi} = \frac{1}{V_\xi} \int_{V_\xi} |\nabla \gamma_i| \, dV. \tag{A.8}$$

In the following often an average interfacial property per unit volume is used, which is defined as

$$\langle \psi^{\text{int}} \rangle = \frac{1}{V_\xi} \int_{V_\xi} \psi^{\text{int}} \, |\nabla \gamma_i| \, dV, \tag{A.9}$$

which according to equation (A.7) can be expressed also as an integral over the interfacial area present in V_ξ as

$$\langle \psi^{\text{int}} \rangle = \frac{a^{\text{int}}}{A_\xi^{\text{int}}} \int_{A_\xi^{\text{int}}} \psi^{\text{int}} \, dA. \tag{A.10}$$

For an observer moving with the interface, the distribution function γ_i does not change in time which means for the total derivative of the distribution function at the interface

$$\left(\frac{d\gamma_i}{dt} \right)^{\text{int}} = \left(\frac{\partial \gamma_i}{\partial t} \right)^{\text{int}} + \vec{u}^{\text{int}} \cdot (\nabla \gamma_i)^{\text{int}} = 0. \tag{A.11}$$

Since the distribution function γ_i is constant everywhere except at the interface, and assuming that the interfacial velocity is $\vec{u}^{\text{int}} = 0$ everywhere except at the interface, equation (A.11) can be generalized for the whole flow domain

$$\frac{d\gamma_i}{dt} = \frac{\partial \gamma_i}{\partial t} + \vec{u}^{\text{int}} \cdot \nabla \gamma_i = 0 \tag{A.12}$$

Integrating equation (A.12) over a space fixed volume V_ξ one obtains

$$\frac{1}{V} \int_{V_\xi} \frac{\partial \gamma_i}{\partial t} \, dV + \frac{1}{V_\xi} \int_{V_\xi} \vec{u}^{\text{int}} \cdot \nabla \gamma_i \, dV = 0. \tag{A.13}$$

Since the control volume V_ξ does not change in time, the time derivative operator can be moved in front of the integral sign. Transforming the second term into an integral over the interfacial area, equation (A.13) can also be written as

$$\frac{\partial}{\partial t}\left[\frac{1}{V_\xi}\int_{V_\xi}\gamma_i V\right]-\frac{1}{V_\xi}\int_{V_\xi}\vec{u}^{\text{int}}\cdot\vec{n}_i^{\text{int}}\left|\nabla\gamma_i\right|dV=0, \tag{A.14}$$

or with the definition of the volume concentration of the phase i (A.2)

$$\boxed{\frac{\partial\alpha_i}{\partial t}=\frac{1}{V_\xi}\int_{A_\xi^{\text{int}}}\vec{u}^{\text{int}}\cdot\vec{n}_i^{\text{int}}\,dA.} \tag{A.15}$$

Equation (A.15) gives a purely kinematic relation between the change in time of the volumetric concentration of the phase i with the movement of the interface within a space fixed control volume, regardless of the underlying physical process (e.g., expansion/compression due to pressure change, evaporation, or condensation).

From the definition of the volumetric concentration (A.2), the corresponding spatial gradient becomes

$$\nabla\alpha_i=\nabla\left[\frac{1}{V_\xi}\int_{V_\xi}\gamma_i\,dV\right] \tag{A.16}$$

or, assuming a space fixed control volume V_ξ

$$\nabla\alpha_i=\frac{1}{V_\xi}\int_{V_\xi}\nabla\gamma_i\,dV. \tag{A.17}$$

Transferring the volume integral into an integral over the interfacial area, the gradient of the volumetric concentration (A.18) can also be expressed as

$$\boxed{\nabla\alpha_i=-\frac{1}{V_\xi}\int_{A_\xi^{\text{int}}}\vec{n}_i^{\text{int}}\,dA.} \tag{A.18}$$

Equations (A.12), (A.15), and (A.18) will be used later for the interpretation of some terms in volume averaged phasic balance equations.

A.1.3 Transport equation for interfacial area

For an observer moving with the interface yields similar to equation (A.12)

$$\frac{\partial}{\partial t}\left(\left|\nabla\gamma_i\right|\right)+\vec{u}^{\text{int}}\cdot\nabla\left(\left|\nabla\gamma_i\right|\right)=0. \tag{A.19}$$

Integrating equation (A.19) over a space fixed control volume V_ξ yields

$$\frac{1}{V_\xi}\int\limits_{V_\xi}\frac{\partial}{\partial t}\left(|\nabla\gamma_i|\right)dV+\frac{1}{V_\xi}\int\limits_{V_\xi}\vec{u}^{\text{int}}\cdot\nabla\left(|\nabla\gamma_i|\right)dV=0 \tag{A.20}$$

or, since the control volume is fixed in time

$$\frac{\partial}{\partial t}\left[\frac{1}{V\xi}\int\limits_{V\xi}\left(|\nabla\gamma_i|\right)dV\right]+\nabla\cdot\left[\frac{1}{V_\xi}\int\limits_{V_\xi}\vec{u}^{\text{int}}\cdot\left(|\nabla\gamma_i|\right)dV\right]$$
$$=\frac{1}{V_\xi}\int\limits_{V_\xi}\left(\nabla\cdot\vec{u}^{\text{int}}\right)|\nabla\gamma_i|\,dV. \tag{A.21}$$

The first term in equation (A.21) represents the time derivative of the interfacial area concentration a^{int}, the other two terms can be transposed into integrals over the interfacial area resulting in

$$\frac{\partial a^{\text{int}}}{\partial t}+\nabla\cdot\left[a^{\text{int}}\frac{1}{A_\xi^{\text{int}}}\int\limits_{A_\xi^{\text{int}}}\vec{u}^{\text{int}}\,dA\right]=\frac{1}{V_\xi}\int\limits_{A_\xi^{\text{int}}}\left(\nabla\cdot\vec{u}^{\text{int}}\right)dA. \tag{A.22}$$

Introducing the average interfacial velocity

$$\langle\vec{u}^{\text{int}}\rangle=\frac{1}{A_\xi^{\text{int}}}\int\limits_{A_\xi^{\text{int}}}\vec{u}^{\text{int}}\,dA \tag{A.23}$$

and the volumetric source term for the interfacial area

$$\sigma_i^A=\frac{1}{V_\xi}\int\limits_{A_\xi^{\text{int}}}\nabla\cdot\vec{u}^{\text{int}}\,dA, \tag{A.24}$$

the final transport equation for the interfacial area becomes

$$\boxed{\frac{\partial a^{\text{int}}}{\partial t}+\nabla\cdot\left[a^{\text{int}}\langle\vec{u}^{\text{int}}\rangle\right]=\sigma_i^A.} \tag{A.25}$$

A.2 Single-phase flow equations

Within the two subdomains where either vapor ($\gamma_g = 1, i = g$) or liquid ($\gamma_l = 1, i = l$) is present, the classical "instantaneous" conservation equations for mass, momentum, and energy are valid which can be written in differential form as follows

mass:

$$\frac{\partial \varrho_i}{\partial t} + \nabla \cdot (\varrho_i \vec{u}_i) = 0 \tag{A.26}$$

momentum:

$$\frac{\partial}{\partial t}(\varrho_i \vec{u}_i) + \nabla \cdot (\varrho_i \vec{u}_i \vec{u}_i) + \nabla \cdot \bar{\mathbf{P}}_i = \vec{F}_i \tag{A.27}$$

with the pressure tensor

$$\bar{\mathbf{P}}_i = p_i \bar{\mathbf{I}} - \bar{\mathbf{T}}_i \tag{A.28}$$

energy

$$\frac{\partial}{\partial t}\left[\varrho_i\left(e_i + \frac{u_i^2}{2}\right)\right] + \nabla\cdot\left[\varrho_i \vec{u}_i\left(e_i + \frac{u_i^2}{2}\right)\right] + \nabla\cdot(\vec{q}_i + \bar{\mathbf{P}}_i\cdot\vec{u}_i) = \varrho_i \vec{F}_i\cdot\vec{u}_i + \varrho_i Q_i \tag{A.29}$$

A.3 Two-phase balance equations

A.3.1 Balance equation for mass

Multiplying the single phase mass conservation equation (A.26)

$$\frac{\partial}{\partial t}(\varrho_i) + \nabla \cdot (\varrho_i \vec{u}) = 0, \tag{A.30}$$

with the corresponding distribution function γ_i and integrating over a small space-fixed volume, one obtains

$$\int\limits_{V_\xi} \gamma_i \left[\frac{\partial}{\partial t}(\varrho_i) + \nabla \cdot (\varrho_i \vec{u}_i)\right] dV = 0, \tag{A.31}$$

or

$$\int\limits_{V_\xi} \left[\frac{\partial}{\partial t}(\gamma_i \varrho_i) + \nabla \cdot (\gamma_i \varrho_i \vec{u}_i)\right] dV - \int\limits_{V_\xi} \varrho_i \left[\frac{\partial \gamma_i}{\partial t} + \vec{u}_i \nabla \cdot \gamma\right] = 0. \tag{A.32}$$

Introducing the interfacial velocity $\vec{u}^{\text{int}}$ equation (A.32) becomes

$$\int\limits_{V_\xi} \left[\frac{\partial}{\partial t}(\gamma_i\, \varrho) + \nabla \cdot (\gamma_i\, \varrho_i\, \vec{u}_i)\right] dV - \int\limits_{V_\xi} \varrho_i \underbrace{\left[\frac{\partial \gamma_i}{\partial t} + \vec{u}^{\text{int}} \cdot \nabla \gamma_i\right] dV}_{=0}$$

$$+ \int\limits_{V_\xi} \left[\varrho_i \left(\vec{u}^{\text{int}} - \vec{u}_i\right) \cdot \nabla \gamma_i\right] dV = 0 \tag{A.33}$$

With the condition at the interface as given by relations (A.6) and (A.12) the balance equation (A.33) can be written as

$$\int_{V_\xi} \left[\frac{\partial}{\partial t} (\gamma_i\, \varrho_i) + \nabla \cdot (\gamma_i\, \varrho_i\, \vec{u}_i) \right] dV = \int_{V_\xi} \left[\varrho_i \left(\vec{u}^{\text{int}} - \vec{u}_i \right) \cdot \vec{n}_i^{\text{int}} \, |\nabla \gamma_i| \right] dV, \qquad \text{(A.34)}$$

which can be further simplified. Since the integration is performed over a space-fixed volume, the time and space differential operators on the l.h.s. of equation (A.34) can be moved in front of the integration

$$\int_{V_\xi} \left[\frac{\partial}{\partial t} (\gamma_i\, \varrho_i) + \nabla \cdot (\gamma_i\, \varrho_i\, \vec{u}_i) \right] dV = \frac{\partial}{\partial t} \int_{V_\xi} (\gamma_i\, \varrho_i)\; dV + \nabla \cdot \int_{V_\xi} (\gamma_i\, \varrho_i\, \vec{u}_i)\, dV. \quad \text{(A.35)}$$

As shown by equations (A.9) and (A.10) the volume integration on the right-hand side of equation (A.34) can be transformed into a integration over the interfaces present in V_ξ

$$\int_{V_\xi} \left[\varrho_i \left(\vec{u}^{\text{int}} - \vec{u}_i \right) \cdot \vec{n}_i^{\text{int}} \, |\nabla \gamma_i| \right] dV = \int_{A_\xi^{\text{int}}} \varrho_i \left(\vec{u}^{\text{int}} - \vec{u}_i \right) \cdot \vec{n}_i^{\text{int}} \, dA. \qquad \text{(A.36)}$$

With equations (A.34) and (A.35) the phasic mass balance equation becomes

$$\underbrace{\frac{\partial}{\partial t} \left[\frac{1}{V_\xi} \int_{V_\xi} (\gamma_i\, \varrho_i)\; dV \right]}_{(1)} + \underbrace{\nabla \cdot \left[\frac{1}{V_\xi} \int_{V_\xi} (\gamma_i\, \varrho_i \vec{u}_i)\; dV \right]}_{(2)}$$
$$= \underbrace{\frac{1}{V_\xi} \int_{A_\xi^{\text{int}}} \varrho_i \left(\vec{u}^{\text{int}} - \vec{u}_i \right) \cdot \vec{n}_i^{\text{int}} \, dA,}_{(3)} \qquad \text{(A.37)}$$

where the inverse of the volume V_ξ is used as constant factor.

The terms (1) and (2) on the left-hand side of equation (A.37) can be easily identified as volume averaged quantities for density

$$\frac{1}{V_\xi} \int_{V_\xi} (\gamma_i\, \varrho_i)\; dV = \langle \varrho_i \rangle = \alpha_i \, \langle \varrho_i \rangle^i \qquad \text{(A.38)}$$

and mass flux

$$\frac{1}{V_\xi} \int_{V_\xi} (\gamma_i\, \varrho_i)\; dV = \langle \varrho_i\, \vec{u}_i \rangle = \alpha_i \, \langle \varrho_i\, \vec{u}_i \rangle^i . \qquad \text{(A.39)}$$

The term (3) on the right-hand side represents a volumetric source term for mass characterized by the mass flux at the interface $\varrho_i \left(\vec{u}^{\text{int}} - \vec{u}_i\right)$

$$\sigma_i^M = \frac{1}{V_\xi} \int\limits_{A_\xi^{\text{int}}} \varrho_i \left(\vec{u}^{\text{int}} - \vec{u}_i\right) \vec{n}_i^{\text{int}} \, dA, \tag{A.40}$$

or, introducing the interfacial area concentration $a^{\text{int}} = A_\xi^{\text{int}}/V_\xi$ as defined by equation A.8

$$\boxed{\sigma_i^M = a^{\text{int}} \frac{1}{A_\xi} \int\limits_{A_\xi^{\text{int}}} \varrho_i \left(\vec{u}^{\text{int}} - \vec{u}_i\right) \vec{n}_i^{\text{int}} \, dA.} \tag{A.41}$$

Finally, the mass balance equation can be written in short form as

$$\frac{\partial}{\partial t} \langle \varrho_i \rangle + \nabla \cdot \langle \varrho_i \vec{u}_i \rangle = \sigma_i^M \tag{A.42}$$

or, if averaging is done over the volume occupied by the phase $i = g, l$

$$\boxed{\frac{\partial}{\partial t} \left[\alpha_i \langle \varrho_i \rangle^i\right] + \nabla \cdot \left[\alpha_i \langle \varrho_i \vec{u}_i \rangle^i\right] = \sigma_i^M.} \tag{A.43}$$

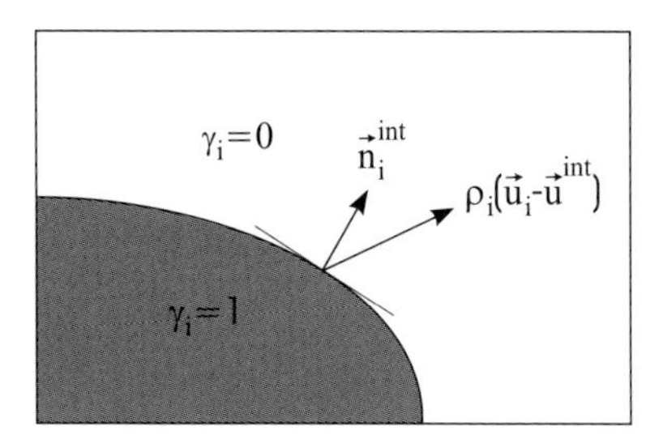

Fig. A.4: Mass transfer at the interface

A.3.2 Balance equation for momentum

Introducing the expression for the pressure tensor (A.27) the balance equation for momentum (A.28) becomes

$$\frac{\partial}{\partial t}(\varrho_i \vec{u}_i) + \nabla \cdot (\varrho_i \vec{u}_i \vec{u}_i) + \nabla p \bar{\mathbf{I}} - \nabla \cdot \bar{\mathbf{T}}_i = \vec{F}_i. \tag{A.44}$$

Multiplying equation (A.44) with γ_i and integrating over a small space-fixed control volume results in

$$\int\limits_{V_\xi} \gamma_i \left[\frac{\partial}{\partial t}(\varrho_i \vec{u}_i) + \nabla \cdot (\varrho_i \vec{u}_i \vec{u}_i) + \nabla(p \bar{\mathbf{I}}) - \nabla \cdot \bar{\mathbf{T}}_i =\right] dV = \int\limits_{V_\xi} \gamma_i \vec{F}_i \, dV \tag{A.45}$$

or

$$\int\limits_{V_\xi} \left[\frac{\partial}{\partial t}(\gamma_i \varrho_i \vec{u}_i) + \nabla \cdot (\gamma_i \varrho_i \vec{u}_i \vec{u}_i) + \nabla\left(\gamma_i p \bar{\mathbf{I}}\right) - \nabla \cdot \left(\gamma_i \bar{\mathbf{T}}_i\right)\right] dV - \int\limits_{V_\xi} p \nabla \gamma_i \, dV$$
$$- \int\limits_{V_\xi} \varrho_i \vec{u}_i \left[\left(\frac{\partial \gamma_i}{\partial t} + \vec{u}_i \cdot \nabla \gamma_i\right)\right] dV + \int\limits_{V_\xi} \bar{\mathbf{T}}_i \cdot \nabla \gamma_i \, dV = \int\limits_{V_\xi} \gamma_i \vec{F}_i \, dV. \tag{A.46}$$

Introducing the interfacial velocity, equation (A.46) changes to

$$\int\limits_{V_\xi} \left[\frac{\partial}{\partial t}(\gamma_i \varrho_i \vec{u}_i) + \nabla \cdot (\gamma_i \varrho_i \vec{u}_i \vec{u}_i) + \nabla\left(\gamma_i p \bar{\mathbf{I}}\right) - \nabla \cdot \left(\gamma_i \bar{\mathbf{T}}_i\right)\right] dV$$
$$+ \int\limits_{V_\xi} \bar{\mathbf{T}}_i \cdot \nabla \gamma_i \, dV - \int\limits_{V_\xi} p \nabla \gamma_i \, dV - \int\limits_{V_\xi} \varrho_i \vec{u}_i \varrho_i \vec{u}_i \underbrace{\left(\frac{\partial \gamma_i}{\partial t} + \vec{u}^{\text{int}} \cdot \nabla \gamma_i\right)}_{=0} dV$$
$$+ \int\limits_{V_\xi} \varrho_i \vec{u}_i (\vec{u}^{\text{int}} - \vec{u}_i) \cdot \nabla \gamma_i \, dV = \int\limits_{V_\xi} \gamma_i \vec{F}_i \, dV \tag{A.47}$$

or, with condition (A.12) at the moving interface to

$$\frac{1}{V_\xi} \int\limits_{V_\xi} \left[\frac{\partial}{\partial t}(\gamma_i \varrho_i \vec{u}_i) + \nabla \cdot (\gamma_i \varrho_i \vec{u}_i \vec{u}_i) + \nabla\left(\gamma_i p \bar{\mathbf{I}}\right) - \nabla \cdot \left(\gamma_i \bar{\mathbf{T}}_i\right)\right] dV - \frac{1}{V_\xi} \int\limits_{V_\xi} p_i \nabla \gamma_i \, dV$$
$$= \frac{1}{V_\xi} \int\limits_{V_\xi} \bar{\mathbf{T}}_{\mathbf{i}} \cdot \nabla \gamma_i \, dV - \frac{1}{V_\xi} \int\limits_{V_\xi} \varrho_i \vec{u}_i (\vec{u}^{\text{int}} - \vec{u}_i) \cdot \nabla \gamma_i \, dV + \int\limits_{V_\xi} \gamma_i \vec{F}_i \, dV. \tag{A.48}$$

Moving the differential operators on the l.h.s. in front of the integrals and replacing the first two volume integrals with equivalent integrals over the interfacial area A_ξ^{int}, and transferring the volume integrals containing $\nabla \gamma_i$ into integrals over the interfacial area, equation (A.48) becomes

$$\underbrace{\frac{\partial}{\partial t}\left[\frac{1}{V_\xi} \int\limits_{V_\xi} (\gamma_i \varrho_i \vec{u}_i) \, dV\right]}_{1} + \nabla \cdot \underbrace{\left[\frac{1}{V_\xi} \int\limits_{V_\xi} \gamma_i (\varrho_i \vec{u}_i \vec{u}_i + p_i \bar{\mathbf{I}} + \bar{\mathbf{T}}_i) \, dV\right]}_{(2)} - \underbrace{\frac{1}{V_\xi} \int\limits_{V_\xi} p_i \nabla \gamma_i \, dV}_{(3)}$$
$$= \underbrace{-\frac{1}{V_\xi} \int\limits_{A_\xi^{\text{int}}} \bar{\mathbf{T}}_i \cdot \vec{n}_i^{\text{int}} \, dA}_{(4)} + \underbrace{\frac{1}{V_\xi} \int\limits_{A_\xi^{\text{int}}} \varrho_i \vec{u}_i (\vec{u}^{int} - \vec{u}_i) \cdot \vec{n}_i^{int} \, dA_i}_{(5)} + \underbrace{\int\limits_{V_\xi} \gamma_i \vec{F}_i \, dV}_{(6)}. \tag{A.49}$$

The first two terms (1) and (2) on the l.h.s. of equation (A.49) can be identified as time or space-averaged quantities, namely

$$\frac{1}{V_\xi} \int\limits_{V_\xi} \gamma_i (\varrho_i \vec{u}_i) = \langle \varrho_i \vec{u}_i \rangle \tag{A.50}$$

and

$$\frac{1}{V_\xi}\int\limits_{V_\xi}\gamma_i(\varrho_i\vec{u}_i\vec{u}_i + p_i\bar{\mathbf{I}} + \bar{\mathbf{T}}_\mathbf{i})\,dV = \left\langle\varrho_i\vec{u}_i\vec{u}_i + p_i\bar{\mathbf{I}} + \bar{\mathbf{T}}_i\right\rangle. \tag{A.51}$$

The third term (3) on the l.h.s. of (A.49) can be transformed into an integral over the interfacial area

$$-\frac{1}{V_\xi}\int\limits_{V_\xi}p_i\nabla\gamma_i\,dV = \frac{1}{V_\xi}\int\limits_{A_\xi^{\text{int}}}p_i\vec{n}_i^{\text{int}}\,dA. \tag{A.52}$$

Introducing the spatial gradient of the void fraction α_i as defined by equation (A.18) the relation (A.52) becomes

$$-\frac{1}{V_\xi}\int\limits_{V_\xi}p_i\nabla\gamma_i\,dV = -\left\langle p_i^{\text{int}}\right\rangle\nabla\alpha_i, \tag{A.53}$$

where the average interfacial pressure is defined as

$$\left\langle p_i^{\text{int}}\right\rangle = \frac{\displaystyle\frac{1}{V_\xi}\int\limits_{A_\xi^{\text{int}}}p_i\,\vec{n}_i^{\text{int}}\,dA}{\displaystyle\frac{1}{V_\xi}\int\limits_{A_\xi^{\text{int}}}\vec{n}_i^{\text{int}}\,dA}. \tag{A.54}$$

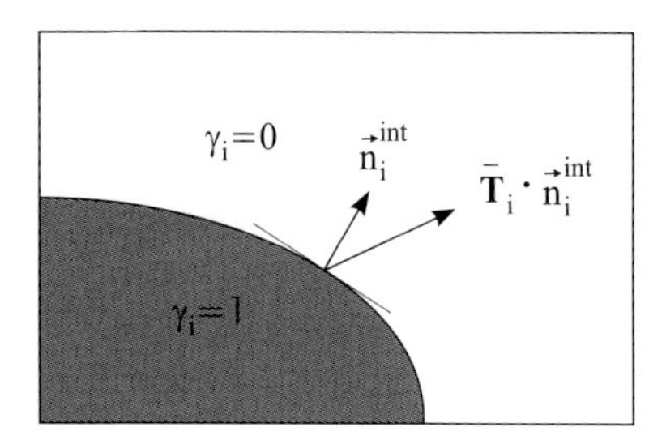

Fig. A.5: Viscous forces at the interface

The fourth term of equation (A.49) can be seen as an integral over all viscous forces acting on the interfacial area which will be abbreviated by $\left\langle\vec{F}_i^{\text{v}}\right\rangle$

$$\left\langle\vec{F}_i^{\text{v}}\right\rangle = \frac{1}{V_\xi}\int\limits_{A_\xi^{\text{int}}}\bar{\mathbf{T}}_i\cdot\vec{n}_i^{\text{int}}\,dA. \tag{A.55}$$

With the interfacial area concentration, the viscous forces (A.55) become

$$\boxed{\left\langle\vec{F}_i^{\text{v}}\right\rangle = a^{\text{int}}\frac{1}{A_\xi^{\text{int}}}\int\limits_{A_\xi^{\text{int}}}\bar{\mathbf{T}}_i\cdot\vec{n}_i^{\text{int}}\,dA.} \tag{A.56}$$

The fifth term of equation (A.49) represents the momentum transfer at the interface associated with the mass transfer

$$\frac{1}{V_\xi}\int\limits_{A_\xi^{\text{int}}} \varrho_i \vec{u}_i(\vec{u}^{\text{int}} - \vec{u}_i)\cdot\vec{n}_i^{\text{int}}\, dA_i = \langle \vec{u}_i^{\text{ex}}\rangle\, \sigma_i^M, \tag{A.57}$$

with u_i^{ex} as the average of the velocity of the mass exchanged between the two phases

$$\langle u_i^{\text{ex}}\rangle = \frac{\dfrac{1}{V_\xi}\displaystyle\int\limits_{A_\xi^{\text{int}}} \varrho_i \vec{u}_i(\vec{u}^{\text{int}} - \vec{u}_i)\cdot\vec{n}_i^{\text{int}}\, dA_i}{\dfrac{1}{V_\xi}\displaystyle\int\limits_{A_\xi^{\text{int}}} \varrho_i(\vec{u}^{\text{int}} - \vec{u}_i)\cdot\vec{n}_i^{\text{int}}\, dA_i}. \tag{A.58}$$

The last term in equation (A.49) represents the average of the external forces acting on the phase i

$$\boxed{\left\langle \vec{F}_i\right\rangle = \frac{1}{V_\xi}\int\limits_{V_\xi} \gamma_i \vec{F}_i\, dV.} \tag{A.59}$$

Introducing equations (A.50), (A.51), (A.54), (A.55), (A.57) and (A.59) into equation (A.49) one obtains the momentum balance for the phase i

$$\frac{\partial}{\partial t}\langle\varrho_i\vec{u}_i\rangle + \nabla\cdot\langle\varrho_i\vec{u}_i\vec{u}_i + p_i\bar{\mathbf{I}} + \bar{\mathbf{T}}_i\rangle - \langle p_i^{\text{int}}\rangle\,\bar{\mathbf{I}}\nabla\alpha_i = \sigma_i^M\,\langle\vec{u}_i^{\text{ex}}\rangle + \left\langle\vec{F}_i^{\text{vis}}\right\rangle + \left\langle\vec{F}_i\right\rangle \tag{A.60}$$

or changing to space averaging related to the volume occupied by the phase i

$$\boxed{\frac{\partial}{\partial t}\left[\alpha_i\,\langle\varrho_i\vec{u}_i\rangle^i\right] + \nabla\cdot\left[\alpha_i\,\langle\varrho_i\vec{u}_i\vec{u}_i + p_i + \bar{\mathbf{T}}_i\rangle^i\right] - \langle p_i^{\text{int}}\rangle\,\bar{\mathbf{I}}\,\nabla\alpha_i = \vec{\sigma}_i^J + \left\langle\vec{F}_i\right\rangle,} \tag{A.61}$$

with the interfacial momentum source term

$$\vec{\sigma}_i^J = \sigma_i^M\,\langle\vec{u}_i^{\text{ex}}\rangle + \left\langle\vec{F}_i^v\right\rangle. \tag{A.62}$$

A.3.3 Balance equation for energy

The starting point is the single phase energy equation (A.29)

$$\frac{\partial}{\partial t}\left[\varrho_i\left(e_i + \frac{u_i^2}{2}\right)\right] + \nabla\cdot\left[\varrho_i\vec{u}_i\left(e_i + \frac{u_i^2}{2}\right)\right] + \nabla\cdot(\vec{q}_i + p_i\vec{u}_i - \bar{\mathbf{T}}_i\cdot\vec{u}_i) = \vec{F}_i\cdot\vec{u}_i + Q_i. \tag{A.63}$$

Multiplying (A.63) with the distribution function γ_i and integrating over a small control volume V_ξ yields

$$\int\limits_{V_\xi} \gamma_i \frac{\partial}{\partial t}\left[\varrho_i\left(e_i+\frac{u_i^2}{2}\right)\right] dV + \int\limits_{V_\xi} \gamma_i \nabla\cdot\left[\varrho_i \vec{u}_i\left(e_i+\frac{u_i^2}{2}\right)\right] dV$$
$$+\int\limits_{V_\xi} \gamma_i \nabla\cdot\left(\vec{q}_i + p_i\vec{u}_i - \bar{\mathbf{T}}_i\cdot\vec{u}_i\right) dV = \int\limits_{V_\xi} \gamma_i \vec{F}_i\cdot\vec{u}_i\, dV + \int\limits_{V_\xi}\gamma_i Q_i\, dV, \qquad (A.64)$$

or respectively

$$\int\limits_{V_\xi} \frac{\partial}{\partial t}\left[\gamma_i\varrho_i\left(e_i+\frac{u_i^2}{2}\right)\right] dV + \int\limits_{V_\xi} \nabla\cdot\left[\gamma_i\varrho_i \vec{u}_i\left(e_i+\frac{u_i^2}{2}\right)\right] dV$$
$$+\int\limits_{V_\xi} \nabla\cdot\left[\gamma_i\left(\vec{q}_i + p_i\vec{u}_i - \bar{\mathbf{T}}_i\cdot\vec{u}_i\right)\right] dV - \int\limits_{V_\xi} \varrho_i\left(e_i+\frac{u_i^2}{2}\right)\left(\frac{\partial\gamma_i}{\partial t}+\vec{u}_i\cdot\nabla\gamma_i\right) dV$$
$$-\int\limits_{V_\xi}\left(\vec{q}_i + p_i\vec{u}_i - \bar{\mathbf{T}}_i\cdot\vec{u}_i\right)\cdot\nabla\gamma_i\, dV = \int\limits_{V_\xi} \gamma_i \vec{F}_i\cdot\vec{u}_i\, dV + \int\limits_{V_\xi}\gamma_i Q_i\, dV. \qquad (A.65)$$

Moving the differential operators outside the integrals over a space fixed volume, and introducing the phasic enthalpy $h_i = e_i + p_i/\varrho_i$ and the interfacial velocity $\vec{u}^{\mathrm{int}}$, equation (A.65) becomes

$$\frac{\partial}{\partial t}\left[\int\limits_{V_\xi}\gamma_i\varrho_i\left(e_i+\frac{u_i^2}{2}\right) dV\right] + \nabla\cdot\left[\int\limits_{V_\xi}\left[\gamma_i\varrho_i\vec{u}_i\left(h_i+\frac{u_i^2}{2}\right)\right] dV\right]$$
$$+\nabla\cdot\left[\int\limits_{V_\xi}\nabla\cdot\left[\gamma_i\left(\vec{q}_i - \bar{\mathbf{T}}_i\cdot\vec{u}_i\right)\right] dV\right] - \int\limits_{V_\xi}\varrho_i\left(e_i+\frac{u_i^2}{2}\right)\underbrace{\left(\frac{\partial\gamma_i}{\partial t}+\vec{u}^{\mathrm{int}}\cdot\nabla\gamma^{\mathrm{int}}\right)}_{=0} dV$$
$$+\int\limits_{V_\xi}\varrho_i\left(h_i+\frac{u_i^2}{2}\right)\left(\vec{u}^{\mathrm{int}}-\vec{u}_i\right)\cdot\nabla\gamma_i\, dV - \int\limits_{V_\xi}\left(\vec{q}_i - \bar{\mathbf{T}}_i\cdot\vec{u}_i\right)\cdot\nabla\gamma_i\, dV$$
$$-\int\limits_{V_\xi} p_i\vec{u}_i^{\mathrm{int}}\cdot\nabla\gamma\, dV = \int\limits_{V_\xi}\gamma_i\vec{F}_i\cdot\vec{u}_i\, dV + \int\limits_{V_\xi}\gamma_i Q_i\, dV \qquad (A.66)$$

or dividing by V_ξ and rearranging some terms

$$\underbrace{\frac{\partial}{\partial t}\left[\frac{1}{V_\xi}\int\limits_{V_\xi}\gamma_i\varrho_i\left(e_i+\frac{u_i^2}{2}\right)dV\right]}_{(1)}+\underbrace{\nabla\cdot\left[\frac{1}{V_\xi}\int\limits_{V_\xi}\nabla\cdot\left[\gamma_i\varrho_i\vec{u}_i\left(h_i+\frac{u_i^2}{2}\right)\right]dV\right]}_{(2)}$$

$$+\underbrace{\nabla\cdot\left[\frac{1}{V_\xi}\int\limits_{V_\xi}\nabla\cdot\left[\gamma_i(\vec{q}_i-\bar{\mathbf{T}}_i\cdot\vec{u}_i)\right]dV\right]}_{(3)}\underbrace{-\frac{1}{V_\xi}\int\limits_{V_\xi}p_i\vec{u}_i^{\,\mathrm{int}}\cdot\nabla\gamma\,dV}_{(4)}$$

$$=\underbrace{-\frac{1}{V_\xi}\int\limits_{V_\xi}\varrho_i\left(h_i+\frac{u_i^2}{2}\right)(\vec{u}^{\,\mathrm{int}}-\vec{u}_i)\cdot\nabla\gamma_i dV}_{(5)}+\underbrace{\frac{1}{V_\xi}\int\limits_{V_\xi}(\vec{q}_i-\bar{\mathbf{T}}_i\cdot\vec{u}_i)\cdot\nabla\gamma_i\,dV}_{(6)}$$

$$+\underbrace{\frac{1}{V_\xi}\int\limits_{V_\xi}\gamma_i\vec{F}_i\cdot\vec{u}_i d}_{(7)}+\underbrace{\frac{1}{V_\xi}\int\limits_{V_\xi}\gamma_i Q_i\,dV}_{(8)}. \tag{A.67}$$

The first three terms of equation (A.67) can be easily interpreted as time and spatial derivatives of volume averaged quantities

$$\frac{1}{V_\xi}\int\limits_{V_\xi}\gamma_i\varrho_i\left(e_i+\frac{u_i^2}{2}\right)dV=\left\langle\varrho_i\left(e_i+\frac{u_i^2}{2}\right)\right\rangle \tag{A.68}$$

$$\frac{1}{V_\xi}\int\limits_{V_\xi}\left[\gamma_i\varrho_i\vec{u}_i\left(h_i+\frac{u_i^2}{2}\right)\right]dV=\left\langle\varrho_i\vec{u}_i\left(h_i+\frac{u_i^2}{2}\right)\right\rangle \tag{A.69}$$

$$\frac{1}{V_\xi}\int\limits_{V_\xi}\gamma_i(\vec{q}_i-\bar{\mathbf{T}}_i\cdot\vec{u}_i)\,dV=\langle\vec{q}_i\rangle-\langle\bar{\mathbf{T}}_i\cdot\vec{u}_i\rangle\,. \tag{A.70}$$

The terms (4) to (6) of equation (A.67) can be transformed into integrals over the interfacial area resulting in

$$-\frac{1}{V_\xi}\int\limits_{V_\xi}p_i\vec{u}_i^{\,\mathrm{int}}\cdot\nabla\gamma\,dV=\frac{1}{V_\xi}\int\limits_{A_\xi^{\mathrm{int}}}p_i\vec{u}_i^{\,\mathrm{int}}\cdot\vec{n}_i^{\,\mathrm{int}}\,dA, \tag{A.71}$$

$$
-\frac{1}{V_\xi}\int\limits_{V_\xi}\left(h_i+\frac{u_i^2}{2}\right)\varrho_i\left(\vec{u}^{\text{int}}-\vec{u}_i\right)\cdot\nabla\gamma\,dV
$$

$$
=\frac{1}{V_\xi}\int\limits_{A_\xi^{\text{int}}}\left(h_i+\frac{u_i^2}{2}\right)\varrho_i\left(\vec{u}^{\text{int}}-\vec{u}_i\right)\cdot\vec{n}_i^{\text{int}}\,dA, \quad (A.72)
$$

and

$$
\frac{1}{V_\xi}\int\limits_{V_\xi}\gamma_i(\vec{q}_i-\bar{\mathbf{T}}_i\cdot\vec{u}_i)\,dV=-\frac{1}{V_\xi}\int\limits_{A_\xi^{\text{int}}}(\vec{q}_i-\bar{\mathbf{T}}_i\cdot\vec{u}_i)\cdot\vec{n}_i^{\text{int}}\,dA. \quad (A.73)
$$

The two terms (7) and (8) on r.h.s. of equation (A.67) represent volume averaged quantities for the work of external forces and the external heat sources, respectively, acting on the phase i

$$
\frac{1}{V_\xi}\int\limits_{V_\xi}\gamma_i\vec{F}_i\cdot\vec{u}_i\,dV=\left\langle\vec{F}_i\cdot\vec{u}_i\right\rangle, \quad (A.74)
$$

and

$$
\boxed{\frac{1}{V_\xi}\int\limits_{V_\xi}\gamma_i Q_i\,dV=\langle Q_i\rangle.} \quad (A.75)
$$

Introducing equations (A.68) to (A.75) into equations (A.67) yields an intermediate form of the energy equation

$$
\frac{\partial}{\partial t}\left\langle\varrho_i\left(e_i+\frac{u_i^2}{2}\right)\right\rangle+\nabla\cdot\left\langle\varrho_i\vec{u}_i\left(h_i+\frac{u_i^2}{2}\right)\right\rangle+\nabla\cdot\langle\vec{q}_i\rangle-\langle\bar{\mathbf{T}}_i\cdot\vec{u}_i\rangle
$$

$$
+\frac{1}{V_\xi}\int\limits_{A_\xi^{\text{int}}}p_i\vec{u}_i^{\text{int}}\vec{n}_i^{\text{int}}\,dA=\frac{1}{V_\xi}\int\limits_{A_\xi^{\text{int}}}\left(h_i+\frac{u_i^2}{2}\right)\varrho_i\left(\vec{u}^{\text{int}}-\vec{u}_i\right)\cdot\vec{n}_i^{\text{int}}\,dA
$$

$$
-\frac{1}{V_\xi}\int\limits_{A_\xi^{\text{int}}}\vec{q}_i\cdot\vec{n}_i^{\text{int}}\,dA+\frac{1}{V_\xi}\int\limits_{A_\xi^{\text{int}}}(\bar{\mathbf{T}}_i\cdot\vec{u}_i)\cdot\vec{n}_i^{\text{int}}\,dA+\left\langle\vec{F}_i\cdot\vec{u}_i\right\rangle+\langle Q_i\rangle. \quad (A.76)
$$

where the interfacial area terms need some further interpretation.

Introducing the time derivative of the volume fraction from equation (A.18), the pressure term on the l.h.s. of equation (A.76) can be written as

$$
\frac{1}{V_\xi}\int\limits_{A_\xi^{\text{int}}}p_i\vec{u}^{\text{int}}\cdot\vec{n}_i^{\text{int}}\,dA=\left\langle p_i^{\text{int}}\right\rangle\frac{\partial\alpha_i}{\partial t}, \quad (A.77)
$$

with the definition of an averaged interfacial pressure as was already introduced for the momentum balance equation (A.54)

$$\langle p_i^{\text{int}} \rangle = \frac{\int\limits_{A_\xi^{\text{int}}} p_i \vec{u}^{\text{int}} \cdot \vec{n}_i^{\text{int}} \, dA}{\int\limits_{A_\xi^{\text{int}}} \vec{u}^{\text{int}} \cdot \vec{n}_i^{\text{int}} \, dA}. \tag{A.78}$$

The first term on the right-hand side of equation (A.76) can be interpreted as the energy transfer between the phases associated with the mass transfer

$$\frac{1}{V_\xi} \int\limits_{A_\xi^{\text{int}}} \left(h_i + \frac{u_i^2}{2} \right) \varrho_i \left(\vec{u}^{\text{int}} - \vec{u}_i \right) \cdot \vec{n}_i^{\text{int}} \, dA = \sigma_i^M \langle h_i^{\text{ex}} \rangle, \tag{A.79}$$

with the total (thermal and kinetic) energy exchanged

$$\boxed{\langle h_i^{\text{ex}} \rangle = \frac{\int\limits_{A_\xi^{\text{int}}} \left(h_i + \frac{u_i^2}{2} \right) \varrho_i \left(\vec{u}^{\text{int}} - \vec{u}_i \right) \cdot \vec{n}_i^{\text{int}} \, dA}{\int\limits_{A_\xi^{\text{int}}} \varrho_i \left(\vec{u}^{\text{int}} - \vec{u}_i \right) \cdot \vec{n}_i^{\text{int}} \, dA},} \tag{A.80}$$

and the mass transfer term as was already defined by equation (A.40)

$$\sigma_i^M = \frac{1}{V_\xi} \int\limits_{A_\xi^{\text{int}}} \varrho_i (\vec{u}^{\text{int}} - \vec{u}_i) \cdot \vec{n}_i^{\text{int}} \, dA. \tag{A.81}$$

The last term on the l.h.s. of equation (A.76) represents the interfacial heat transfer and the work of interfacial viscous forces. For the interfacial heat transfer one obtains

$$\sigma_i^Q = -\frac{1}{V_\xi} \int\limits_{A_\xi^{\text{int}}} \vec{q}_i \cdot \vec{n}_i^{\text{int}} \, dA, \tag{A.82}$$

or

$$\boxed{\sigma_i^Q = -a^{\text{int}} \frac{1}{A_\xi^{\text{int}}} \int\limits_{A_\xi^{\text{int}}} \vec{q}_i \cdot \vec{n}_i^{\text{int}} \, dA.} \tag{A.83}$$

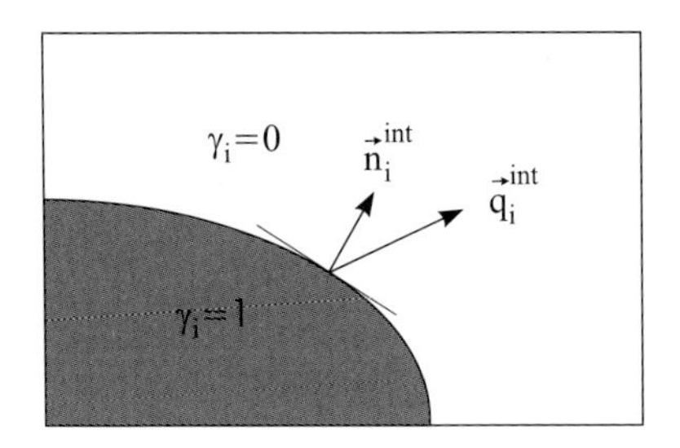

Fig. A.6: Heat transfer at the interface

The work of interfacial viscous forces becomes

$$-\frac{1}{V_\xi}\int\limits_{A_\xi^{\mathrm{int}}}(\bar{\mathbf{T}}_i\cdot\vec{u}_i)\cdot\vec{n}_i^{\mathrm{int}}\,dA=\left\langle\vec{F}_i^v\right\rangle\cdot\langle\vec{u}_i^v\rangle\,, \tag{A.84}$$

with the interfacial viscous force term as was already defined for the momentum balance equation by equation (A.55)

$$\langle F_i^{\mathrm{v}}\rangle=\frac{1}{V_\xi}\int\limits_{A_\xi^{\mathrm{int}}}\bar{\mathbf{T}}_i\cdot\vec{n}_i^{\mathrm{int}}\,dA \tag{A.85}$$

and an associated velocity

$$\boxed{\langle\vec{u}_i^{\mathrm{v}}\rangle=\frac{\int\limits_{A_\xi^{\mathrm{int}}}(\bar{\mathbf{T}}_i\cdot\vec{u}_i)\cdot\vec{n}_i^{\mathrm{int}}\,dA}{\int\limits_{A_\xi^{\mathrm{int}}}\bar{\mathbf{T}}_i\cdot\vec{n}_i^{\mathrm{int}}\,dA}\,.} \tag{A.86}$$

With expressions (A.77), (A.79), (A.82), and (A.84) the balance equation for energy (A.76) finally becomes

$$\begin{aligned}&\frac{\partial}{\partial t}\left\langle\varrho_i\left(e_i+\frac{u_i^2}{2}\right)\right\rangle+\nabla\cdot\left\langle\varrho_i\vec{u}_i\left(h_i+\frac{u_i^2}{2}\right)\right\rangle+\nabla\cdot\langle\vec{q}_i\rangle-\langle\bar{\mathbf{T}}_i\cdot\vec{u}_i\rangle\\&+\langle p_i^{\mathrm{int}}\rangle\frac{\partial\alpha_i}{\partial t}=\sigma_i^Q+\sigma_i^M\langle h_i^{\mathrm{ex}}\rangle+\langle F_i^v\rangle\cdot\langle\vec{u}_i^v\rangle+\left\langle\vec{F}_i\right\rangle\cdot\langle\vec{u}_i\rangle+\langle Q_i\rangle\,,\end{aligned} \tag{A.87}$$

or introducing the volumetric concentration α_i

$$\begin{aligned}&\frac{\partial}{\partial t}\left[\alpha_i\left\langle\varrho_i\left(e_i+\frac{u_i^2}{2}\right)\right\rangle^i\right]+\nabla\cdot\left[\alpha_i\left\langle\varrho_i\vec{u}_i\left(h_i+\frac{u_i^2}{2}\right)\right\rangle^i\right]+\langle p_i^{\mathrm{int}}\rangle\frac{\partial\alpha_i}{\partial t}\\&+\nabla\cdot\left[\alpha_i\langle\vec{q}_i\rangle^i\right]-\left[\alpha_i\langle\bar{\mathbf{T}}_i\cdot\vec{u}_i\rangle^i\right]=\sigma_i^E+\left\langle\vec{F}_i\right\rangle\cdot\langle\vec{u}_i\rangle+\alpha_i\,\langle Q_i\rangle^i\,,\end{aligned} \tag{A.88}$$

with the interfacial energy source term

$$\sigma_i^E=\sigma_i^Q+\sigma_i^M\langle h_i^{\mathrm{ex}}\rangle+\left\langle\vec{F}_i^{\mathrm{v}}\right\rangle\cdot\langle\vec{u}_i^{\mathrm{v}}\rangle \tag{A.89}$$

A.3.4 Summary of two-phase balance equations

Assuming that the averaging is performed only on the volume where the phase i is present, the balance equations for mass (A.43), momentum (A.61), and energy (A.88) can be written as

mass:

$$\frac{\partial}{\partial t}\left[\alpha_i \left\langle \varrho_i \right\rangle^i\right] + \nabla \cdot \left[\alpha_i \left\langle \varrho_i \, \vec{u}_i \right\rangle^i\right] = \sigma_i^M . \tag{A.90}$$

momentum:

$$\frac{\partial}{\partial t}\left[\alpha_i \left\langle \varrho_i \vec{u}_i \right\rangle^i\right] + \nabla \cdot \left[\left\langle \alpha_i \varrho_i \vec{u}_i \vec{u}_i + \alpha_i p_i + \alpha_i \bar{\mathbf{T}}_i \right\rangle\right] - \left\langle p_i^{\text{int}} \right\rangle \nabla \alpha_i$$
$$= \left\langle \vec{F}_i^{\text{v}} \right\rangle + \sigma_i^M \left\langle u_i^{\text{ex}} \right\rangle + \left\langle \vec{F}_i \right\rangle . \tag{A.91}$$

energy:

$$\frac{\partial}{\partial t}\left[\alpha_i \left\langle \varrho_i \left(e_i + \frac{u_i^2}{2}\right)\right\rangle^i\right] + \nabla \cdot \left[\alpha_i \left\langle \varrho_i \vec{u}_i \left(h_i + \frac{u_i^2}{2}\right)\right\rangle^i\right] + \nabla \cdot \left[\alpha_i \left\langle \vec{q}_i \right\rangle\right]$$
$$-\alpha_i \left\langle \bar{\mathbf{T}}_i \cdot \vec{u}_i \right\rangle^i + \left\langle p_i^{\text{int}} \right\rangle \frac{\partial \alpha_i}{\partial t} = \sigma_i^Q + \sigma_i^M \left\langle h_i^{\text{ex}} \right\rangle$$
$$+ \left\langle F_i^{\text{v}} \right\rangle \cdot \left\langle \vec{u}_i^{\text{v}} \right\rangle + \alpha_i \left\langle \vec{F}_i \right\rangle \cdot \left\langle \vec{u}_i \right\rangle + \alpha_i \left\langle Q_i \right\rangle^i . \tag{A.92}$$

Assuming that the average of a product can be approximated by the product of the averages of the parameters involved, e.g.,

$$\left\langle \varrho_i \, \vec{u}_i \right\rangle^i = \left\langle \varrho_i \right\rangle^i \left\langle \vec{u}_i \right\rangle^i \tag{A.93}$$

and dropping the parentheses for volume averages, the basic conservation laws (A.90) to (A.92) can be written as

mass:

$$\frac{\partial}{\partial t}\left(\alpha_i \varrho_i\right) + \nabla \cdot \left(\alpha_i \varrho_i \, \vec{u}_i^i\right) = \sigma_i^M . \tag{A.94}$$

momentum:

$$\frac{\partial}{\partial t}(\alpha_i \varrho_i \vec{u}_i) + \nabla \cdot (\alpha_i \varrho_i \vec{u}_i \vec{u}_i + \alpha_i p_i \bar{\mathbf{I}} + \alpha_i \bar{\mathbf{T}}_i)$$
$$-p_i^{\text{int}} \nabla \alpha_i = \vec{F}_i^{\text{v}} + u_i^{\text{ex}} \sigma_i^M + \vec{F}_i . \tag{A.95}$$

energy:

$$\frac{\partial}{\partial t}\left[\alpha_i \varrho_i \left(e_i + \frac{u_i^2}{2}\right)\right] + \nabla \cdot \left[\alpha_i \varrho_i \vec{u}_i \left(h_i + \frac{u_i^2}{2}\right)\right] + \nabla \cdot (\alpha_i \vec{q}_i) - \nabla \cdot \left(\alpha_i \bar{\mathbf{T}}_i \cdot \vec{u}_i^i\right)$$
$$+p_i^{\text{int}} \frac{\partial \alpha_i}{\partial t} = \sigma_i^Q + \sigma_i^M h_i^{\text{ex}} + F_i^{\text{v}} \cdot \vec{u}_i^{\text{v}} + \alpha_i \vec{F}_i \cdot \vec{u}_i + \alpha_i Q_i . \tag{A.96}$$

The source terms and abbreviations used in equations (A.90) to (A.96) are defined as:

interfacial source term for mass:

$$\sigma_i^M = a^{\text{int}} \frac{1}{A_\xi^{\text{int}}} \int\limits_{A_\xi^{\text{int}}} \varrho_i \left(\vec{u}^{\text{int}} - \vec{u}_i\right) \vec{n}_i^{\text{int}} \, dA \tag{A.97}$$

viscous contributions to interfacial forces:

$$\vec{F}_i^{\text{v}} = a^{\text{int}} \frac{1}{A_\xi^{\text{int}}} \int\limits_{A_\xi^{\text{int}}} \bar{\mathbf{T}}_i \cdot \vec{n}_i^{\text{int}} \, dA \tag{A.98}$$

interfacial pressure

$$p_i^{\text{int}} = \frac{\int\limits_{A_\xi^{\text{int}}} p_i \, \vec{n}_i^{\text{int}} \, dA}{\int\limits_{A_\xi^{\text{int}}} \vec{n}_i^{\text{int}} \, dA}. \tag{A.99}$$

external forces

$$\vec{F}_i = \frac{1}{V_\xi} \int\limits_{V_\xi} \gamma_i \vec{F}_i \tag{A.100}$$

velocity associated with interfacial mass transfer:

$$u_i^{\text{ex}} = \frac{\int\limits_{A_\xi^{\text{int}}} \varrho_i \vec{u}_i (\vec{u}^{\text{int}} - \vec{u}_i) \cdot \vec{n}_i^{\text{int}} \, dA_i}{\int\limits_{A_\xi^{\text{int}}} \varrho_i (\vec{u}^{\text{int}} - \vec{u}_i) \cdot \vec{n}_i^{\text{int}} \, dA_i} \tag{A.101}$$

interfacial source term for heat:

$$\sigma_i^Q = -a^{\text{int}} \frac{1}{A_\xi^{\text{int}}} \int\limits_{A_\xi^{\text{int}}} \vec{q}_i \cdot \vec{n}_i^{\text{int}} \, dA \tag{A.102}$$

total enthalpy associated with mass transfer:

$$h_i^{\text{ex}} = \frac{\int\limits_{A_\xi^{\text{int}}} \left(h_i + \frac{u_i^2}{2}\right) \varrho_i \left(\vec{u}^{\text{int}} - \vec{u}_i\right) \cdot \vec{n}_i^{\text{int}} \, dA}{\int\limits_{A_\xi^{\text{int}}} \varrho_i \left(\vec{u}^{\text{int}} - \vec{u}_i\right) \cdot \vec{n}_i^{\text{int}} \, dA} \tag{A.103}$$

velocity associated with the work of the viscous interfacial forces:

$$\vec{u}_i^{\mathrm{v}} = \frac{\int\limits_{A_\xi^{\mathrm{int}}} (\bar{\mathbf{T}}_i \cdot \vec{u}_i) \cdot \vec{n}_i^{\mathrm{int}}\, dA}{\int\limits_{A_\xi^{\mathrm{int}}} \bar{\mathbf{T}}_i \cdot \vec{n}_i^{\mathrm{int}}\, dA} \tag{A.104}$$

external heat sources:

$$Q_i = \frac{1}{V_\xi} \int\limits_{V_\xi} \gamma_i Q_i \, dV. \tag{A.105}$$

B Characteristic Analysis of Flow Equations: Vectors and Matrices

B.1 Single-phase gas flow, one-dimensional conditions

$$\boxed{\frac{\partial \mathbf{U}}{\partial t} + \mathbf{G}\frac{\partial \mathbf{U}}{\partial x} = \mathbf{D}}$$

$$\mathbf{U} = \begin{bmatrix} p \\ u \\ s \end{bmatrix}, \quad \mathbf{G} = \begin{bmatrix} u & \varrho a^2 & 0 \\ \frac{1}{\varrho} & u & 0 \\ 0 & 0 & u \end{bmatrix}, \quad \mathbf{D} = \begin{bmatrix} a^2 \left(\frac{\partial \varrho}{\partial s}\right)_p \frac{q}{T} \\ f \\ \frac{q}{T} \end{bmatrix}$$

Table B.1: Coefficient matrix, and state and source term vector

$$\mathbf{V}_R = \begin{bmatrix} \frac{1}{2} & \frac{1}{2a\varrho} & 0 \\ \frac{1}{2} & \frac{-1}{2a\varrho} & 0 \\ 0 & 0 & 1 \end{bmatrix}, \quad \mathbf{V}_L = \mathbf{T} = \begin{bmatrix} 1 & +a\,\varrho & 0 \\ 1 & -a\,\varrho & 0 \\ 0 & 0 & 1 \end{bmatrix}$$

Table B.2: Matrix of right and left eigenvectors

Gasdynamic Aspects of Two-Phase Flow. Herbert Städtke

ISBN: 3-527-40578-X

$$\mathbf{\Lambda} = \mathbf{T}^{-1}\mathbf{G}\,\mathbf{T} = \begin{bmatrix} u+a & 0 & 0 \\ 1 & u-a & 0 \\ 0 & 0 & 1 \end{bmatrix}$$

Table B.3: Diagonal matrix of eigenvalues $\mathbf{\Lambda}$

$$\mathbf{G}_1 = (u+a)\begin{bmatrix} \frac{1}{2} & \frac{\varrho a}{2} & 0 \\ \frac{1}{2\varrho a} & \frac{1}{2} & 0 \\ 0 & 0 & 0 \end{bmatrix}, \qquad \mathbf{G}_1 = (u-a)\begin{bmatrix} \frac{1}{2} & \frac{\varrho a}{2} & 0 \\ -\frac{1}{2\varrho a} & \frac{1}{2} & 0 \\ 0 & 0 & 0 \end{bmatrix}$$

$$\mathbf{G}_3 = u\begin{bmatrix} 0 & 0 & 0 \\ 0 & 0 & 0 \\ 0 & 0 & 1 \end{bmatrix}, \qquad \sum \mathbf{G}_k = \mathbf{G} = \begin{bmatrix} u & \varrho a^2 & 0 \\ \frac{1}{\varrho} & u & 0 \\ 0 & 0 & 0 \end{bmatrix}$$

Table B.4: Split matrices $\mathbf{G}_k$, for eigenvalues λ_k with $k = 1, 2, 3$

$$\boxed{\frac{\partial \mathbf{V}}{\partial t} + \mathbf{H}\frac{\partial \mathbf{V}}{\partial x} = \mathbf{C}} \quad \text{with} \quad \mathbf{H} = \frac{\partial \mathbf{F}}{\partial \mathbf{V}}$$

$$\mathbf{V} = \begin{bmatrix} \varrho \\ \varrho u \\ \varrho(e + \frac{1}{2}u^2) \end{bmatrix}, \quad \mathbf{F} = \begin{bmatrix} \varrho u \\ \varrho u^2 + p \\ \varrho(h + \frac{1}{2}u^2) \end{bmatrix}, \quad \mathbf{C} = \begin{bmatrix} 0 \\ \varrho f \\ \varrho q + \varrho f u \end{bmatrix}$$

Table B.5: State vector for conserved variables **F**, flux vector **F**, and source term vector **C**

$$\mathbf{H} = \begin{bmatrix} 0 & 1 & 0 \\ a^2 - u^2 & 2u & 0 \\ ua^2 - u(h + \frac{1}{2}u^2) & (h + \frac{1}{2}u^2) & u \end{bmatrix}$$

$$+ \left(\frac{\partial \varrho}{\partial s}\right)_p \frac{a^2}{T\varrho} \begin{bmatrix} 0 & 0 & 0 \\ (h - \frac{1}{2}u^2) & u & -1 \\ u(h - \frac{1}{2}u^2) & u^2 & 0 \end{bmatrix}$$

Table B.6: Coefficient matrix **H** for conservative form of flow equations

B.2 Single-phase gas flow, two-dimensional conditions

$$\boxed{\frac{\partial \mathbf{U}}{\partial t} + \mathbf{G}_n \frac{\partial \mathbf{U}}{\partial x} = \mathbf{D}}$$

$$\mathbf{U} = \begin{bmatrix} p \\ u_x \\ u_y \\ s \end{bmatrix} \qquad \mathbf{G}_n = \begin{bmatrix} u & \varrho a^2 & 0 & 0 \\ \frac{1}{\varrho} & u & 0 & 0 \\ & 0 & u & 0 \\ 0 & 0 & 0 & u \end{bmatrix} \qquad \mathbf{D} = \begin{bmatrix} \frac{\varrho \beta a^2}{C^p} q \\ f_x \\ f_y \\ \frac{q}{T} \end{bmatrix}$$

Table B.7: State coefficient matrix and source term vector for two-dimensional flow

$$\mathbf{V}_R = \begin{bmatrix} \frac{1}{2} & +\frac{1}{2\varrho a} & 0 & 0 \\ \frac{1}{2} & -\frac{1}{2\varrho a} & 0 & 0 \\ 0 & 0 & 1 & 0 \\ 0 & 0 & 0 & 1 \end{bmatrix}, \qquad \mathbf{V}_L = \begin{bmatrix} 1 & +\varrho a & 0 & 0 \\ 1 & -\varrho a & 0 & 0 \\ 0 & 0 & 1 & 0 \\ 0 & 0 & 0 & 1 \end{bmatrix}$$

Table B.8: Matrix of right and left eigenvectors

$$\mathbf{\Lambda} = \mathbf{T}^{-1}\mathbf{G}\,\mathbf{T} = \begin{bmatrix} u_n + a & 0 & 0 & 0 \\ 0 & u_n - a & 0 & 0 \\ 0 & 0 & u_n & 0 \\ 0 & 0 & 0 & u_n \end{bmatrix} \qquad \text{with} \qquad \mathbf{T} = \mathbf{V}_R^T$$

Table B.9: Diagonal matrix of eigenvalues $\mathbf{\Lambda}$

$$\mathbf{G}_{n,1} = (u_n + a) \begin{bmatrix} \frac{1}{2} & \frac{\varrho a}{2} & 0 & 0 \\ \frac{1}{2\varrho a} & \frac{1}{2} & 0 & 0 \\ 0 & 0 & 0 & 0 \\ 0 & 0 & 0 & 0 \end{bmatrix}$$

Table B.10: Split matrix $\mathbf{G}_{n,1}$, for eigenvalue $\lambda_1 = u_n + a$

$$\mathbf{G}_{n,2} = (u_n - a) \begin{bmatrix} \frac{1}{2} & \frac{\varrho a}{2} & 0 & 0 \\ -\frac{1}{2\varrho a} & \frac{1}{2} & 0 & 0 \\ 0 & 0 & 0 & 0 \\ 0 & 0 & 0 & 0 \end{bmatrix}$$

Table B.11: Split matrix $\mathbf{G}_{n,2}$, for eigenvalue $\lambda_2 = u_n - a$

$$\mathbf{G}_{n,3} = u_n \begin{bmatrix} 0 & 0 & 0 & 0 \\ 0 & 0 & 0 & 0 \\ 0 & 0 & 1 & 0 \\ 0 & 0 & 0 & 0 \end{bmatrix}, \qquad \mathbf{G}_{n,4} = u_n \begin{bmatrix} 0 & 0 & 0 & 0 \\ 0 & 0 & 0 & 0 \\ 0 & 0 & 0 & 0 \\ 0 & 0 & 0 & 1 \end{bmatrix}$$

Table B.12: Split matrix $\mathbf{G}_{n,3,4}$, for eigenvalue $\lambda_{3,4} = u_n$

B.3 Homogeneous nonequilibrium two-phase flow

$$\boxed{\mathbf{A}\frac{\partial \mathbf{U}}{\partial t} + \mathbf{B}\frac{\partial \mathbf{U}}{\partial x} = \mathbf{C}}$$

$$\mathbf{U} = \begin{bmatrix} p \\ u \\ \alpha_g \\ s_g \\ s_l \end{bmatrix}, \qquad \mathbf{C} = \begin{bmatrix} \sigma_g^M \\ \sigma_l^M \\ F \\ \sigma_g^M \left(h^{\text{ex}} - h_g\right) + \dfrac{\sigma_g^M}{T_g} + \dfrac{Q_g}{T_g} \\ \sigma_l^M \left(h^{\text{ex}} - h_l\right) + \dfrac{\sigma_l^M}{T_l} + \dfrac{Q_l}{T_l} \end{bmatrix}$$

Table B.13: State vector of primitive variables $\mathbf{U}$ and source term vector $\mathbf{C}$

$$\mathbf{A} = \begin{bmatrix} \dfrac{\alpha_g}{a_g^2} & 0 & \varrho_g & \alpha_g \left(\dfrac{\partial \varrho_g}{\partial s_g}\right)_p & 0 \\ \dfrac{\alpha_l}{a_l^2} & 0 & -\varrho_l & 0 & \alpha_l \left(\dfrac{\partial \varrho_l}{\partial s_l}\right)_p \\ 0 & \varrho & 0 & 0 & 0 \\ 0 & 0 & 0 & \alpha_g \varrho_g & 0 \\ 0 & 0 & 0 & 0 & \alpha_l \varrho_l \end{bmatrix}$$

Table B.14: Coefficient matrix $\mathbf{A}$

$$\mathbf{B} = \begin{bmatrix} \dfrac{\alpha_g u}{a_g^2} & \alpha_g \varrho_g & \varrho_g u & \alpha_g u \left(\dfrac{\partial \varrho_g}{\partial s_g}\right)_p & 0 \\ \dfrac{\alpha_l u}{a_l^2} & \alpha_l \varrho_l & -\varrho_l u & 0 & \alpha_l u \left(\dfrac{\partial \varrho_l}{\partial s_l}\right)_p \\ 1 & \varrho v & 0 & 0 & 0 \\ 0 & 0 & 0 & \alpha_g \varrho_g u & 0 \\ 0 & 0 & 0 & 0 & \alpha_l \varrho_l u \end{bmatrix}$$

Table B.15: Coefficient matrix $\mathbf{B}$

$$\boxed{\frac{\partial \mathbf{U}}{\partial t} + \mathbf{G}\frac{\partial \mathbf{U}}{\partial x} = \mathbf{D}} \quad \text{with} \quad \mathbf{G} = \mathbf{A}^{-1}\mathbf{B}$$

$$\mathbf{G} = \begin{bmatrix} u & \varrho a^2 & 0 & 0 & 0 \\ \dfrac{1}{\varrho} & u & 0 & 0 & 0 \\ 0 & -\alpha_g \alpha_l \left(\dfrac{\varrho a^2}{\varrho_g a_g^2} - \dfrac{\varrho a^2}{\varrho_l a_l^2}\right) & u & 0 & 0 \\ 0 & 0 & 0 & u & 0 \\ 0 & 0 & 0 & 0 & u \end{bmatrix} = \sum_{k=1}^{5} \mathbf{G}_k$$

with the homogeneous sound velocity $a = \sqrt{\dfrac{1}{\dfrac{\alpha_l \varrho}{\varrho_l a_l^2} + \dfrac{\alpha_g \varrho}{\varrho_g a_g^2}}}$

Table B.16: Coefficient matrix $\mathbf{G}$

$$
\mathbf{D} = \begin{bmatrix}
\frac{\varrho a^2}{\varrho_g}\left[\sigma_g^M - \frac{1}{\varrho_g}\left(\frac{\partial \varrho_g}{\partial s_g}\right)\sigma_g^S\right] + \frac{\varrho a^2}{\varrho_l}\left[\sigma_l^M - \frac{1}{\varrho_l}\left(\frac{\partial \varrho_l}{\partial s_l}\right)_p \sigma_l^S\right] \\
\frac{F}{\varrho} \\
\frac{\alpha_l \varrho a^2}{\varrho_l \varrho_g a_l^2}\left[\sigma_g^M - \frac{1}{\varrho_g}\left(\frac{\partial \varrho_g}{\partial s_g}\right)_p \sigma_l^S\right] - \frac{\alpha_g \varrho a^2}{\varrho_g \varrho_g a_g^2}\left[\sigma_l^M - \frac{1}{\varrho_l}\left(\frac{\partial \varrho_l}{\partial s_l}\right)_p \sigma_l^S\right] \\
\sigma_g^S \frac{1}{\alpha_g \varrho_g} \\
\sigma_l^S \frac{1}{\alpha_l \varrho_l}
\end{bmatrix}
$$

Table B.17: Source term vector $\mathbf{D}$

$$
\mathbf{G}_1 = (u+a)\begin{bmatrix}
\frac{1}{2} & \frac{\varrho a}{2} & 0 & 0 & 0 \\
\frac{1}{2\varrho a} & \frac{1}{2} & 0 & 0 & 0 \\
-\frac{\alpha_g \alpha_l}{2}\left(\frac{1}{\varrho_g a_g^2} - \frac{1}{\varrho_l a_l^2}\right) & -\frac{\alpha_g \alpha_l}{2}\varrho a\left(\frac{1}{\varrho_g a_g^2} - \frac{1}{\varrho_l a_l^2}\right) & 0 & 0 & 0 \\
0 & 0 & 0 & 0 & 0 \\
0 & 0 & 0 & 0 & 0
\end{bmatrix}
$$

Table B.18: Coefficient matrix $\mathbf{G}_{n,1}$ for the pressure wave $\lambda_1 = u + a$

$$\mathbf{G}_2 = (u-a)\begin{bmatrix} \frac{1}{2} & -\frac{\varrho a}{2} & 0 & 0 & 0 \\ -\frac{1}{2\varrho a} & \frac{1}{2} & 0 & 0 & 0 \\ -\frac{\alpha_g \alpha_l}{2}\left(\frac{1}{\varrho_g a_g^2} - \frac{1}{\varrho_l a_l^2}\right) & +\frac{\alpha_g \alpha_l}{2}\varrho a\left(\frac{1}{\varrho_g a_g^2} - \frac{1}{\varrho_l a_l^2}\right) & 0 & 0 & 0 \\ 0 & 0 & 0 & 0 & 0 \\ 0 & 0 & 0 & 0 & 0 \end{bmatrix}$$

Table B.19: Split coefficient matrix $\mathbf{G}_2$ for the pressure wave $\lambda_2 = u - a$

$$\mathbf{G}_3 = u\begin{bmatrix} 0 & 0 & 0 & 0 & 0 \\ 0 & 0 & 0 & 0 & 0 \\ \frac{\alpha_g \alpha_l}{2}\left(\frac{1}{\varrho_g a_g^2} - \frac{1}{\varrho_l a_l^2}\right) & 0 & 1 & 0 & 0 \\ 0 & 0 & 0 & 0 & 0 \\ 0 & 0 & 0 & 0 & 0 \end{bmatrix}$$

Table B.20: Split coefficient matrix $\mathbf{G}_3$ for the void wave $\lambda_3 = u$

$$\mathbf{G}_4 = u\begin{bmatrix} 0 & 0 & 0 & 0 & 0 \\ 0 & 0 & 0 & 0 & 0 \\ 0 & 0 & 0 & 0 & 0 \\ 0 & 0 & 0 & 1 & 0 \\ 0 & 0 & 0 & 0 & 0 \end{bmatrix}, \qquad \mathbf{G}_5 = u\begin{bmatrix} 0 & 0 & 0 & 0 & 0 \\ 0 & 0 & 0 & 0 & 0 \\ 0 & 0 & 0 & 0 & 0 \\ 0 & 0 & 0 & 0 & 0 \\ 0 & 0 & 0 & 0 & 1 \end{bmatrix}$$

Table B.21: Split coefficient matrices $\mathbf{G}_4$ and $\mathbf{G}_5$ for the entropy waves: $\lambda_{3,4} = u$

B.4 Wallis model

$$\boxed{\mathbf{A}\frac{\partial \mathbf{U}}{\partial t} + \mathbf{B}\frac{\partial \mathbf{U}}{\partial x} = \mathbf{C}}$$

$$\mathbf{U} = \begin{bmatrix} p \\ u_g \\ u_l \\ \alpha \\ s_g \\ s_l \end{bmatrix}, \qquad \mathbf{C} = \begin{bmatrix} \sigma_g^M \\ \sigma_l^M \\ F_g^{\mathrm{v}} + \sigma_g^M (u^{ex} - u_g) + F_g \\ F_l^{\mathrm{v}} + \sigma_l^M (u^{ex} - u_l) + F_l \\ \sigma_g^S \\ \sigma_l^S \end{bmatrix}$$

Table B.22: State vector for primititive variables $\mathbf{U}$ and source term vector $\mathbf{C}$

$$\mathbf{A} = \begin{bmatrix} \dfrac{\alpha_g}{a_g^2} & 0 & 0 & \varrho_g & \alpha_g \left(\dfrac{\partial \varrho_g}{\partial s_g}\right)_p & 0 \\ \dfrac{\alpha_l}{a_l^2} & 0 & 0 & -\varrho_l & 0 & \alpha_l \left(\dfrac{\partial \varrho_l}{\partial s_l}\right)_p \\ 0 & \alpha_g \varrho_g & 0 & 0 & 0 & 0 \\ 0 & 0 & \alpha_l \varrho_l & 0 & 0 & 0 \\ 0 & 0 & 0 & 0 & \alpha_g \varrho_g & 0 \\ 0 & 0 & 0 & 0 & 0 & \alpha_l \varrho_l \end{bmatrix}$$

Table B.23: Coefficient matrix $\mathbf{A}$

$$\mathbf{B} = \begin{bmatrix} \dfrac{\alpha_g u_g}{a_g^2} & \alpha_g \varrho_g & 0 & \varrho_g u_g & \alpha_g u_g \left(\dfrac{\partial \varrho_g}{\partial s_g}\right)_p & 0 \\ \dfrac{\alpha_l u_l}{a_l^2} & 0 & \alpha_l \varrho_l & -\varrho_f u_l & 0 & \alpha_l u_l \left(\dfrac{\partial \varrho_l}{\partial s_l}\right)_p \\ \alpha_g & \alpha_g \varrho_g u_g & 0 & 0 & 0 & 0 \\ \alpha_l & 0 & \alpha_l \varrho_l u_l & 0 & 0 & 0 \\ 0 & 0 & 0 & 0 & \alpha_g \varrho_g u_g & 0 \\ 0 & 0 & 0 & 0 & 0 & \alpha_l \varrho_l u_l \end{bmatrix}$$

Table B.24: Coefficient matrix $\mathbf{B}$

$$\boxed{\frac{\partial \mathbf{U}}{\partial t} + \mathbf{G}\frac{\partial \mathbf{U}}{\partial x} = \mathbf{D}} \quad \text{with} \quad \mathbf{G} = \mathbf{A}^{-1}\mathbf{B}$$

$$\mathbf{G} = \begin{bmatrix} u_g - \alpha_l \Delta u \dfrac{\varrho_g a_0^2}{\varrho_s a_l^2} & \alpha_g a_0^2 \dfrac{\varrho_g \varrho_l}{\varrho_s} & \alpha_l a_0^2 \dfrac{\varrho_g \varrho_l}{\varrho_s} & a_0^2 \Delta u \dfrac{\varrho_g \varrho_l}{\varrho_s} & 0 & 0 \\ \dfrac{1}{\varrho_g} & u_g & 0 & 0 & 0 & 0 \\ \dfrac{1}{\varrho_l} & 0 & u_l & 0 & 0 & 0 \\ \alpha_g \alpha_l \dfrac{\Delta u}{\varrho_s} \dfrac{a_0^2}{a_g^2 a_l^2} & \alpha_g \alpha_l \dfrac{\varrho_g}{\varrho_s} \dfrac{a_0^2}{a_l^2} & -\alpha_g \alpha_l \dfrac{\varrho_l}{\varrho_s} \dfrac{a_0^2}{a_g^2} & u_l + \alpha_l \Delta u \dfrac{\varrho_g a_0^2}{\varrho_s a_l^2} & 0 & 0 \\ 0 & 0 & 0 & 0 & u_g & 0 \\ 0 & 0 & 0 & 0 & 0 & u_l \end{bmatrix}$$

Table B.25: Coefficient matrix $\mathbf{G}$

B.5 Hyperbolic two-phase flow model – one-dimensional conditions

$$\boxed{\mathbf{A}\frac{\partial \mathbf{U}}{\partial t} + \mathbf{B}\frac{\partial \mathbf{U}}{\partial x} = \mathbf{C}}$$

$$\mathbf{U} = \begin{bmatrix} p \\ u_g \\ u_l \\ \alpha_g \\ s_g \\ s_l \end{bmatrix}, \qquad \mathbf{C} = \begin{bmatrix} \sigma_g^M \\ \sigma_l^M \\ F_g^{\mathrm{v}} + \sigma_g^M (u^{\mathrm{ex}} - u_g) + F_g \\ F_l^{\mathrm{v}} + \sigma_l^M (u^{\mathrm{ex}} - u_l) + F_l \\ \sigma_g^S \\ \sigma_l^S \end{bmatrix}$$

with entropy source terms

$$\sigma_g^S = \frac{\sigma_g^M \left[(h^{\mathrm{ex}} - h_g) + \frac{1}{2} (u^{\mathrm{ex}} - u_g)^2 \right]}{T_g} + \frac{F_g^{\mathrm{v}} \left(u^{\mathrm{int}} - u_g \right)}{T_g} + \frac{\sigma_g^Q}{T_g} + \frac{Q_g}{T_g}$$

$$\sigma_l^S = \frac{\sigma_l^M \left[(h^{\mathrm{ex}} - h_l) + \frac{1}{2} (u^{ex} - u_l)^2 \right]}{T_l} + \frac{F_l^{\mathrm{v}} \left(u^{\mathrm{int}} - u_l \right)}{T_l} + \frac{\sigma_l^Q}{T_l} + \frac{Q_l}{T_l}$$

Table B.26: State vector for primitive variables **U** and source term vector **C**

$$\boxed{\mathbf{A}\frac{\partial \mathbf{U}}{\partial t} + \frac{\partial \mathbf{U}}{\partial x} = \mathbf{C}}$$

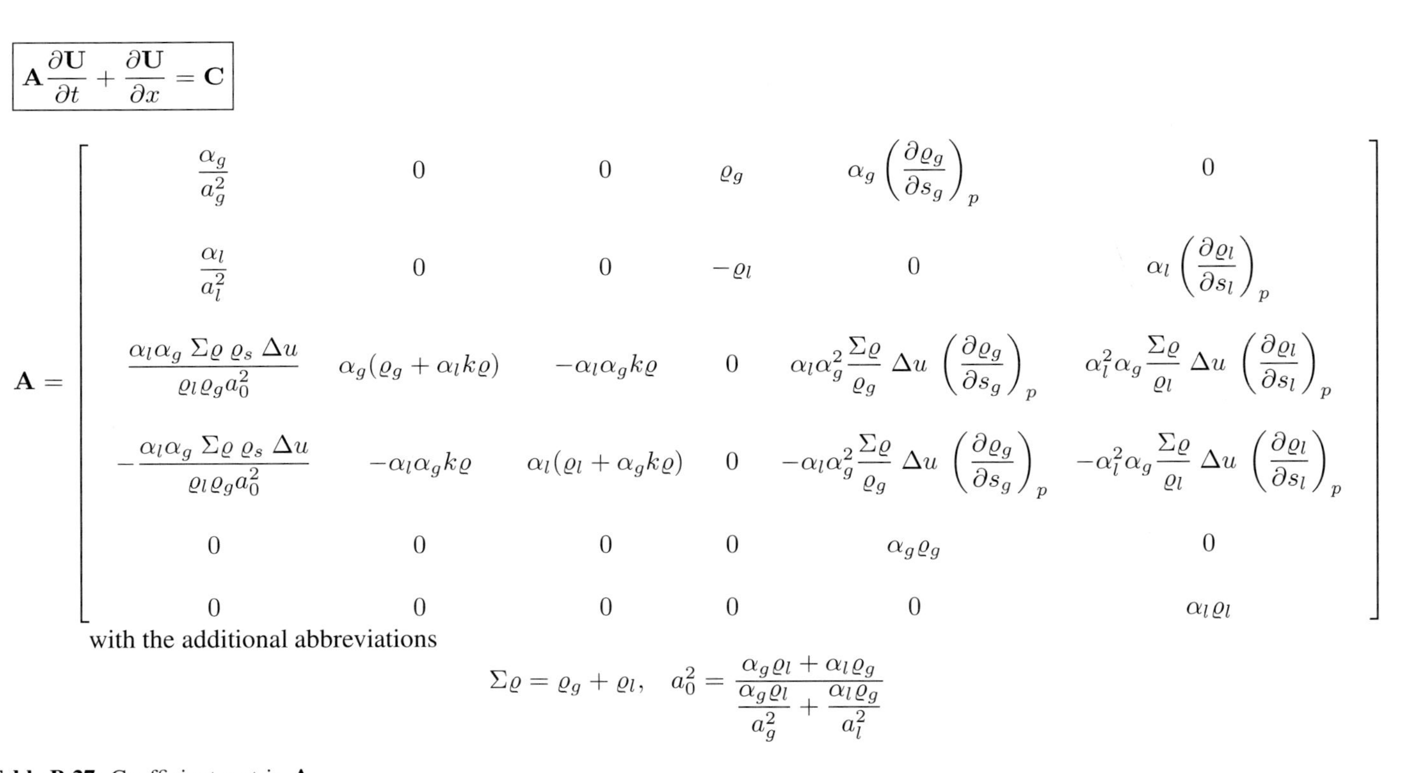

$$\mathbf{A} = \begin{bmatrix} \frac{\alpha_g}{a_g^2} & 0 & 0 & \varrho_g & \alpha_g \left(\frac{\partial \varrho_g}{\partial s_g}\right)_p & 0 \\ \frac{\alpha_l}{a_l^2} & 0 & 0 & -\varrho_l & 0 & \alpha_l \left(\frac{\partial \varrho_l}{\partial s_l}\right)_p \\ \frac{\alpha_l \alpha_g \, \Sigma\varrho \, \varrho_s \, \Delta u}{\varrho_l \varrho_g a_0^2} & \alpha_g(\varrho_g + \alpha_l k \varrho) & -\alpha_l \alpha_g k \varrho & 0 & \alpha_l \alpha_g^2 \frac{\Sigma\varrho}{\varrho_g} \, \Delta u \left(\frac{\partial \varrho_g}{\partial s_g}\right)_p & \alpha_l^2 \alpha_g \frac{\Sigma\varrho}{\varrho_l} \, \Delta u \left(\frac{\partial \varrho_l}{\partial s_l}\right)_p \\ -\frac{\alpha_l \alpha_g \, \Sigma\varrho \, \varrho_s \, \Delta u}{\varrho_l \varrho_g a_0^2} & -\alpha_l \alpha_g k \varrho & \alpha_l(\varrho_l + \alpha_g k \varrho) & 0 & -\alpha_l \alpha_g^2 \frac{\Sigma\varrho}{\varrho_g} \, \Delta u \left(\frac{\partial \varrho_g}{\partial s_g}\right)_p & -\alpha_l^2 \alpha_g \frac{\Sigma\varrho}{\varrho_l} \, \Delta u \left(\frac{\partial \varrho_l}{\partial s_l}\right)_p \\ 0 & 0 & 0 & 0 & \alpha_g \varrho_g & 0 \\ 0 & 0 & 0 & 0 & 0 & \alpha_l \varrho_l \end{bmatrix}$$

with the additional abbreviations

$$\Sigma\varrho = \varrho_g + \varrho_l, \quad a_0^2 = \frac{\alpha_g \varrho_l + \alpha_l \varrho_g}{\frac{\alpha_g \varrho_l}{a_g^2} + \frac{\alpha_l \varrho_g}{a_l^2}}$$

Table B.27: Coefficient matrix $\mathbf{A}$

$$\boxed{\mathbf{A}\frac{\partial \mathbf{U}}{\partial t} + \frac{\partial \mathbf{U}}{\partial x} = \mathbf{C}}$$

$$\mathbf{B} = \begin{bmatrix} \dfrac{\alpha_g u_g}{a_g^2} & \alpha_g \varrho_g & 0 & \varrho_g u_g & \alpha_g u_g \left(\dfrac{\partial \varrho_g}{\partial s_g}\right)_p & 0 \\ \dfrac{\alpha_l u_l}{a_l^2} & 0 & \alpha_l \varrho_l & -\varrho_l u_l & 0 & \alpha_l u_l \left(\dfrac{\partial \varrho_l}{\partial sl}\right)_p \\ \begin{array}{c} \alpha_g \\ +\dfrac{\alpha_l \alpha_g \Sigma\varrho u_a \Delta u}{\varrho_l \varrho_g} \end{array} & \begin{array}{c} \alpha_g \varrho_g u_g \\ +\alpha_l \alpha_g k \varrho u_l \\ -\alpha_l \alpha_g \, \Delta\varrho \, \Delta u \end{array} & \begin{array}{c} -\alpha_l \alpha_g k \varrho u_g \\ \alpha_l \alpha_g \, \Delta\varrho \, \Delta u \end{array} & \alpha_l \alpha_g \, \Sigma\varrho \, (\Delta u)^2 & \begin{array}{c} \alpha_l \alpha_g^2 \, \Delta u \, \dfrac{\Sigma\varrho}{\varrho_g} \\ \times \left(\dfrac{\partial \varrho_g}{\partial s_g}\right)_p u_g \end{array} & \begin{array}{c} \alpha_l^2 \alpha_g \, \Delta u \, \dfrac{\Sigma\varrho}{\varrho_l} \\ \times \left(\dfrac{\partial \varrho_l}{\partial s_l}\right)_p u_l \end{array} \\ \begin{array}{c} \alpha_l \\ +\dfrac{\alpha_l \alpha_g \Sigma\varrho u_a \Delta u}{\varrho_l \varrho_g} \end{array} & \begin{array}{c} -\alpha_l \alpha_g k \varrho u_l \\ +\alpha_l \alpha_g \, \Delta\varrho \, \Delta u \end{array} & \begin{array}{c} \alpha_l \varrho_l u_l \\ +\alpha_l \alpha_g k \varrho u_l \\ -\alpha_l \alpha_g \, \Delta\varrho \, \Delta u \end{array} & -\alpha_l \alpha_g \, \Sigma\varrho \, (\Delta u)^2 & \begin{array}{c} -\alpha_l \alpha_g^2 \, \Delta u \, \dfrac{\Sigma\varrho}{\varrho_g} \\ \times \left(\dfrac{\partial \varrho_g}{\partial s_g}\right)_p u_g \end{array} & \begin{array}{c} -\alpha_l^2 \alpha_g \, \Delta u \, \dfrac{\Sigma\varrho}{\varrho l} \\ \times \left(\dfrac{\partial \varrho_l}{\partial s_l}\right)_p u_l \end{array} \\ 0 & 0 & 0 & 0 & \varrho_g u_g & 0 \\ 0 & 0 & 0 & 0 & 0 & \varrho_l u_l \end{bmatrix}$$

with the additional abbreviations:

$$\Delta\varrho = \alpha_l \varrho_g - \alpha_g \varrho_l, \quad \Sigma\varrho = \varrho_g + \varrho_l, \quad u_a = \frac{\alpha_g \varrho_l u_g}{a_l^2} + \frac{\alpha_l \varrho_g u_l}{a_l^2}$$

Table B.28: Coefficient matrix $\mathbf{B}$

$$\boxed{\frac{\partial \mathbf{U}}{\partial t} + \mathbf{G}\frac{\partial \mathbf{U}}{\partial x} = \mathbf{D}} \qquad \text{with} \qquad \mathbf{G} = \mathbf{A}^{-1}\mathbf{B}$$

$$\mathbf{G} = \begin{bmatrix} \alpha_l u_l \frac{\varrho_g}{\varrho_s}\frac{a_0^2}{a_l^2} + \alpha_g u_g \frac{\varrho_l}{\varrho_s}\frac{a_0^2}{a_g^2} & \alpha_g a_0^2 \frac{\varrho_l \varrho_g}{\varrho_s} & \alpha_l a_0^2 \frac{\varrho_l \varrho_g}{\varrho_s} & a_0^2 \Delta u \frac{\varrho_l \varrho_g}{\varrho_s} & 0 & 0 \\ \frac{\hat{\varrho}_l}{\hat{\varrho}^2} & u_1 \frac{\varrho_l \varrho_g}{\hat{\varrho}^2} + u_2 \frac{k\varrho^2}{\hat{\varrho}^2} & -\alpha_l \Delta u \frac{\varrho_l \hat{\varrho}_l}{\hat{\varrho}^2} & 0 & 0 & 0 \\ \frac{\hat{\varrho}_g}{\hat{\varrho}^2} & \alpha_g \Delta u \frac{\varrho_g \hat{\varrho}_g}{\hat{\varrho}^2} & u_1 \frac{\varrho_l \varrho_g}{\hat{\varrho}^2} + u_2 \frac{k\varrho^2}{\hat{\varrho}^2} & 0 & 0 & 0 \\ \alpha_l \alpha_g \frac{\Delta u}{\varrho_s}\frac{a_0^2}{a_l^2 a_g^2} & \alpha_l \alpha_g \frac{\varrho_g}{\varrho_s}\frac{a_0^2}{a_l^2} & -\alpha_l \alpha_g \frac{\varrho_l}{\varrho_s}\frac{a_0^2}{a_g^2} & \alpha_l u_g \frac{\varrho_g}{\varrho_s}\frac{a_0^2}{a_l^2} + \alpha_g u_l \frac{\varrho_l}{\varrho_s}\frac{a_0^2}{a_g^2} & 0 & 0 \\ 0 & 0 & 0 & 0 & u_g & 0 \\ 0 & 0 & 0 & 0 & 0 & u_l \end{bmatrix}$$

with the additional abbreviations:

$$\hat{\varrho} = \sqrt{\varrho_g \varrho_l + k\varrho^2}, \quad u_1 = \alpha_g u_g + \alpha_l u_l, \quad u_2 = \frac{\alpha_g \varrho_g u_g + \alpha_l \varrho_l u_l}{\alpha_g \varrho_g + \alpha_l \varrho_l}, \quad a_0^2 = \frac{\alpha_g \varrho_l + \alpha_l \varrho_g}{\frac{\alpha_g \varrho_l}{a_g^2} + \frac{\alpha_l \varrho_g}{a_l^2}}$$

Table B.29: Coefficient matrix **G**

$$
\mathbf{D} = \begin{bmatrix}
\frac{\varrho_l a_0^2}{\varrho_s}\left[\sigma_g^M - \frac{1}{\varrho_g}\left(\frac{\partial \varrho_g}{\partial s_g}\right)_p \sigma_g^S\right] + \frac{\varrho_g a_0^2}{\varrho_s}\left[\sigma_l^M - \frac{1}{\varrho_l}\left(\frac{\partial \varrho_l}{\partial s_l}\right)_p \sigma_l^S\right] \\
\frac{\varrho_l}{\alpha_g \hat{\varrho}^2}\left[F_g^{\mathrm{v}} + \sigma_g^M (u^{\mathrm{ex}} - u_g)\right] - \sigma_g^M \frac{k\varrho\varrho_g + \alpha_l(\varrho_l^2 - \varrho_g^2)}{\varrho_g \hat{\varrho}^2}\,\Delta u + \frac{F_g(\varrho_l + \alpha_g k\varrho) + F_l \alpha_g k\varrho}{\alpha_g \hat{\varrho}_2} \\
-\frac{\varrho_g}{\alpha_l \hat{\varrho}^2}\left[F_g^{\mathrm{v}} + \sigma_g^M (u^{\mathrm{ex}} - u_l)\right] - \sigma_g^M \frac{k\varrho\varrho_l + \alpha_g(\varrho_g^2 - \varrho_l^2)}{\varrho_l \hat{\varrho}^2}\,\Delta u + \frac{F_g \alpha_l k\varrho + F_l(\varrho_g + \alpha_l k\varrho)}{\alpha_l \hat{\varrho}^2} \\
\frac{\alpha_l a_0^2}{\varrho_s a_l^2}\left[\sigma_g^M - \frac{1}{\varrho_g}\left(\frac{\partial \varrho_g}{\partial s_g}\right)_p \sigma_g^S\right] - \frac{\alpha_g a_0^2}{\varrho_s a_g^2}\left[\sigma_l^M - \frac{1}{\varrho_l}\left(\frac{\partial \varrho_l}{\partial s_l}\right)_p \sigma_l^S\right] \\
\frac{1}{\alpha_g \varrho_g}\sigma_g^S \\
\frac{1}{\alpha_l \varrho_l}\sigma_l^S
\end{bmatrix}
$$

with

$$
\sigma_g^S = \frac{\sigma_g^M\left[(h^{\mathrm{ex}} - h_g) + \frac{1}{2}\left(u^{\mathrm{ex}} - u_g\right)^2\right]}{T_g} + \frac{F_g^v\left(u^{\mathrm{int}} - u_g\right)}{T_g} + \frac{\sigma_g^Q}{T_g} + \frac{Q_g}{T_g}
$$

$$
\sigma_l^S = \frac{\sigma_l^M\left[(h^{\mathrm{ex}} - h_l) + \frac{1}{2}\left(u^{\mathrm{ex}} - u_l\right)^2\right]}{T_l} + \frac{F_l^v\left(u^{\mathrm{int}} - u_l\right)}{T_l} + \frac{\sigma_l^Q}{T_l} + \frac{Q_l}{T_l}
$$

Table B.30: Source term vector $\mathbf{D}$

$$\mathbf{V}_R = \begin{bmatrix} \alpha_l \varrho_l (\Delta u)^2 & 0 & \Delta u & -\alpha_l \left(1 - \alpha_l \dfrac{(\Delta u)^2}{a_l^2}\right) & 0 & 0 \\ \alpha_g \varrho_g (\Delta u)^2 & -\Delta u & 0 & \alpha_g \left(1 - \alpha_g \dfrac{(\Delta u)^2}{a_g^2}\right) & 0 & 0 \\ \hat{\varrho}^2 \tilde{a}^2 & \hat{\varrho}_l \left(a - \dfrac{\alpha_l \varrho_l \hat{\varrho}_g}{\hat{\varrho}^2} \Delta u\right) & \hat{\varrho}_g \left(a + \dfrac{\alpha_g \varrho_g \hat{\varrho}_l}{\hat{\varrho}^2} \Delta u\right) & -\alpha_g \alpha_l \dfrac{\hat{\varrho}^2}{\hat{\varrho}_s} \left(\dfrac{\hat{\varrho}_g \tilde{a}^2}{\varrho_g a_g^2} - \dfrac{\hat{\varrho}_l \tilde{a}^2}{\varrho_l a_l^2}\right) & 0 & 0 \\ \hat{\varrho}^2 \tilde{a}^2 & \hat{\varrho}_l \left(-a - \dfrac{\alpha_l \varrho_l \hat{\varrho}_g}{\hat{\varrho}^2} \Delta u\right) & \hat{\varrho}_g \left(-a + \dfrac{\alpha_g \varrho_g \hat{\varrho}_l}{\hat{\varrho}^2} \Delta u\right) & -\alpha_g \alpha_l \dfrac{\hat{\varrho}^2}{\hat{\varrho}_s} \left(\dfrac{\hat{\varrho}_g \tilde{a}^2}{\varrho_g a_g^2} - \dfrac{\hat{\varrho}_l \tilde{a}^2}{\varrho_l a_l^2}\right) & 0 & 0 \\ 0 & 0 & 0 & 0 & 1 & 0 \\ 0 & 0 & 0 & 0 & 0 & 1 \end{bmatrix}$$

Table B.31: Matrix of right-hand side eigenvectors $\mathbf{V}_R$ of the coefficient matrix $\mathbf{G}$

$$
\mathbf{V}_L = \begin{bmatrix}
\alpha_g \dfrac{\Delta u}{a_g^2} & \alpha_g \varrho_g & -\alpha_g \varrho_g \dfrac{\hat{\varrho}_l}{\hat{\varrho}_g} & \varrho_g \,\Delta u & 0 & 0 \\
\alpha_l \dfrac{\Delta u}{a_l^2} & \alpha_l \varrho_l \dfrac{\hat{\varrho}_g}{\hat{\varrho}_l} & -\alpha_l \varrho_l & -\varrho_l \,\Delta u & 0 & 0 \\
a\hat{\varrho}_s + \hat{\varrho}\Delta u \,\Delta a & \alpha_g \hat{\varrho}^2 \left(a_2^2 + a\,\Delta u \dfrac{\varrho_g \hat{\varrho}_s}{\hat{\varrho}^2} \right) & \alpha_l \hat{\varrho}^2 \left(a_1^2 - a\,\Delta u \dfrac{\varrho_l \hat{\varrho}_s}{\hat{\varrho}^2} \right) & \hat{\varrho}^2 \,\Delta u \,\tilde{a}^2 & 0 & 0 \\
-a\hat{\varrho}_s + \hat{\varrho}\Delta u \,\Delta a & \alpha_g \hat{\varrho}^2 \left(a_2^2 - a\,\Delta u \dfrac{\varrho_g \hat{\varrho}_s}{\hat{\varrho}^2} \right) & \alpha_l \hat{\varrho}^2 \left(a_1^2 + a\,\Delta u \dfrac{\varrho_l \hat{\varrho}_s}{\hat{\varrho}^2} \right) & \hat{\varrho}^2 \,\Delta u \,\tilde{a}^2 & 0 & 0 \\
0 & 0 & 0 & 0 & 1 & 0 \\
0 & 0 & 0 & 0 & 0 & 1
\end{bmatrix}
$$

with

$$
\Delta u = u_g - u_l, \quad a_1 = \sqrt{a^2 - \frac{{\alpha_g}^2 \varrho_l \varrho_g \hat{\varrho}_l}{\hat{\varrho}^4} (\Delta u)^2}, \quad a_2 = \sqrt{a^2 - \frac{{\alpha_l}^2 \varrho_l \varrho_g \hat{\varrho}_g^2}{\hat{\varrho}^4} (\Delta u)^2}, \quad \Delta a = {\alpha_g}^2 \frac{\hat{\varrho}_l \tilde{a}^2}{\hat{\varrho} a_g^2} - {\alpha_l}^2 \frac{\hat{\varrho}_g \tilde{a}^2}{\hat{\varrho} a_l^2}
$$

Table B.32: Matrix of left-hand side eigenvectors $\mathbf{V}_L$ of the coefficient matrix $\mathbf{G}$

$$\boxed{\mathbf{T}^{-1}\frac{\partial \mathbf{U}}{\partial t}+\mathbf{\Lambda}\mathbf{T}^{-1}\frac{\partial \mathbf{U}}{\partial x} = \mathbf{E}}$$

$$E_1 = \sigma_g^{MS} + \left[F_g^{\mathrm{v}} + \sigma_g^M(u^{\mathrm{ex}} - u^m)\right] \frac{\varrho_g \varrho_l}{\alpha_l \hat{\varrho}_g} - \sigma_g^M \Delta u \frac{\alpha_l \alpha_g (\varrho_l^2 - \varrho_g^2)}{\alpha_l \varrho_l \hat{\varrho}_g} + \alpha_g \frac{\varrho_g}{\hat{\varrho}_g}(\varrho_g - \varrho_l) g$$

$$E_2 = \sigma_l^{MS} + \left[F_g^{\mathrm{v}} + \sigma_g^M(u^{\mathrm{ex}} - u^m)\right] \frac{\varrho_g \varrho_l}{\alpha_g \hat{\varrho}_l} - \sigma_g^M \Delta u \frac{\alpha_l \alpha_g (\varrho_l^2 - \varrho_g^2)}{\alpha_g \varrho_g \hat{\varrho}_g} + \alpha_l \frac{\varrho_l}{\hat{\varrho}_l}(\varrho_g - \varrho_l) g$$

$$E_3 = \sigma_g^{MS} \frac{a_0^2 \Delta u}{a_l^2 \varrho_s} (\tilde{a}^2 \alpha_l \varrho_l^2 \Delta u + a a_l^2 \varrho_l \hat{\varrho}_s + a_l^2 \Delta a \varrho_l \hat{\varrho}) + \sigma_l^{MS} \frac{a_0^2 \Delta u}{a_g^2 \varrho_s} (-\tilde{a}^2 \alpha_g \hat{\varrho}^2 + a a_g^2 \varrho_g \hat{\varrho}_s + a_g^2 \Delta a \varrho_g \hat{\varrho})$$

$$+ \left[F_g^{\mathrm{v}} + \sigma_g^M(u^{\mathrm{ex}} - u^m)\right] \left[a_2^2 \varrho_l - a_1^2 \varrho_g + 2\frac{a \varrho_l \varrho_g}{\hat{\varrho}^2} \hat{\varrho}_s^2 \Delta u\right] + \left[a_1^2 \alpha_l \hat{\varrho}^2 + a_2^2 \alpha_g \hat{\varrho}^2 - a\, \Delta u\, (\alpha_l \varrho_l - \alpha_g \varrho_g) \hat{\varrho}_s\right] g$$

$$E_4 = \sigma_g^{MS} \frac{a_0^2 \Delta u}{a_l^2 \varrho_s} \left[\tilde{a}^2 \alpha_l \hat{\varrho}^2 - a a_l^2 \varrho_l \hat{\varrho}_s + a_l^2 \Delta a \varrho_l \hat{\varrho}\right] + \sigma_l^{MS} \frac{a_0^2 \Delta u}{a_g^2 \varrho_s} (-\tilde{a}^2 \alpha_g \hat{\varrho}^2 - a a_g^2 \varrho_g \hat{\varrho}_s + a_g^2 \Delta a u \varrho_g \hat{\varrho})$$

$$+ \left[F_g^{\mathrm{v}} + \sigma_g^M(u^{\mathrm{ex}} - u^m)\right] \left[a_2^2 \varrho_l - a_1^2 \varrho_g - 2\frac{a \varrho_l \varrho_g}{\hat{\varrho}^2} \hat{\varrho}_s^2 \Delta u\right] + \left[a_1^2 \alpha_l \hat{\varrho}^2 + a_2^2 \alpha_g \hat{\varrho}^2 + a\, \Delta u\, (\alpha_l \varrho_l - \alpha_g \varrho_g) \hat{\varrho}_s\right] g$$

$$E_5 = \sigma_g^S \frac{1}{\alpha_g \varrho_g}, \qquad E_6 = \sigma_l^S \frac{1}{\alpha_l \varrho_l}$$

with $\sigma_g^{MS} = \sigma_g^M - \sigma_g^S \frac{1}{\varrho_g} \left(\frac{\partial \varrho_g}{\partial s_g}\right)_p$, $\quad \sigma_l^{MS} = \sigma_l^M - \sigma_l^S \frac{1}{\varrho_l} \left(\frac{\partial \varrho_l}{\partial s_l}\right)_p$, $\quad u^m = \alpha_g u_l + \alpha_l u_g$

Table B.33: Elements of new state vector as used in characteristic form of flow equations

$$
\mathbf{G}_1 = -u_g \frac{\hat{\varrho}_g}{\Delta u \left(\varrho_g \hat{\varrho}_s - \alpha_l \hat{\varrho}_g \varrho_s \frac{(\Delta u)^2}{a_0^2} \right)}
$$

$$
\times \begin{bmatrix}
\alpha_l \alpha_g \varrho_l \Delta u \frac{\Delta u^2}{a_g^2} & \alpha_l \alpha_g \varrho_l \varrho_g \Delta u^2 & -\alpha_l \alpha_g \varrho_l \varrho_g \frac{\hat{\varrho}_l}{\hat{\varrho}_g} \Delta u^2 & \alpha_l \varrho_l \varrho_g \Delta u^3 & 0 & 0 \\
0 & 0 & 0 & 0 & 0 & 0 \\
\alpha_g \frac{\Delta u^2}{a_g^2} & \alpha_g \varrho_g \Delta u & -\alpha_g \varrho_g \frac{\hat{\varrho}_l}{\hat{\varrho}_g} \Delta u & \varrho_g \Delta u^2 & 0 & 0 \\
-\alpha_l \alpha_g g_l \frac{\Delta u^2}{a_g^2} & -\alpha_l \alpha_g \varrho_g g_l & \alpha_l \alpha_g \varrho_g g_l \frac{\hat{\varrho}_l}{\hat{\varrho}_g} & -\alpha_l \varrho_g g_l \Delta u & 0 & 0 \\
0 & 0 & 0 & 0 & 0 & 0 \\
0 & 0 & 0 & 0 & 0 & 0
\end{bmatrix}
$$

with $\quad g_l = 1 - \alpha_l \frac{\Delta u^2}{a_l^2}$

Table B.34: Split coefficient matrix $\mathbf{G}_1$ for the eigenvalue $\lambda_1 = u_g$

$$
\mathbf{G}_2 = -u_l \frac{\hat{\varrho}_l}{\Delta u \left(\varrho_l \hat{\varrho}_s - \alpha_g \hat{\varrho}_l \varrho_s \dfrac{(\Delta u)^2}{a_0^2} \right)}
$$

$$
\times \begin{bmatrix}
\alpha_l \alpha_g \varrho_g \Delta u \dfrac{\Delta u^3}{a_l^2} & \alpha_l \alpha_g \varrho_l \varrho_g \dfrac{\hat{\varrho}_g}{\hat{\varrho}_l} \Delta u^2 & -\alpha_l \alpha_g \varrho_l \varrho_g \Delta u^2 & \alpha_g \varrho_l \varrho_g \Delta u^3 & 0 & 0 \\
-\alpha_l \dfrac{\Delta u^2}{a_l^2} & -\alpha_l \varrho_l \dfrac{\hat{\varrho}_g}{\hat{\varrho}_l} \Delta u & \alpha_l \varrho_l \Delta u & \varrho_l \Delta u^2 & 0 & 0 \\
0 & 0 & 0 & 0 & 0 & 0 \\
\alpha_l \alpha_g g_g \dfrac{\Delta u}{a_l^2} & \alpha_l \alpha_g \varrho_l g_g \dfrac{\hat{\varrho}_g}{\hat{\varrho}_l} & -\alpha_l \alpha_g \varrho_l g_g & -\alpha_g \varrho_g g_g \Delta u & 0 & 0 \\
0 & 0 & 0 & 0 & 0 & 0 \\
0 & 0 & 0 & 0 & 0 & 0
\end{bmatrix}
$$

$$
\text{with} \qquad g_g = 1 - \alpha_g \frac{\Delta u^2}{a_g^2}
$$

Table B.35: Split coefficient matrix $\mathbf{G}_2$ for the eigenvalue $\lambda_2 = u_l$

$$\mathbf{G}_{12}^{\text{hom}} = \lim_{\Delta u \to 0} [\mathbf{G}_1 + \mathbf{G}_2]$$

$$\mathbf{G}_{12}^{\text{hom}} = \begin{bmatrix} 0 & 0 & 0 & 0 & 0 & 0 \\ 0 & \alpha_l \dfrac{\hat{\varrho}_g}{\hat{\varrho}_s} u & -\alpha_l \dfrac{\hat{\varrho}_l}{\hat{\varrho}_s} u & 0 & 0 & 0 \\ 0 & \alpha_g \dfrac{\hat{\varrho}_g}{\hat{\varrho}_s} u & \alpha_l \dfrac{\hat{\varrho}_l}{\hat{\varrho}_s} u & 0 & 0 & 0 \\ \dfrac{\alpha_l \alpha_g}{\varrho_l \varrho_g \hat{\varrho}_s} \left[\dfrac{\varrho_l \hat{\varrho}_g}{a_g^2} - \dfrac{\varrho_g \hat{\varrho}_l}{a_l^2} \right] & \alpha_l \alpha_g \varrho_l \dfrac{\hat{\varrho}_g}{\hat{\varrho}_l} & -\alpha_l \alpha_g \varrho_l & u & 0 & 0 \\ 0 & 0 & 0 & 0 & 0 & 0 \\ 0 & 0 & 0 & 0 & 0 & 0 \end{bmatrix}$$

Table B.36: Split coefficient matrix $\mathbf{G}_{12}^{\text{hom}}$ for homogeneous conditions $\lambda_1 = \lambda_2 = u$

$$\mathbf{G}_3 = (u+a)\frac{\tilde{a}^2 + a\,\Delta u\,\Delta\alpha - 2\alpha_l\alpha_g\dfrac{\varrho_l\varrho_g\hat{\varrho}_l\hat{\varrho}_g}{\hat{\varrho}^4}(\Delta u)^2}{2a\hat{\varrho}^2\hat{\varrho}_s a_3^2 a_4^2}$$

$$\begin{bmatrix} \hat{\varrho}^2\tilde{a}^2 \\ \hat{\varrho}_l\left(a - \dfrac{\alpha_l\varrho_l\hat{\varrho}_g}{\hat{\varrho}^2}\Delta u\right) \\ \hat{\varrho}_g\left(a + \dfrac{\alpha_g\varrho_g\hat{\varrho}_l}{\hat{\varrho}^2}\Delta u\right) \\ -\alpha_g\alpha_l\dfrac{\hat{\varrho}^2}{\hat{\varrho}_s}\left(\dfrac{\hat{\varrho}_g\tilde{a}^2}{\varrho_g a_g^2} - \dfrac{\hat{\varrho}_l\tilde{a}^2}{\varrho_l a_l^2}\right) \\ 0 \\ 0 \end{bmatrix} \times \begin{bmatrix} a\hat{\varrho}_s + \hat{\varrho}\Delta u\,\Delta a \\ \alpha_g\hat{\varrho}^2\left(a_2^2 + a\,\Delta u\dfrac{\varrho_g\hat{\varrho}_s}{\hat{\varrho}^2}\right) \\ \alpha_l\hat{\varrho}^2\left(a_1^2 - a\,\Delta u\dfrac{\varrho_l\hat{\varrho}_s}{\hat{\varrho}^2}\right) \\ \hat{\varrho}^2\,\Delta u\,\tilde{a}^2 \\ 0 \\ 0 \end{bmatrix}^T$$

with

$$\Delta\alpha = \frac{\alpha_l\varrho_l\hat{\varrho}_g}{\hat{\varrho}^2} - \frac{\alpha_g\varrho_g\hat{\varrho}_l}{\hat{\varrho}^2}$$

$$a_3^2 = \tilde{a}^2 - \frac{\alpha_l\varrho_l\hat{\varrho}_g}{\hat{\varrho}^2}(\Delta u)^2$$

$$a_4^2 = \tilde{a}^2 - \frac{\alpha_g\varrho_g\hat{\varrho}_l}{\hat{\varrho}^2}(\Delta u)^2$$

Table B.37: Split coefficient matrix $\mathbf{G}_3$ for the eigenvalues $\lambda_3 = u + a$

$$\mathbf{G}_4 = -(u-a)\frac{\tilde{a}^2 - a\,\Delta u\,\Delta\alpha - 2\alpha_f\alpha_g\dfrac{\varrho_l\varrho_g\hat{\varrho}_l\hat{\varrho}_g}{\hat{\varrho}^4}(\Delta u)^2}{2a\hat{\varrho}^2\hat{\varrho}_s a_3^2 a_4^2}$$

$$\begin{bmatrix} \hat{\varrho}^2\tilde{a}^2 \\ \hat{\varrho}_l\left(-a - \dfrac{\alpha_l\varrho_l\hat{\varrho}_g}{\hat{\varrho}^2}\Delta u\right) \\ \hat{\varrho}_g\left(-a + \dfrac{\alpha_g\varrho_g\hat{\varrho}_l}{\hat{\varrho}^2}\Delta u\right) \\ -\alpha_g\alpha_l\dfrac{\hat{\varrho}^2}{\hat{\varrho}_s}\left(\dfrac{\hat{\varrho}_g\tilde{a}^2}{\varrho_g a_g^2} - \dfrac{\hat{\varrho}_l\tilde{a}^2}{\varrho_l a_l^2}\right) \\ 0 \\ 0 \end{bmatrix} \times \begin{bmatrix} -a\hat{\varrho}_s + \hat{\varrho}\Delta u\,\Delta a \\ \alpha_g\hat{\varrho}^2\left(a_2^2 - a\,\Delta u\dfrac{\varrho_g\hat{\varrho}_s}{\hat{\varrho}^2}\right) \\ \alpha_l\hat{\varrho}^2\left(a_1^2 + a\,\Delta u\dfrac{\varrho_l\hat{\varrho}_s}{\hat{\varrho}^2}\right) \\ \hat{\varrho}^2\,\Delta u\,\tilde{a}^2 \\ 0 \\ 0 \end{bmatrix}^T$$

with

$$\Delta\alpha = \frac{\alpha_l\varrho_l\hat{\varrho}_g}{\hat{\varrho}^2} - \frac{\alpha_g\varrho_g\hat{\varrho}_l}{\hat{\varrho}^2}$$

$$a_3^2 = \tilde{a}^2 - \frac{\alpha_l\varrho_l\hat{\varrho}_g}{\hat{\varrho}^2}(\Delta u)^2$$

$$a_4^2 = \tilde{a}^2 - \frac{\alpha_g\varrho_g\hat{\varrho}_l}{\hat{\varrho}^2}(\Delta u)^2$$

Table B.38: Split coefficient matrix $\mathbf{G}_3$ for the eigenvalues $\lambda_4 = u - a$

$$\mathbf{G}_5 = u_g \begin{bmatrix} 0 & 0 & 0 & 0 & 0 & 0 \\ 0 & 0 & 0 & 0 & 0 & 0 \\ 0 & 0 & 0 & 0 & 0 & 0 \\ 0 & 0 & 0 & 0 & 0 & 0 \\ 0 & 0 & 0 & 0 & 1 & 0 \\ 0 & 0 & 0 & 0 & 0 & 0 \end{bmatrix}, \quad \mathbf{G}_6 = u_l \begin{bmatrix} 0 & 0 & 0 & 0 & 0 & 0 \\ 0 & 0 & 0 & 0 & 0 & 0 \\ 0 & 0 & 0 & 0 & 0 & 0 \\ 0 & 0 & 0 & 0 & 0 & 0 \\ 0 & 0 & 0 & 0 & 0 & 0 \\ 0 & 0 & 0 & 0 & 0 & 1 \end{bmatrix}$$

Table B.39: Split coefficient matrices $\mathbf{G}_5$ and $\mathbf{G}_6$ for the entropy waves: $\lambda_5 = u_g$ and $\lambda_6 = u_l$

$$\boxed{\frac{\partial \mathbf{V}}{\partial t} + \mathbf{H}\frac{\partial \mathbf{V}}{\partial x} = \mathbf{D}} \quad \text{with} \quad \mathbf{H} = \mathbf{J}\,\mathbf{G}\,\mathbf{J}^{-1}$$

$$\mathbf{U} = \begin{bmatrix} p \\ u_g \\ u_l \\ \alpha_g \\ s_g \\ s_l \end{bmatrix}, \quad \mathbf{V(U)} = \begin{bmatrix} \alpha_g \varrho_g \\ \alpha_l \varrho_l \\ \alpha_g \varrho_g u_g \\ \alpha_l \varrho_l u_l \\ \alpha_g \varrho_g s_g \\ \alpha_l \varrho_l s_l \end{bmatrix}, \quad \mathbf{F(U)} = \begin{bmatrix} \alpha_g \varrho_g u_g \\ \alpha_l \varrho_l u_l \\ \alpha_g (\varrho_g u_g^2 + p) \\ \alpha_l (\varrho_l u_l^2 + p) \\ \alpha_g \varrho_g u_g s_g \\ \alpha_l \varrho_l u_l s_l \end{bmatrix}$$

$$\mathbf{J} = \frac{\partial \mathbf{V}}{\partial \mathbf{U}} = \begin{bmatrix} \frac{\alpha_g}{a_g^2} & 0 & 0 & \varrho_g & \alpha_g \left(\frac{\partial \varrho_g}{\partial s_g}\right)_p & 0 \\ \frac{\alpha_l}{a_l^2} & 0 & 0 & -\varrho_l & 0 & \alpha_l \left(\frac{\partial \varrho_l}{\partial s_l}\right)_p \\ \frac{\alpha_g u_g}{a_g^2} & \alpha_g \varrho_g & 0 & \varrho_g u_g & \alpha_g u_g \left(\frac{\partial \varrho_g}{\partial s_g}\right)_p & 0 \\ \frac{\alpha_l u_l}{a_l^2} & 0 & \alpha_l \varrho_l & -\varrho_l u_l & 0 & \alpha_l u_l \left(\frac{\partial \varrho_l}{\partial s_l}\right)_p \\ \frac{\alpha_g s_g}{a_g^2} & 0 & 0 & \varrho_g s_g & \alpha_g \varrho_g f_g & 0 \\ \frac{\alpha_l s_l}{a_l^2} & 0 & 0 & -\varrho_l s_l & 0 & \alpha_l \varrho_l f_l \end{bmatrix}$$

$$\text{with} \quad f_g = 1 + \frac{1}{\varrho_g}\left(\frac{\partial \varrho_g}{\partial s_g}\right)_p \quad \text{and} \quad f_l = 1 + \frac{1}{\varrho_l}\left(\frac{\partial \varrho_l}{\partial s_l}\right)_p$$

Table B.40: Jacobian matrix **J** used in conservative form of flow equations

$$\mathbf{K} = \frac{\partial \mathbf{F}}{\partial \mathbf{U}} = \begin{bmatrix} \dfrac{\alpha_g u_g}{a_g^2} & \alpha_g \varrho_g & 0 & \varrho_g u_g & \alpha_g u_g \left(\dfrac{\partial \varrho_g}{\partial s_g}\right)_p & 0 \\ \dfrac{\alpha_l u_l}{a_l^2} & 0 & \alpha_l \varrho_l & -\varrho_l u_l & 0 & \alpha_l u_l \left(\dfrac{\partial \varrho_l}{\partial s_l}\right)_p \\ \alpha_g \left(1 + \dfrac{u_g^2}{a_g^2}\right) & 2\alpha_g \varrho_g u_g & 0 & p + \varrho_g u_g^2 & \alpha_g u_g^2 \left(\dfrac{\partial \varrho_g}{\partial s_g}\right)_p & 0 \\ \alpha_l \left(1 + \dfrac{u_l^2}{a_l^2}\right) & 0 & 2\alpha_l \varrho_l u_l & -p - \varrho_l u_l^2 & 0 & \alpha_l u_l^2 \left(\dfrac{\partial \varrho_l}{\partial s_l}\right)_p \\ \dfrac{\alpha_g u_g s_g}{a_g^2} & \alpha_g u_g s_g & 0 & \varrho_g u_g s_g & \alpha_g \varrho_g u_g f_g & 0 \\ \dfrac{\alpha_l u_l s_l}{a_l^2} & 0 & \alpha_l u_l s_l & -\varrho_l u_l s_l & 0 & \alpha_l \varrho_l u_l f_l \end{bmatrix}$$

$$\text{with} \quad f_g = 1 + \frac{1}{\varrho_g}\left(\frac{\partial \varrho_g}{\partial s_g}\right)_p \quad \text{and} \quad f_l = 1 + \frac{1}{\varrho_l}\left(\frac{\partial \varrho_l}{\partial s_l}\right)_p$$

Table B.41: Jacobian matrix **K** used in conservative form of flow equations

$$\boxed{\mathbf{H} = \mathbf{H}_1 + \mathbf{H}_2}$$

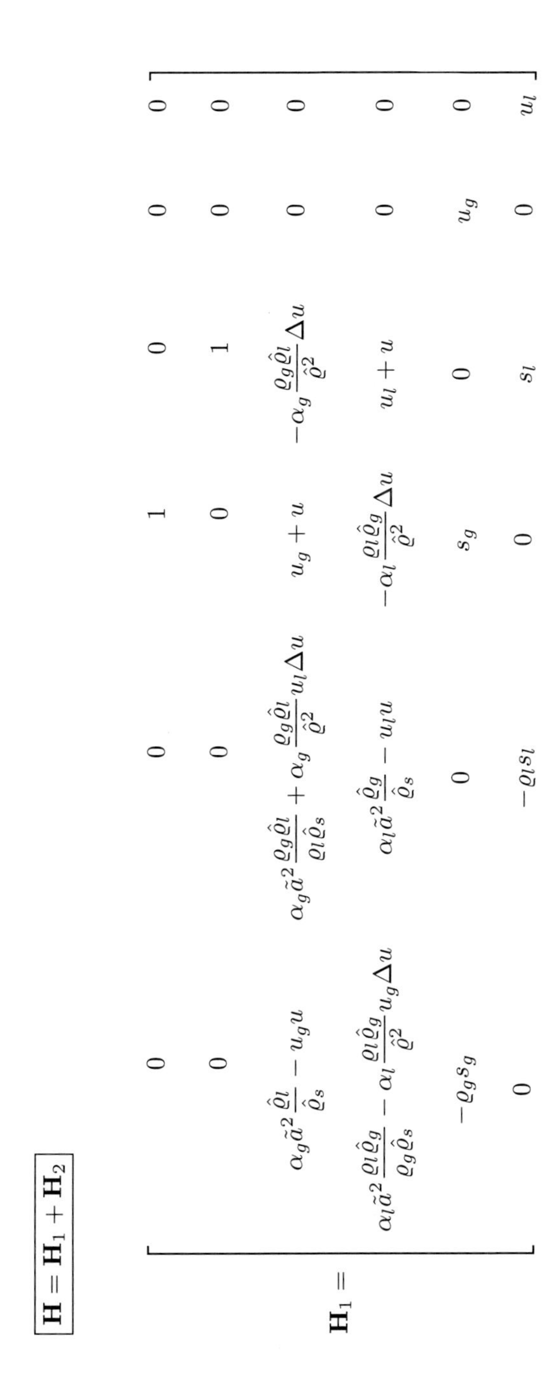

$$\mathbf{H}_1 = \begin{bmatrix} 0 & 0 & 1 & 0 & 0 & 0 \\ 0 & 0 & 0 & 1 & 0 & 0 \\ \alpha_g \tilde{a}^2 \dfrac{\hat{\varrho}_l}{\hat{\varrho}_s} - u_g u & \alpha_g \tilde{a}^2 \dfrac{\varrho_g \hat{\varrho}_l}{\varrho_l \hat{\varrho}_s} + \alpha_g \dfrac{\varrho_g \hat{\varrho}_l}{\hat{\varrho}^2} u_l \Delta u & u_g + u & -\alpha_g \dfrac{\varrho_g \hat{\varrho}_l}{\hat{\varrho}^2} \Delta u & 0 & 0 \\ \alpha_l \tilde{a}^2 \dfrac{\varrho_l \hat{\varrho}_g}{\varrho_g \hat{\varrho}_s} - \alpha_l \dfrac{\varrho_l \hat{\varrho}_g}{\hat{\varrho}^2} u_g \Delta u & \alpha_l \tilde{a}^2 \dfrac{\hat{\varrho}_g}{\hat{\varrho}_s} - u_l u & -\alpha_l \dfrac{\varrho_l \hat{\varrho}_g}{\hat{\varrho}^2} \Delta u & u_l + u & 0 & 0 \\ -\varrho_g s_g & 0 & s_g & 0 & u_g & 0 \\ 0 & -\varrho_l s_l & 0 & s_l & 0 & u_l \end{bmatrix}$$

Table B.42: Coefficient matrix $\mathbf{H}_1$ for the conservative form of flow equations

$$\boxed{\mathbf{H} = \mathbf{H}_1 + \mathbf{H}_2}$$

$$\mathbf{H}_2 = \begin{bmatrix} 0 & 0 & 0 & 0 & 0 & 0 \\ 0 & 0 & 0 & 0 & 0 & 0 \\ \alpha_g \tilde{a}^2 \dfrac{\hat{\varrho}_l}{\hat{\varrho}_s} f_g & \alpha_g \tilde{a}^2 \dfrac{\varrho_g \hat{\varrho}_l}{\varrho_l \hat{\varrho}_s} f_l & 0 & 0 & -\alpha_g \tilde{a}^2 \dfrac{\hat{\varrho}_l}{\hat{\varrho}_s s_g} f_g & -\alpha_g \tilde{a}^2 \dfrac{\varrho_g \hat{\varrho}_l}{\varrho_l \hat{\varrho}_s s_l} f_l \\ \alpha_l \tilde{a}^2 \dfrac{\varrho_l \hat{\varrho}_g}{\varrho_g \hat{\varrho}_s} f_g & \alpha_l \tilde{a}^2 \dfrac{\hat{\varrho}_g}{\hat{\varrho}_s} f_l & 0 & 0 & -\alpha_l \tilde{a}^2 \dfrac{\varrho_l \hat{\varrho}_g}{\varrho_g \hat{\varrho}_s s_g} f_g & -\alpha_l \tilde{a}^2 \dfrac{\hat{\varrho}_g}{\hat{\varrho}_s s_l} f_l \\ 0 & 0 & 0 & 0 & 0 & 0 \\ 0 & 0 & 0 & 0 & 0 & 0 \end{bmatrix}$$

$$f_g = \frac{s_g}{\varrho_g} \left(\frac{\partial \varrho_g}{\partial s_g} \right)_p \qquad f_l = \frac{s_l}{\varrho_l} \left(\frac{\partial \varrho_l}{\partial s_l} \right)_p$$

Table B.43: Coefficient matrix $\mathbf{H}_2$ for the conservative form of flow equations

$$\boxed{\frac{\partial \mathbf{V}}{\partial t} + \frac{\partial \mathbf{F}}{\partial x} + \mathbf{H}^{\mathrm{nc}} \frac{\partial \mathbf{F}}{\partial x} = \mathbf{D}} \qquad \text{with} \qquad \mathbf{H}^{\mathrm{nc}} = \mathbf{X}^{\mathrm{nc}}\,\mathbf{K}^{-1}$$

$$\mathbf{X}^{\mathrm{nc}} = \begin{bmatrix} 0 & 0 & 0 & 0 & 0 & 0 \\ 0 & 0 & 0 & 0 & 0 & 0 \\ -\alpha_g \alpha_l \frac{k\varrho}{\hat{\varrho}^2} (\varrho_l - \varrho_g) & -\alpha_g \alpha_l \frac{k \varrho_l \varrho_g \hat{\varrho}_g}{\hat{\varrho}^2} (u_g - u_l) & -\alpha_g \alpha_l \frac{k \varrho_l \varrho_g \hat{\varrho}_l}{\hat{\varrho}^2} (u_g - u_l) & 0 & 0 & 0 \\ +\alpha_g \alpha_l \frac{k\varrho}{\hat{\varrho}^2} (\varrho_l - \varrho_g) & k\alpha_g \alpha_l \frac{k \varrho_l \varrho_g \hat{\varrho}_g}{\hat{\varrho}^2} (u_g - u_l) & -\alpha_g \alpha_l \frac{k \varrho_l \varrho_g \hat{\varrho}_l}{\hat{\varrho}^2} (u_g - u_l) & 0 & 0 & 0 \\ 0 & 0 & 0 & 0 & 0 & 0 \\ 0 & 0 & 0 & 0 & 0 & 0 \end{bmatrix}$$

Table B.44: Nonconservative contribution to the coefficient matrix $\mathbf{H}^{\mathrm{nc}}$

$$\boxed{\frac{\partial \mathbf{V}}{\partial t} + \frac{\partial \mathbf{F}}{\partial x} + \mathbf{H}^{\mathrm{nc}} \frac{\partial \mathbf{F}}{\partial x} = \mathbf{D}}$$

$$\mathbf{D} = \begin{bmatrix} \sigma_g^M \\ \sigma_f^M \\ \frac{\varrho_g \varrho_l}{\hat{\varrho}^2}\left(F_g^{\mathrm{v}} + \sigma_g^M u^{\mathrm{ex}}\right) + \frac{\sigma_g^M}{\hat{\varrho}^2}\left[k\varrho(\alpha_g \varrho_g u_l + \alpha_l \varrho_l u_g) - \alpha_l \alpha_g (\varrho_l^2 - \varrho_g^2)(u_g - u_l)\right] + \frac{\varrho_g}{\hat{\varrho}^2}\left[F_g(\varrho_l + \alpha_g k \varrho) + F_l \alpha_g k \varrho\right] \\ \frac{\varrho_g \varrho_l}{\hat{\varrho}^2}\left(F_l^{\mathrm{v}} + \sigma_l^M u^{\mathrm{ex}}\right) + \frac{\sigma_l^M}{\hat{\varrho}^2}\left[k\varrho(\alpha_g \varrho_g u_l + \alpha_l \varrho_l u_g) - \alpha_l \alpha_g (\varrho_l^2 - \varrho_g^2)(u_g - u_l)\right] + \frac{\varrho_l}{\hat{\varrho}^2}\left[F_g \alpha_l k \varrho + F_l(\varrho_g + \alpha_l k \varrho)\right] \\ \sigma_g^S + \sigma_g^M s_g \\ \sigma_l^S + \sigma_l^M s_l \end{bmatrix}$$

Table B.45: Source term vector **E** for the conservative form of flow equations

$$\boxed{\frac{\partial \mathbf{U}}{\partial t} + \mathbf{G}\frac{\partial \mathbf{U}}{\partial x} = \mathbf{D}}$$

$$\mathbf{D} = \begin{pmatrix} \dfrac{\varrho_l a_0^2}{\varrho_s}\left[\sigma_g^M - \dfrac{1}{\varrho_g}\left(\dfrac{\partial \varrho_g}{\partial s_g}\right)_p \sigma_g^S\right] + \dfrac{\varrho_g a_0^2}{\varrho_s}\left[\sigma_l^M - \dfrac{1}{\varrho_l}\left(\dfrac{\partial \varrho_l}{\partial s_l}\right)_p \sigma_l^S\right] - \dfrac{\varrho_l \varrho_g}{\varrho_s}(\alpha_g u_g + \alpha_l u_l) a_0^2 \dfrac{1}{A}\dfrac{\partial A}{\partial x} \\ \dfrac{\varrho_l}{\alpha_g \hat{\varrho}^2}\left[F_g^{\mathrm{v}} + \sigma_g^M (u^{\mathrm{ex}} - u_g)\right] - \sigma_g^M \dfrac{k\varrho\varrho_g + \alpha_l(\varrho_l^2 - \varrho_g^2)}{\varrho_g \hat{\varrho}^2}\Delta u + \dfrac{F_g(\varrho_l + \alpha_g k\varrho) + F_l \alpha_g k\varrho}{\alpha_g \hat{\varrho}^2} \\ -\dfrac{\varrho_g}{\alpha_l \hat{\varrho}^2}\left[F_g^{\mathrm{v}} + \sigma_g^M (u^{\mathrm{ex}} - u_l)\right] - \sigma_g^M \dfrac{k\varrho\varrho_l + \alpha_g(\varrho_g^2 - \varrho_l^2)}{\varrho_l \hat{\varrho}^2}\Delta u + \dfrac{F_g \alpha_l k\varrho + F_l(\varrho_g + \alpha_l k\varrho)}{\alpha_l \hat{\varrho}^2} \\ \dfrac{\alpha_l a_0^2}{\varrho_s a_l^2}\left[\sigma_g^M - \dfrac{1}{\varrho_g}\left(\dfrac{\partial \varrho_g}{\partial s_g}\right)_p \sigma_g^S\right] - \dfrac{\alpha_g a_0^2}{\varrho_s a_g^2}\left[\sigma_l^M - \dfrac{1}{\varrho_l}\left(\dfrac{\partial \varrho_l}{\partial s_l}\right)_p \sigma_l^S\right] + \alpha_l \alpha_g \dfrac{a_0^2}{\varrho_s}\left(\dfrac{\varrho_l u_l}{a_g^2} - \dfrac{\varrho_g u_g}{a_l^2}\right)\dfrac{1}{A}\dfrac{\partial A}{\partial x} \\ \sigma_g^S \dfrac{1}{\alpha_g \varrho_g} \\ \sigma_l^S \dfrac{1}{\alpha_l \varrho_l} \end{pmatrix}$$

Table B.46: Vector of source terms $\mathbf{D}$ for quasi-one-dimensional flow

B.6 Hyperbolic two-phase flow model – two-dimensional conditions

$$\boxed{\frac{\partial \mathbf{U}}{\partial t} + \mathbf{G}_x \frac{\partial \mathbf{U}}{\partial x} + \mathbf{G}_x \frac{\partial \mathbf{U}}{\partial y} = \mathbf{D}}$$

$$\mathbf{U} = \begin{bmatrix} p \\ u_{g,x} \\ u_{l,x} \\ u_{g,y} \\ u_{l,y} \\ \alpha_g \\ s_g \\ s_l \end{bmatrix}, \quad \mathbf{V} = \begin{bmatrix} \alpha_g \varrho_g \\ \alpha_l \varrho_l \\ \alpha_g \varrho_g u_{g,x} \\ \alpha_l \varrho_l u_{l,x} \\ \alpha_g \varrho_g u_{l,y} \\ \alpha_l \varrho_l u_{l,y} \\ \alpha_g s_g \\ \alpha_l s_l \end{bmatrix}$$

$$\mathbf{F}_x = \begin{bmatrix} \alpha_g \varrho_g u_{g,x} \\ \alpha_l \varrho_l \left(u_l\right)_x \\ \alpha_g \varrho_g u_{g,x}^2 + \alpha_g p \\ \alpha_l \varrho_l u_{l,x}^2 + \alpha_l p \\ \alpha_g \varrho_g u_{g,x} u_{l,y} \\ \alpha_g \varrho_g u_{l,x} u_{l,y} \\ \alpha_g s_g u_{g,x} \\ \alpha_l s_l u_{l,x} \end{bmatrix}, \quad \mathbf{F}_y = \begin{bmatrix} \alpha_g \varrho_g u_{g,y} \\ \alpha_l \varrho_l u_{l,y} \\ \alpha_g \varrho_g u_{g,x} u_{l,y} \\ \alpha_g \varrho_g u_{l,x} u_{l,y} \\ \alpha_g \varrho_g u_{g,y}^2 + \alpha_g p \\ \alpha_l \varrho_l u_{l,y}^2 + \alpha_l p \\ \alpha_g s_g u_{g,y} \\ \alpha_l s_l u_{l,y} \end{bmatrix}$$

Table B.47: State vectors for primitive and conserved state parameters and related flux vectors for two-dimensional flow conditions

$$\boxed{\frac{\partial \mathbf{U}}{\partial t} + \mathbf{G}_x \frac{\partial \mathbf{U}}{\partial x} + \mathbf{G}_x \frac{\partial \mathbf{U}}{\partial x} = \mathbf{D}}$$

$$\mathbf{G}_x = \begin{bmatrix}
u_{g,x} + \overline{\Delta u_x} & \alpha_g a_0^2 \frac{\varrho_l \varrho_g}{\varrho_s} & \alpha_l a_0^2 \frac{\varrho_l \varrho_g}{\varrho_s} & 0 & 0 & a_0^2 \Delta u_x \frac{\varrho_l \varrho_g}{\varrho_s} & 0 & 0 \\
 & & & & & & & 0 \\
\frac{\hat{\varrho}_l}{\hat{\varrho}^2} & \hat{\alpha}_g u_{g,x} + \hat{\alpha}_l u_{l,x} & -\hat{\alpha}_l \frac{\hat{\varrho}_l}{\hat{\varrho}_g} \Delta u_x & 0 & 0 & 0 & 0 & 0 \\
\frac{\hat{\varrho}_g}{\hat{\varrho}^2} & \hat{\alpha}_g \frac{\hat{\varrho}_g}{\hat{\varrho}_l} \Delta u_x & \hat{\alpha}_g u_{g,x} + \hat{\alpha}_l u_{l,x} & 0 & 0 & 0 & 0 & 0 \\
0 & -\hat{\alpha}_l \Delta u_y & -\hat{\alpha}_l \frac{\hat{\varrho}_l}{\hat{\varrho}_g} \Delta u_y & u_{g,x} & 0 & 0 & 0 & 0 \\
0 & +\hat{\alpha}_g \frac{\hat{\varrho}_g}{\hat{\varrho}_l} \Delta u_y & +\hat{\alpha}_g \Delta u_y & 0 & u_{l,x} & 0 & 0 & 0 \\
\alpha_l \alpha_g \frac{\Delta u_x}{\varrho_s} \frac{a_0^2}{a_l^2 a_g^2} & \alpha_l \alpha_g \frac{\varrho_g}{\varrho_s} \frac{a_0^2}{a_l^2} & \alpha_l \alpha_g \frac{\varrho_l}{\varrho_s} \frac{a_0^2}{a_g^2} & 0 & 0 & u_{l,x} - \overline{\Delta u_x} & 0 & 0 \\
0 & 0 & 0 & 0 & 0 & 0 & u_{g,x} & 0 \\
0 & 0 & 0 & 0 & 0 & 0 & 0 & u_{l,x}
\end{bmatrix}$$

with $\hat{\alpha}_g = \alpha_g \frac{\varrho_g \hat{\varrho}_l}{\hat{\varrho}^2}$, $\hat{\alpha}_l = \alpha_l \frac{\varrho_l \hat{\varrho}_g}{\hat{\varrho}^2}$, $\overline{\Delta u_x} = \alpha_l \Delta u_x \frac{\varrho_l}{\varrho_s} \frac{a_0^2}{a_l^2}$

Table B.48: Coefficient matrix $\mathbf{G}_x$for two-dimensional flow

$$\boxed{\frac{\partial \mathbf{U}}{\partial t} + \mathbf{G}_x \frac{\partial \mathbf{U}}{\partial x} + \mathbf{G}_x \frac{\partial \mathbf{U}}{\partial x} = \mathbf{D}}$$

$$\mathbf{G}_y = \begin{bmatrix}
u_{g,y} + \overline{\Delta u_y} & 0 & 0 & \alpha_g a_0^2 \frac{\varrho_l \varrho_g}{\varrho_s} & \alpha_l a_0^2 \frac{\varrho_l \varrho_g}{\varrho_s} & a_0^2 \Delta u_y \frac{\varrho_l \varrho_g}{\varrho_s} & 0 & 0 \\
 & & & & & & & 0 \\
0 & u_{g,y} & 0 & -\hat{\alpha}_l \Delta u_x & -\hat{\alpha}_l \frac{\hat{\varrho}_l}{\hat{\varrho}_g} \Delta u_x & 0 & 0 & 0 \\
0 & 0 & u_{l,y} & +\hat{\alpha}_g \frac{\hat{\varrho}_g}{\hat{\varrho}_l} \Delta u_x & +\hat{\alpha}_g \Delta u_x & 0 & 0 & 0 \\
\frac{\hat{\varrho}_l}{\hat{\varrho}^2} & 0 & 0 & \hat{\alpha}_g u_{g,y} + \hat{\alpha}_l u_{l,y} & -\hat{\alpha}_l \frac{\hat{\varrho}_l}{\hat{\varrho}_g} \Delta u_y & 0 & 0 & 0 \\
\frac{\hat{\varrho}_g}{\hat{\varrho}^2} & 0 & 0 & \hat{\alpha}_g \frac{\hat{\varrho}_g}{\hat{\varrho}_l} \Delta u_y & \hat{\alpha}_g u_{g,y} + \hat{\alpha}_l u_{l,y} & 0 & 0 & 0 \\
\alpha_l \alpha_g \frac{\Delta u_y}{\varrho_s} \frac{a_0^2}{a_l^2 a_g^2} & 0 & 0 & \alpha_l \alpha_g \frac{\varrho_g}{\varrho_s} \frac{a_0^2}{a_l^2} & \alpha_l \alpha_g \frac{\varrho_l}{\varrho_s} \frac{a_0^2}{a_g^2} & u_{l,y} - \Delta u_y & 0 & 0 \\
0 & 0 & 0 & 0 & 0 & 0 & u_{gy} & 0 \\
0 & 0 & 0 & 0 & 0 & 0 & 0 & u_{l,y}
\end{bmatrix}$$

with $\quad \hat{\alpha}_g = \alpha_g \frac{\varrho_g \hat{\varrho}_l}{\hat{\varrho}^2}, \quad \hat{\alpha}_l = \alpha_l \frac{\varrho_l \hat{\varrho}_g}{\hat{\varrho}^2}, \quad \overline{\Delta u_y} = \alpha_l \Delta u_y \frac{\varrho_l}{\varrho_s} \frac{a_0^2}{a_l^2}$

Table B.49: Coefficient matrix $\mathbf{G}_y$ for two-dimensional flow

$$\boxed{\frac{\partial \mathbf{U}}{\partial t} + \mathbf{G}_n = \mathbf{D}_n}$$

$$\mathbf{G}_n = \begin{bmatrix} u_{g,n} + \overline{\Delta u_n} & \alpha_g a_0^2 \frac{\varrho_l \varrho_g}{\varrho_s} & \alpha_l a_0^2 \frac{\varrho_l \varrho_g}{\varrho_s} & 0 & 0 & a_0^2 \Delta u_n \frac{\varrho_l \varrho_g}{\varrho_s} & 0 & 0 \\ \frac{\hat{\varrho}_l}{\hat{\varrho}^2} & \hat{\alpha}_g u_{g,n} + \hat{\alpha}_l u_{l,n} & -\hat{\alpha}_l \frac{\hat{\varrho}_l}{\hat{\varrho}_g} \Delta u_n & 0 & 0 & 0 & 0 & 0 \\ \frac{\hat{\varrho}_g}{\hat{\varrho}^2} & \hat{\alpha}_g \frac{\hat{\varrho}_g}{\hat{\varrho}_l} \Delta u_n & \hat{\alpha}_g u_{g,n} + \hat{\alpha}_l u_{l,n} & 0 & 0 & 0 & 0 & 0 \\ 0 & -\hat{\alpha}_l \Delta u_t & -\hat{\alpha}_l \frac{\hat{\varrho}_l}{\hat{\varrho}_g} \Delta u_t & u_{g,n} + \epsilon & 0 & 0 & 0 & 0 \\ 0 & \hat{\alpha}_g \frac{\hat{\varrho}_g}{\hat{\varrho}_l} \Delta u_t & \hat{\alpha}_g \Delta u_t & 0 & u_{l,n} - \epsilon & 0 & 0 & 0 \\ \alpha_l \alpha_g \frac{\Delta u_n}{\varrho_s} \frac{a_0^2}{a_l^2 a_g^2} & \alpha_l \alpha_g \frac{\varrho_g}{\varrho_s} \frac{a_0^2}{a_l^2} & \alpha_l \alpha_g \frac{\varrho_l}{\varrho_s} \frac{a_0^2}{a_g^2} & 0 & 0 & u_{l,n} - \overline{\Delta u_n} & 0 & 0 \\ 0 & 0 & 0 & 0 & 0 & 0 & u_{g,n} & 0 \\ 0 & 0 & 0 & 0 & 0 & 0 & 0 & u_{l,n} \end{bmatrix}$$

with $\quad \hat{\alpha}_g = \alpha_g \frac{\varrho_g \hat{\varrho}_l}{\hat{\varrho}^2}, \quad \hat{\alpha}_l = \alpha_l \frac{\varrho_l \hat{\varrho}_g}{\hat{\varrho}^2}, \quad \overline{\Delta u_n} = \alpha_l \Delta u_n \frac{\varrho_l}{\varrho_s} \frac{a_0^2}{a_l^2}$

Table B.50: Projection of coefficient matrix $\mathbf{G}_n$ in the n-direction

Index

Gasdynamic Aspects of Two-Phase Flow. Herbert Städtke

ISBN: 3-527-40578-X

273